FIRES, EXPLOSIONS, AND TOXIC GAS DISPERSIONS

FIRES, EXPLOSIONS, AND TOXIC GAS DISPERSIONS

Effects Calculation and Risk Analysis

Marc J. Assael
Konstantinos E. Kakosimos

CRC Press
Taylor & Francis Group
Boca Raton London New York

CRC Press is an imprint of the
Taylor & Francis Group, an **informa** business

CRC Press
Taylor & Francis Group
6000 Broken Sound Parkway NW, Suite 300
Boca Raton, FL 33487-2742

International Standard Book Number: 978-1-4398-2675-1 (Hardback)

Library of Congress Cataloging-in-Publication Data

Assael, Marc J.
Fires, explosions, and toxic gas dispersions : effects calculation and risk analysis / Marc J. Assael, Konstantinos E. Kakosimos.
p. cm.
Includes bibliographical references and index.
ISBN 978-1-4398-2675-1 (hardcover : alk. paper)
1. Hazardous substances--Risk assessment. 2. Hazardous substances--Accidents. 3. Hazardous substances--Safety measures. 4. Hazardous substances--Health aspects. I. Kakosimos, Konstantinos E. II. Title.

T55.3.H3A847 2010
604.7--dc22 2009047099

Visit the Taylor & Francis Web site at
http://www.taylorandfrancis.com

and the CRC Press Web site at
http://www.crcpress.com

for Theodora and John-Alexander,
for Katerina and Evangelo,

and
our parents

contents

preface

The aim of this book is to provide an overview of the calculation procedures for Risk Analysis, that is, the effects and consequences from fires, explosions and toxic gas dispersion. Starting from the probability of an accident to occur, the reader is introduced to the calculation of a leak and then to the calculation of the heat flux, the overpressure, and the concentrations of toxic clouds. Following that, the consequences to people (injuries or deaths) as well as material damages are described. The book concludes with a discussion on the possible causes of destruction.

The book is targeted to undergraduate students, as well as to professional engineers and researchers whose work calls for predicting the consequences of an accident. The procedures described are practical and attempt to give a very good understanding of the subject. The use of modern simple methods helps to understand the meaning of all the variables involved, in contrast to current complicated computer packages which produce only results. The book is also ideally suited for the basis of an undergraduate course in Risk Analysis. The structure of the book, the step-by-step calculation procedures and the large number of examples* aim to assist both teaching and learning, while also filling an existing gap in the teaching of risk analysis.

The book is presented in four chapters. Chapter A introduces some basic techniques of hazard identification with special emphasis on "What-if" and HAZOP analyses. These are followed by a discussion on event occurrence probability, and the use of fault and event tree analyses are shown through examples. The influence of the "human factor" is also discussed. In Chapter B we begin with calculations for various types of leaks and present specific calculations for leaks of a pressurized gas and a liquid, respectively. Case studies refer to small leaks or full vessel rupture, and the outflow of pressurized liquefied gases is also examined.

Chapter C is the largest chapter in this book. It consists of five sections. In the first section some complementary definitions are given. The calculation of the

* All examples given in this book are also available as Microsoft Excell sheets in ftp://transp.cheng.auth.gr/

heat flux from a fire is carried out in the second section, which covers the pool fire, ball fire, jet fire and flash fire. The probability for injuries or lethality is also calculated. In the third section, calculation procedures for the overpressure of a shock wave resulting from an unconfined vapor cloud are presented. From the change in time and space of the overpressure, the probability of injuries, deaths and material damages is calculated. In the fourth section a brief analysis of a BLEVE (Boiling Liquid Expanding Vapour Explosion) is presented. The continuous or instantaneous dispersion of light or heavy toxic gases or airborne particles (e.g., PM10), is examined in detail in the last section of this book. For every case, temporal and local concentrations are calculated and based on these, safety zones are specified. At the end of every section, extensive tables with major industrial accidents according to the section's subject are presented in chronological order.

In the last chapter of this book, Chapter D, an analysis of various causes of destruction, based on major industrial accidents of the last century, is carried out. The aim of this chapter is to show some common circumstances which can result in an accident. The discussion refers to facilities (storage tanks, stacks, etc.) and equipment (pipes, valves, heat exchangers). Possible ignition sources are also discussed.

We are grateful to Dr Thomas Tsolakis for his ideas and his enthusiastic encouragement in writing this book, as well as to Mr Tiantafyllos Parthenopoulos, whose excellent cooperation in our risk analysis projects formed the basis for this book. We also would like to thank Mr Apostolos Rafaelidis for his help with the human factor section, and Ms Adriane Thrash for editing the book.

Finally, we are indebted to Dr Konstantinos Antoniadis who agreed to read the entire text and provide us with his comments. We found these most useful and we thank him for his dedicated and extremely efficient efforts; he even read and corrected our examples! Any remaining errors (or perhaps we should say "the remaining errors") are of course entirely our fault. Writing this book was a pleasure for us, and we hope that its reading will also be a pleasure for you.

Marc J. Assael & Konstantinos E. Kakosimos

Thessaloniki

The cover photograph is from the destruction of the AZF unit in Toulouse, France, on September 21, 2001 (photographer Emmanuel Grimault).

main symbols list

A	surface (m^2)
b	flame elevation (m)
b	downwind dispersion width or radius (m)
B	second virial coefficient (m^3/mol)
C	third virial coefficient (m^6/mol^2)
C	concentration (μg/m^3)
C_d	drag coefficient (-)
C_P	specific heat capacity under constant pressure (J $kg^{-1}K^{-1}$)
C_V	specific heat capacity under constant volume (J $kg^{-1}K^{-1}$)
d_p	particle diameter, tube diameter (m)
D	diameter (m)
D	thermal radiation dose ($W^{4/3}sm^{-8/3}$)
erf	error function (-)
f_F	Fanning friction factor (-)
F_b	buoyancy parameter (m^4/s^3)
Fr	Froude number (-)
F_{view}	view factor (-)
g	acceleration due to gravity (9.81 m/s^2)
h	height (m)
h_e	plume's rise (m)
h_s	actual stack's height (m)
H	heat, enthalpy (J/kg)
i_s	impulse (MPas)
l_p	length of tube (m)
L	length (m)
m'	burning rate (kg $m^{-2}s^{-1}$)
$\dot{m}$	outflow rate (kg/s)
M	mass (kg)
M_j	Mach number (-)
M_{TNT}	equivalent TNT mass (kg)
N	number of persons (persons)
P	pressure (Pa)
P	probability (-)
P_a	ambient pressure (Pa)
P_c	critical pressure (Pa)
P_s	overpressure (kPa or bar)
P_w	water vapor pressure (Pa)
P_W^O	saturation water vapor pressure (Pa)
Pr	probit function (-)
q'	heat flux (kW/m^2)
Q_c	source emission rate (kg/s)
R	universal gas constant (8.314 J $mol^{-1}K^{-1}$)
R	radius (m)
Re	Reynolds number (-)
Ri	Richardson number (-)
RH	relative humidity (-)
s	flame's surface fraction covered by soot (-)
S	entropy (J $kg^{-1}K^{-1}$)
SEP	surface emitting power (kW/m^2)
SEP_{soot}	surface emitting power of soot (kW/m^2)
t	time (s)
t_p	duration of positive phase of explosion (s)
T	absolute temperature (K)
T_a	ambient temperature (K)
T_b	boiling temperature (K)
T_c	critical temperature (K)
T_f	flame temperature (K)
u	velocity, speed (m/s)
u_w	wind speed (m/s)
V	volume (m^3)
W_g	molecular weight (kg/mol)
x_f	distance of maximum plume rise (m)
X	distance (m)
z_{ref}	reference height (m)
Z	compressibility factor (-)

Greek Symbols

γ	heat capacities ratio (-)
ΔH_c	heat of combustion (J/kg)
ΔH_v	heat of vaporization (J/kg)
η	viscosity (μPas)
Θ	flame's angle of tilt (°)
λ	heat of vaporization (J/kg)
v_s	gas exit speed (m/s)
ρ	density (kg/m^3)
ρ_n	molar density (mol/m^3)
σ	surface tension (Pa m)
σ	Stefan-Boltzmann's constant (5.6703x10^{-8} W m^{-2}K^{-4})
σ_y	lateral dispersion coefficient (m)
σ_z	vertical dispersion coefficient (m)
τ_a	atmospheric transmissivity (-)
ϕ	fill factor, load factor (-)
ω	acentric factor (-)

A hazard identification - event frequency

A1 Hazard and Risk

In everyday conversation the term "hazard" is often used in a general sense and confused with the term "risk." By actual definition, however, the two terms differ substantially:

- **Hazard** is defined as an action that has the potential to cause harm to human health or the environment.
- **Risk** is the *likelihood* of harm, that is, the probability for a certain effect to appear in a specific time under predefined conditions. Risk can be defined as

$$\text{Risk} = \text{Event frequency} \times \text{Event's effects}$$

The hazards that can appear during the operation of a plant / facility where hazardous substances are produced or transported, can be attributed to: (a) equipment failure, (b) operational problems, (c) human factor, (d) external causes or natural phenomena.

Murphy's law states that "the natural course of events is to go from bad to worse." This prediction, is in a way, demonstrated by the second thermodynamic law, which refers to the entropy. Entropy, a measure of the "disturbance" of a system, always increases – a system being the universe or any thermal insulated enclosure or space. In reality, this means that for a system to remain in a static condition, certain energy must be added. All the above, with a strong dose of philosophy, simply translates to the fact that accidents will always happen, unless the correct amount of "energy" is added. Such an energy is the risk analysis of a plant or facility and its management.

The risk management of a plant consists of the following stages:

a) Identification of the hazards involved in the operation of the plant, due to the nature, properties, and conditions of production and handling of dangerous substances. Formulation of potential accident scenarios is a necessary part of this stage.
b) Analysis of the risks involved in emergency situations likely to occur within the environs of the plant which can affect its employees, adjacent plant

personnel, and adjacent community population. Evaluation of the consequences resulting from the identified accident scenarios is a necessary part of this stage.

c) Development of a Safety Management System to prevent accident occurrence and/or mitigation of its consequences.

In the following sections of this chapter we will be discussing:

- Hazard identification techniques (What-if analysis, HAZOP analysis, FMEA analysis and some Qualitative Evaluation techniques).
- Two methods related to the probability of occurrence (frequency) of a hazard (Fault Tree Analysis, and Event Tree Analysis).
- A preliminary approach to the subject of the human factor effect.

Fire and explosion on February 18, 2008, in Big Spring Refinery, TX, U.S.A. (Reproduced by kind permission of Texas Forest Service U.S.A.)

A2
Hazard Identification Techniques

What-if Analysis

HAZOP Analysis

FMEA Analysis

Qualitative Evaluation Techniques

The most common hazard identification techniques applied today are the following:

- "What-if" Analysis (Section A2.1.).
  The "What-if" analysis is the simplest technique used to identify hazards. The analysis is based on the question "What will happen if...", an essential component of a process or plant does not operate according to its design. Depending on the requirements of the analysis, this method may be implemented to all components comprising a process or plant, including the procedures governing its operation.
- HAZard and OPerabiliy Analysis, HAZOP (Section A2.2.).
  This is one of the most structured techniques to identify hazards in a process or plant, and aims to find all possible deviations from the normal function of process parameters. A list of "key-words" is used to define the deviations.
- Failure Modes and Effects Analysis, FMEA (Section A2.3.).
  This is a standard evaluation procedure for systematically identifying potential failures in equipment or system design, and analyzing the effects on the performance of that equipment or system.

In addition to the above three techniques, qualitative evaluation techniques are normally applied to identify any potential hazard as a consequence of the operation of a facility. For existing technology and an experienced evaluation team, a simple qualitative evaluation technique may be sufficient to identify any conceivable hazard. For new technology applications of limited past experience, the hazard evaluation team may brainstorm using techniques like "What-if" analysis. Once a design progresses into the pre-engineering phase, a more-detailed technique like HAZOP or FMEA is preferable for hazard identification and evaluation [TNO 2005, Clemens 1982]. Such qualitative evaluation techniques are:

- Safety Review (Section A2.4.1).
- Checklist Analysis (Section A2.4.2).
- Preliminary Hazard Analysis (Section A2.4.3).
- Criticality Analysis (Section A2.4.4).
- Change Analysis (Section A2.4.5).

Identifying Hazards

procedure

What-if Analysis

HAZOP Analysis

FMEA Analysis

Qualitative Evaluation Techniques

- Critical Incident Technique (Section A2.4.6).
- Energy Analysis (Section A2.4.7).
- Worst-Case Analysis (Section A2.4.8).
- Network Logic Analysis (Section A2.4.9).
- Scenario Analysis (Section A2.4.10).
- Systematic Inspection (Section A2.4.11).

For a more-detailed report, the reader is referred to more specialized literature [TNO 2005, Clemens 1982].

This photograph of the Hellenic Petroleum Plant in Thessaloniki, Greece, clearly shows the difficulties in identifying all possible hazards in complex industrial facilities. (Reproduced by kind permission of Ch. Paschoudis.)

A2.1. "What-If" Analysis

The "What-if" analysis is the simplest technique used to identify hazards. As aforementioned, the analysis is based on the question "What will happen if...", an essential component of a process or plant does not operate according to its design. This method may be applied to all components comprising a process or plant, even including the procedures governing its operation, depending on the analysis requirements. It is a brainstorming approach according to which a group of experienced people familiar with the subject ask questions or voice concerns about possible undesired events. Although it is not as inherently structured as HAZOP or FMEA analyses, it encourages the team to think of questions that begin with "What if...".

Assembling an experienced, knowledgeable team is probably the single most important element in conducting a successful "What-if" analysis. Individuals experienced in the design, operation, and servicing of similar equipment or facilities are essential. Their knowledge of design standards, regulatory codes, past and potential operational errors, as well as maintenance difficulties, brings a practical reality to the review. Team members may include Process or Laboratory Manager, and representatives with specific skills as needed (from maintenance, compressed gas, manufacturing, etc.).

The next most important step is gathering the needed information. The operation or process must be understood by the review team. One important way to gather information on an existing process or piece of equipment is for each team member to visit and walk through the operation site. Additionally, piping and instrument diagrams, design documents, operational procedures, and maintenance procedures are essential information for the review team. If these documents are not available, the first recommendation for the review team becomes clear: *Develop the supporting documentation!* Effective reviews cannot be conducted without up-to-date and reliable documentation. An experienced team can provide an overview analysis, but the nuances of specific issues such as interlocks, pressure relief valves, or code requirements are not likely to be found without documentation.

The great advantage of the "What-if" analysis is its flexibility. In essence it can be applied in any stage of a process or plant using any available information in connection with the available knowledge. The disadvantage of the technique is that it requires personnel with detailed knowledge of the process or plant, who will also be able to conceive and predict deviations from normal operation.

Identifying Hazards

example

What-if Analysis

HAZOP Analysis

FMEA Analysis

Qualitative Evaluation Techniques

EXAMPLE A2.1. *"What-if" Analysis*

A simplified flow diagram for the feed line of a propane-butane separation column system is shown in Figure A2.1. The mixture enters the vessel D-1 at 75°C and 22 bar. The mixture is pumped from the bottom of the vessel to the separation column T-1, by the P-1 pump. An FRC valve controls the flow rate. The mixture is pre-heated to 85°C using steam at the heat exchanger E-1. Perform the What-if analysis (only for the flow parameter).

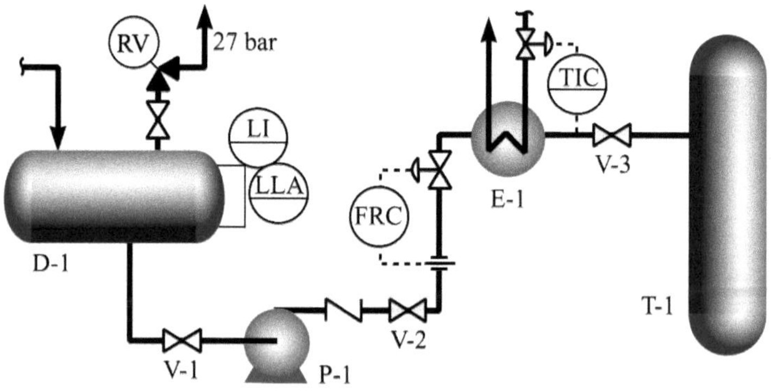

Figure A2.1. Feed line of a propane-butane separation column system.

Symbol interpretation:
- RV : Relief Valve
- LI : Level Indicator
- LLA : Low Level Alarm
- FRC : Flow Recorder Controller
- TIC : Temperature Indicator Controller

According to the aforementioned discussion, the analysis, the consequences and the recommendations for this particular example are shown in Table A2.1.

In the following section, the same example will be examined with the HAZOP analysis, so that the advantages and disadvantages of each technique will become apparent.

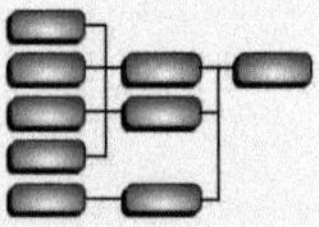

Table A2.1. "What-if" Analysis.

Question "What-if"	Consequences	Recommendations
... the operator accidentally closes the valve V-1?	- Liquid level rises in D-1 vessel. - Operational upset of the T-1 column due to feed interruption.	- There is a level indicator LI. In case LI fails, the relief valve RV will open.
... the pump P-1 shuts down?	As above.	
... the flow control valve FRC is leaking?	- Risk of generating a flammable mixture, and potential fire.	- More frequent valve maintenance. - Consider installation of double-seal systems.
... there is a fire close to the vessel D-1?	- Temperature and pressure increase in the vessel. Possible boiling of the contents.	- Check the capacity of the relief valve to vent the generated vapors. - Install a pressure indicator, PI, on the vessel, along with a high-pressure alarm signal PHA in the control room.
... a crack appears on the tubes of E-1 due to corrosion?	- Hydrocarbon carry-over to the steam network - a hazardous source in other uses of the steam.	- Consider replacing the steam with another heating fluid.

❐

A2.2. Hazard and Operability Analysis (HAZOP)

The HAzard and OPerability study, HAZOP was originally developed by engineers in ICI Chemicals, UK, during the middle of 1970. It is one of the most structured techniques to identify hazards in a process plant, and aims to find all possible deviations from the normal function of process parameters. A list of "key-words," Table A2.2, is used to define the deviations. The HAZOP analysis can be applied to all processes. It is based upon the assumption that any operating problem arising in equipment will be the cause of, or have as a consequence, the deviation from the normal operation of a parameter of one of the lines connected to the equipment concerned.

The primary purpose of the HAZOP analysis is the identification of possible hazard scenarios. The team must not waste time in finding solutions. If the solution is obvious, the team recommends it, otherwise it is referred to the corresponding engineering team. The HAZOP study should preferably be carried out as early in the design phase as possible in order to have influence on the design. On the other hand, to carry out a HAZOP we need a rather complete design. As a compromise, it is usually carried out as a final check when the detailed design has been completed. A HAZOP study may also be conducted on an existing facility to identify modifications that should be implemented to reduce risk and operability problems. HAZOP studies may also be used more extensively, including:

- At the initial concept stage when design drawings are available.
- When the final piping and instrumentation diagrams (P&ID) are available.
- During construction and installation to ensure that recommendations are implemented.
- During commissioning.
- During operation to ensure that plant emergency and operating procedures are regularly reviewed and updated as required.

In recent years HAZOP analysis has been widely accepted as the most preferred technique for hazard identification.

Table A2.2. Keywords.

Keywords	Deviations from normal operation
NO	Complete negation
LESS	Quantitative decrease
MORE	Quantitative increase
PART OF	Qualitative decrease
AS WELL AS	Qualitative increase
REVERSE	Logical opposite
OTHER THAN	Complete substitution

Identifying Hazards
example
What-if Analysis
HAZOP Analysis
FMEA Analysis
Qualitative Evaluation Techniques

EXAMPLE A2.2. *HAZOP Analysis*

This simplified flow diagram in Figure A2.2 shows the mixing of phosphoric acid and ammonia to produce diammonium phosphate, which is not toxic. Perform the HAZOP analysis in Table A2.3 (only for the flow parameter.)

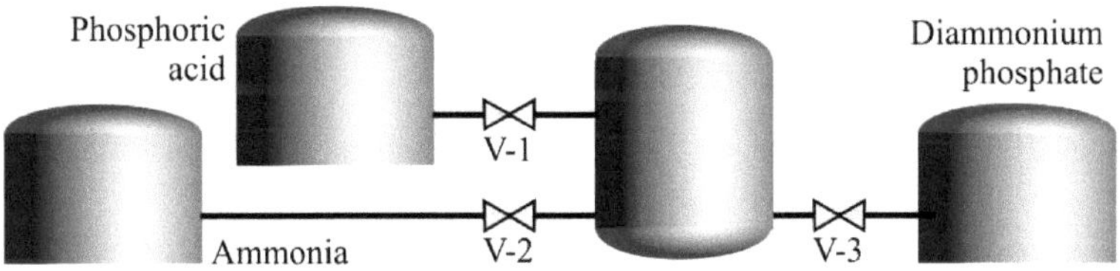

Figure A2.2. Mixing of phosphoric acid and ammonia.

Table A2.3. HAZOP Analysis.

Key-word	Deviation	Possible cause	Consequence	Necessary corrective action
NO	No flow	- Valve V-1 closes. - Phosphoric acid supply exhausted. - Plug in pipe or pipe rupture.	Excess ammonia in reactor. Release to work area.	Automatic closure of valve V-2 on loss of flow from phosphoric acid supply.
LESS	Less flow	- Valve V-1 partially closed. - Partial plug or leak in pipe.	Excess ammonia in reactor. Release to work area, with amount released related to quantitative reduction in supply.	Automatic closure of valve V-2 on reduced flow from phosphoric acid supply.
MORE	More flow		Excess phosphoric acid degrades product.	No hazard to work area.
PART OF	Partial flow	- Delivery of wrong material or wrong concentration. - Incorrect filling of vessel.	Excess ammonia in reactor. Release to work area with amount released related to supply reduction.	Check phosphoric acid supply tank concentration after charging.
AS WELL AS	Flow		*Not applicable*	
REVERSE	Flow		*Not applicable*	
OTHER THAN	Other flow	- Delivery of wrong material.	Depends on flow.	Properly check material before filling.

❐

EXAMPLE A2.3. *HAZOP Analysis*

A simplified flow diagram for the feed line of a propane-butane separation column system is shown in Figure A2.3. The mixture enters the vessel D-1 at 75°C and 22 bar. The mixture is pumped from the bottom of the vessel to the separation column T-1, by the P-1 pump. An FRC valve controls the flow rate. The mixture is pre-heated at 85°C using steam at the heat exchanger E-1. Perform the HAZOP analysis (only for the flow parameter).

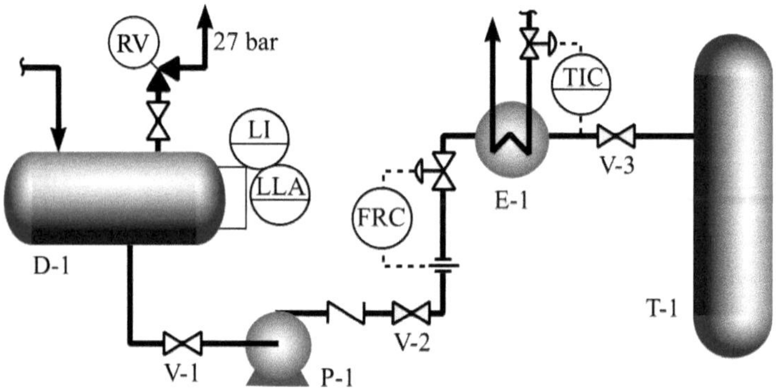

Figure A2.3. Feed line of a propane-butane separation column system.

Symbol interpretation:

RV : Relief Valve
LI : Level Indicator
LLA : Low Level Alarm
FRC : Flow Recorder Controller
TIC : Temperature Indicator Controller

The analysis, consequences and recommendations for this particular example are shown in Table A2.4. In the previous section, the same example was examined with the "What-if" analysis, so that the advantages and disadvantages of each technique become apparent.

Table A2.4. HAZOP Analysis.

Guide Word	Deviation	Possible cause	Consequence	Necessary corrective action
NO	No flow	Loss of suction of the P-1 pump, due to the low liquid level in the D-1 vessel.	a) Pump overheating that may result in leak from the mechanical seal, and possible fire. b) Operational upset in column T-1.	1) A low-level alarm, LLA already exists on the D-1 vessel. 2) Place a low flow alarm LFA on the flow recorder FRC.
		The P-1 pump stops (due to failure or power loss).	c) Liquid level rise in the D-1 vessel. d) As (b) above.	3) There is a safety valve, RV. It is recommended to place a high level alarm, HLA on D-1.
		There is a major leak due to damages to the mechanical seal of the P-1 pump.	e) Increased likelihood of fire. f) As (b) above.	4) More frequent maintenance. 5) Investigate the cause of damage to the mechanical seal. 6) Install a double-seal system.
		The V-1 valve in the suction line is accidentally closed by an operator.	As (a) and (b) above.	7) Point out the error in the operating procedures.
		The FRC valve is closed due to failure (human error, power or instrument-air loss, etc.).	g) As (a) and (b). Furthermore, pressure is rising in the discharge line (until the valve) up to the shut-off pressure of P-1.	8) As (2) above. Furthermore, check if the shut-off pressure of P-1 exceeds the design pressure of the discharge line. 9) Consider modification of the FCV valve so as to remain open in case of power or instrument-air losses.
		The V-2 valve, in the discharge line of P-1, is closed due to human error.	As (g) above.	10) As (2) above. Furthermore, consider installing a recirculation line from the discharge line of P-1 to the vessel D-1.

- Continued

Table A2.4 (cont.) HAZOP Analysis.

Key Word	Deviation	Possible cause	Consequence	Necessary corrective action
No	No flow	V-3 valve is closed.	As (g) above.	11) As (2) above. Furthermore, check if a problem on the shell and tubes of E-1 is expected as the pressure rises up to the shut-off pressure of P-1.
		Mechanical pipe failure and cracking (due to external cause, corrosion, etc.).	h) Significant release of hydrocarbons to the air. Risk of fire or explosion.	
MORE	More flow	Malfunction of the FRC valve.	i) Operational upset in T-1. j) Level decrease in D-1.	12) Preventive actions (more frequent inspection - regulation). As (1) above.
LESS	Less flow	Malfunction of the FRC valve.	As (i) above.. k) Level rise in D-1.	As (12) above. As (3) above.
		Minor leak (from the FRC valve, or the P-1, or flanges).	l) Hydrocarbon release in the air. Risk of fire.	13) More frequent preventive actions. 14) Investigate the causes of damage of the existing seal. 15) Consider installing double-seal systems on the valve and pump, or replacing them with up-to-date equipment. 16) Minimize the use of flanges where possible.
		Leaking tubes on the E-1 heat exchanger (from cracks due to corrosion).	m) Hydrocarbon carry-over to the steam network - a hazardous source in the other potential use of the steam.	17) Consider replacing the steam with other heating fluid.

❒

A2.3. Failure Modes and Effects Analysis (FMEA)

The Failure Modes and Effects Analysis, FMEA, evaluates the ways in which equipment can fail and the effect these failures can have on an installation. These failure descriptions provide analysts with a basis for determining where changes can be made to improve a system design. Single equipment failures are defined by the analysts and the effects of such failures, both locally and on the system as a whole, are investigated. Each individual failure is considered as an independent occurrence with no relation to other failures in the system, except for the subsequent effects which it might produce.

The FMEA analysis is usually applied to systems, subsystems, components, procedures, interfaces etc. The technique is most suited to installations where the danger comes from mechanical equipment and electrical failures, but not from the dynamics of the processes. This is in contrast to the HAZOP technique which is applied to whole processes, whereby the danger comes from hazardous materials in chemical process systems.

In order to determine and define priorities, usually the following three criteria are employed:

S : Severity of the consequences.

P : Probability of occurrence of the event over a period of one year.

b : Difficulty in identifying the particular event.

These three criteria define the Risk Priority Number, *RPN*, as

$$\mathrm{RPN} = \mathrm{S} \times \mathrm{P} \times \mathrm{b}$$

Teams determine the minimum RPN values, as a measure of comparison for further analysis and investigation.

The principles of an FMEA analysis are easy to understand and to learn. It is, however, more important that the analysts are familiar with the components of the system to be analyzed. They must know the failure modes of the components and the effects of those failure modes on the system as a whole. Thus, although the technique is not difficult to apply, it is enormously time-consuming. Although only failure modes (e.g., component faults) are explored, both types of failure modes (those which will, and those which will not result in great harm) must be investigated to fully develop the analysis.

A2.4. Overview of Qualitative Evaluation Techniques

Qualitative evaluation techniques are normally applied to identify any potential hazard as a consequence of the operation of a facility. For the existing technology and an experienced evaluation team, a simple qualitative evaluation technique may be sufficient to identify any conceivable hazard. For new technology applications of limited past experience, the hazard evaluation team may brainstorm using techniques like "What-if" analysis. Once a design progresses into the pre-engineering phase, a more detailed technique like HAZOP or FMEA is certainly preferable for hazard identification and evaluation [TNO 2005, Clemens 1982].

A2.4.1. Safety Review

The Safety Review, also known as Process Safety Review, or Design Safety Review, can be employed at any stage during the life cycle of the plant. It can typically comprise anything from a simple walk-through visual inspection (completed in a day or less) up to a formal examination by a specialized team that can take several weeks. In the case of plants still in the stage of design, the Safety Review can consist of an inspection of documents and drawings.

Safety Reviews intend to identify those operating procedures or plant conditions that could lead to injuries, significant property damage or environmental impacts. A typical Safety Review includes interviews with many people in the plant: operators, engineers, maintenance personnel and others. It should be regarded as a cooperative effort, aiming to improve the overall safety and performance of the plant.

The Safety Review Team must have a lot of experience in applying safety standards and procedures, but also expertise in the evaluation of facilities, electrical systems, pressure vessel inspections and materials characteristics. The plant personnel should be ready to fully cooperate with the team.

A2.4.2. Checklist Analysis

Checklist Analysis uses a written list of objects or procedural steps that must be checked so that the status of a system/facility is verified. The written list includes possible failures and causes of hazardous events. It is based on the personnel experience and it is most useful to identify customarily recognized hazards.

As a minimum, a Checklist Analysis can be employed to ensure that the design is in accordance with standard practices. The Checklist Analysis depends directly upon the experience of those personnel involved in its composition, and it is very simple in its application.

A2.4.3. Preliminary Hazard Analysis

Preliminary Hazard Analysis refers to the effort to identify possible hazards from a very initial stage, preferably at the design stage of the plant or the facility. The technique can be employed in all systems, subsystems, components, procedures, etc., and aims at the identification of possible hazards.

The Preliminary Hazard Analysis is not a discrete technique, but it depends on the expert team, who will apply it based upon its experience. It can incorporate any other technique of hazard identification, as long as it is applied at the design stage of the plant or facility.

A2.4.4. Criticality Analysis

Criticality Analysis ranks the damage potential of system elements according to a scale which represents the harm each element might cause in case of failure. The purpose of the analysis is to rank the criticality of components through unconnected failures, according to

a) their effects (injury, damage, or system degradation, etc.).

b) the probability for this particular failure to occur.

A2.4.5. Change Analysis

Change Analysis is based upon the examination of possible changes of a system/plant/facility. The original system is taken as a base, and on this, possible changes, by themselves or in cooperation with others, are considered as well as the effects they could cause. Usually another hazard identification technique is considered as a base, and on it new possible changes and their effects are examined.

In this case, the full understanding of the physical principles governing the behavior of the system being changed is essential, so that the effects of the change can be determined with an adequate degree of confidence for the analysis.

A2.4.6. Critical Incident Technique

The Critical Incident Technique is based upon the critical evaluation of previous mistakes, failures, hazards and near misses. It identifies dominant high-risk cases. The technique requires interviews and/or distribution of questionnaires to all personnel and uses the collective accumulated experiences. In recent years, there is a tendency to substitute the "What-if" analysis for this technique (Section A2.1).

A2.4.7. Energy Analysis

Energy Analysis refers to the identification of all energy sources within a system, and the examination of the adequacy of barriers to the unwanted flow of that energy to "targets" which might suffer harm. The technique is usually applied to all systems that store, use or incorporate any form of energy.

A2.4.8. Worst-Case Analysis

Worst-Case Analysis technique examines all possible failures that could occur and focus on the worst case of all of them. It subsequently investigates all possible causes that could lead to this worst case.

A2.4.9. Network Logic Analysis

Network Logic Analysis describes the system operation as a network of logic elements, and develops Boolean expressions for proper system functions. Following this, it analyses the network and/or expressions to identify elements of system vulnerability to mishap.

A2.4.10. Scenario Analysis

Scenario Analysis is based upon the examination of possible scenarios proposed by personnel with a great deal of experience in the operation of the plant or facility.

A2.4.11. Systematic Inspection

Systematic Inspection uses checklists, codes, regulations, industrial standards and guidelines, prior mishap experiences, and common sense to methodically examine a design/system/process and identify discrepancies representing hazards.

Event Frequency

procedure

A3
Event Frequency Techniques

Fault Tree Analysis

Event Tree Analysis

The hazard identification techniques discussed in the previous section were primarily based upon constructive thinking, as well as the collective experience of a plant's personnel. A different approach would be to employ techniques which are based on strict logic and analysis. Such techniques include Fault Tree Analysis and Event Tree Analysis. Although both these techniques look similar from a mathematical point of view, the way they are structured makes them appear different. In this section we will examine both techniques. They are usually employed as a compliment or a continuation of the previously discussed techniques.

Two workers were fatally injured on August 28, 2008, when a waste tank containing the pesticide methomyl violently exploded, damaging a process unit at the Bayer CropScience chemical plant in Institute, WV, U.S.A. (Reproduced by kind permission of U.S. Chemical Safety Board.)

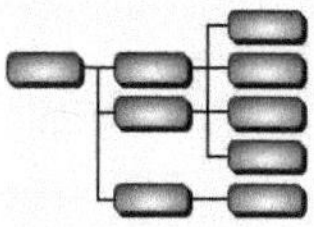

A3.1. Fault Tree Analysis (FTA)

This method systematically defines, with a logical tree, the exact sequence of primary and intermediate event failures likely to lead to a top event failure. The advantage of this method is that it recognizes situations that may lead to hazardous consequences if they are combined with other undesired events.

The essential components of a fault tree analysis, in simplified form, are the following:

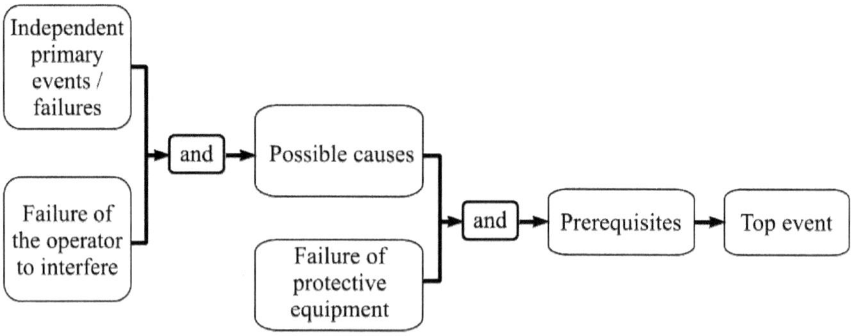

Figure A3.1. Fault tree analysis.

The fault tree technique can be described as an analytical technique, whereby an undesired state of the system is specified, and the system is then analyzed in the context of its environment and operation to find all credible ways in which the undesired event can occur.

Fault tree analysis is a deductive failure analysis which focuses on one particular undesired event and which provides a method for determining causes of this event. The undesired event constitutes the top event in a fault tree diagram constructed for the system, and generally consists of a complete, or catastrophic failure of the system under consideration. Careful formulation of the top event is important for the success of the analysis. The fault tree itself is a graphic model of the various parallel and sequential combinations of faults that will result in the occurrence of the predefined undesired event. The faults can be events that are associated with component hardware failures, human errors, or any other pertinent event which can contribute to the top event.

It is important to understand that a fault tree is not a model of all possible system failures or all possible causes of system failure. A fault tree is tailored to its top event, which corresponds to some particular system failure mode, and the fault tree thus includes only those faults that contribute to this top event. These are connected with "and" or "or" gates as seen in Figure A3.1. For more information the reader is referred to the corresponding literature.

EXAMPLE A3.1. *Fault Tree Analysis*

A flammable liquid storage system is shown in Figure A3.2. The tank T-1 has been designed to keep the liquid under slight positive nitrogen pressure. The pressure in the tank is controlled by PIC-1, which sends an alarm signal to the control room when the pressure exceeds a certain limit. Furthermore, there is a relief valve, RV-1 that opens to the atmosphere in case of emergency. The liquid is fed to the tank by tanker trucks. The pump P-1 sends the liquid to the production line when necessary.
Perform the fault tree analysis for the top event of the tank rupture due to overpressure.

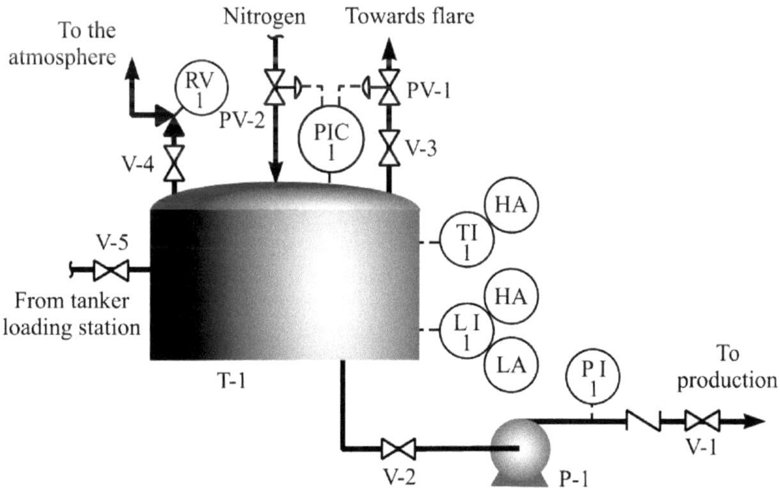

Figure A3.2. Flammable liquid storage system.

Symbol Interpretation

- P - Pressure
- T - Temperature
- L - Level
- I - Indicator
- C - Controller
- A - Aware
- HA - High Alarm...
- LA - Low Alarm ...

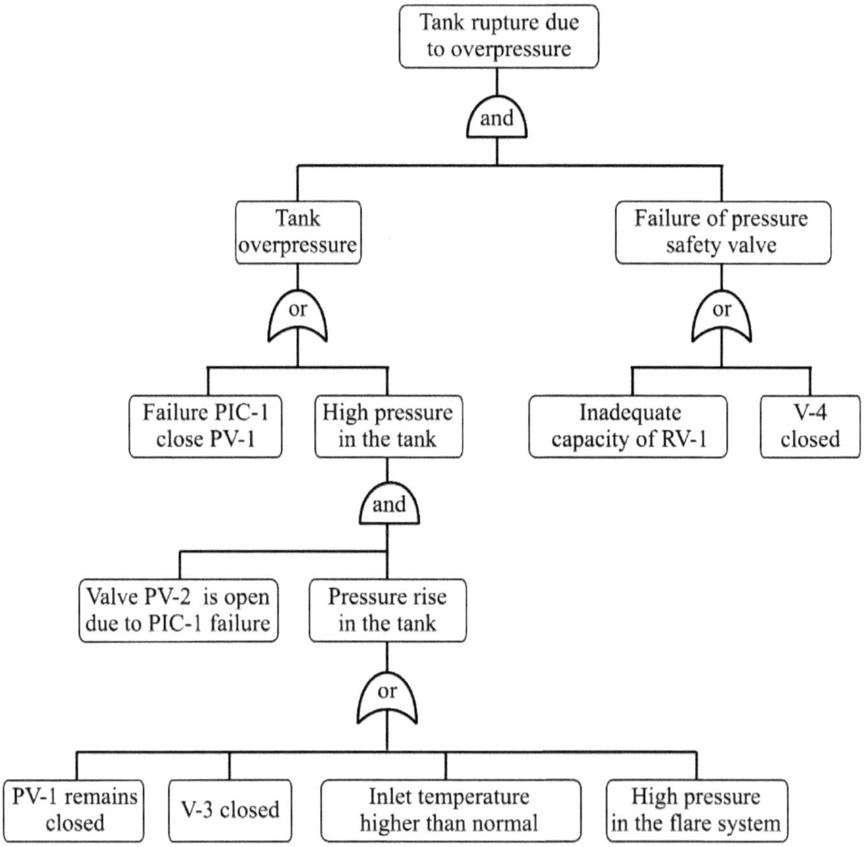

Figure A3.3. Fault tree analysis.

The fault tree analysis which can lead to the tank rupture because of overpressure is shown in Figure A3.3. Notice the use of "and" and "or."

As an example we note that tank rupture because of overpressure can occur if the pressure inside the tank is increased and at the same time the pressure safety valve fails. For the pressure safety valve to fail, RV-1 must be of inadequate capacity or valve V-4 must be closed.

❒

A3.2. Event Tree Analysis (ETA)

The event tree analysis, ETA, evaluates potential accident outcomes that might result following an equipment or process failure (initiating event). Unlike fault tree analysis, ETA records in a precise way the accident sequences, and defines the relationship between the initiating events and the subsequent events that combine to result in an accident.

The aim of an ETA is to examine the possible consequences of an initial undesired event, while fault tree analysis explains how an undesired event can be the result of other events.

The typical structure of an ETA is shown in Figure A3.4.

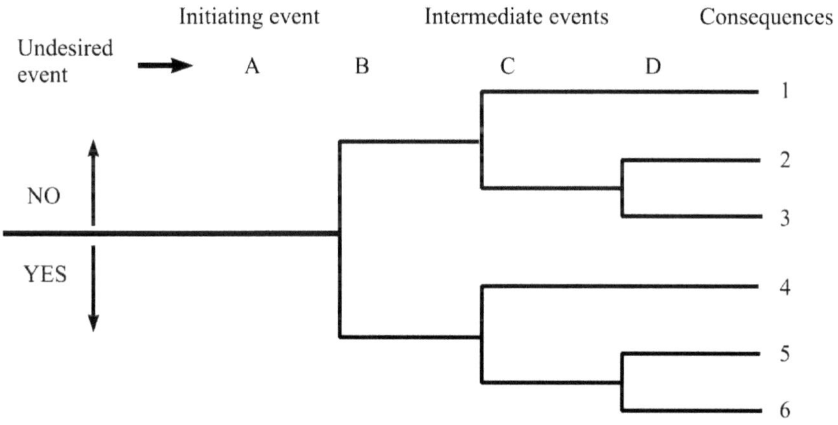

Figure A3.4. Event tree analysis.

The technique is universally applicable to systems of all kinds, with the limitation that unwanted events (as well as wanted events) must be anticipated to produce meaningful analytical results. The technique can be exhaustively thorough, with only two theoretical limits, i.e., the presumptions that:

- all system events have been anticipated, and
- all consequences of those events have been explored.

The technique is among the more difficult in practice. Successful application to complex systems cannot be undertaken without formal study over a period of time, combined with practical experience and diligence in recording of data.

EXAMPLE A3.2. *Event Tree Analysis*

Perform the ETA for the undesired event of a pipe rupture at point K, shown in Figure A3.5. If P_A is the pipe rupture probability, P_B the EFV failure probability, and P_C the RCV failure probability, calculate the probability of

a) Continuous leak.

b) Minor leak (drainage of the fluid in the line).

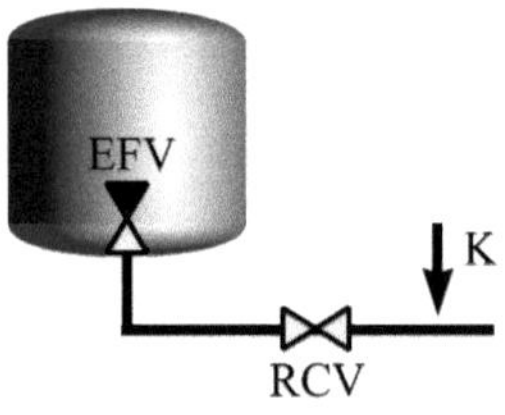

Symbol interpretation

EFV Excessive Flow Valve

RCV Remote Controller isolation Valve

Figure A3.5. Pipe connected to vessel.

Undesired event A : Pipe rupture

Intermediate event B : EFV failure

Intermediate event C : RCV failure

The event tree is shown in Figure A3.6.

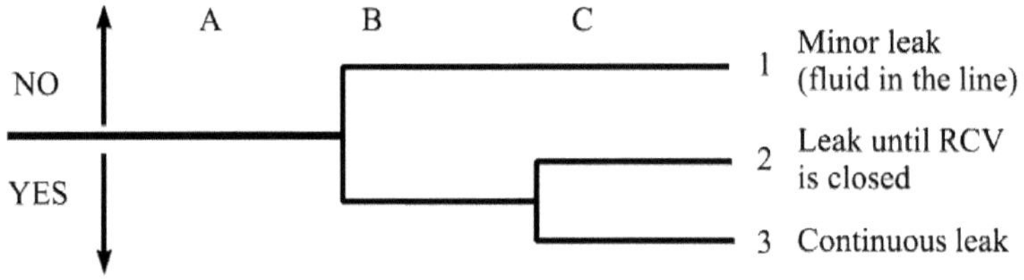

Figure A3.6. Event tree for Example A3.2.

Based on this event tree, we obtain

a) $P = P_A \times P_B \times P_C$.

b) $P = P_A \times (1 - P_B)$.

❒

Human Factor

procedure

Definitions
↓
Incidents at Petroleum Refineries
↓
Checklists

A4 Human Factor

The human factor deals with how effectively and safely personnel interact with processes. It can cover a variety of issues, from facility design, to procedures, to overall management systems. To minimize human factors as potential causes for incidents, the causal relationship of human factors to safety incidents must be understood. A look back into previous incidents and incident investigations can provide valuable insight into the types of human factors that most often contribute to human error and, ultimately, to loss events.

This section refers to a qualitative approach to the identification of the contribution of the human factor to failures or accidents. Until 1992, some industries included a quantitative evaluation of human errors in their risk analyses. In 1992, the U.S. Occupational Safety and Health Administration [OSHA 1992], with the Process Safety Management Rule 29 CFR 1910.119, proposed that a quantitative evaluation of human errors was no longer necessary. Since there is no further guidance from OSHA on how to include human factors in the required Process Hazard Analyses (PHAs), practitioners prefer a more qualitative identification of the human factor.

A4.1. Some Definitions

Historically, statistical analysis of the basic causes of industrial incidents show that for up to 50% of incidents, the prime cause is human error. Over recent years, this fact has led to the design of more automatic or computer-controlled systems. Unfortunately, this philosophy has not resulted in a large improvement in safety record of the Chemical and Petroleum Industry.

In the next sections, we first present some definitions and then focus our discussion on the influence of the human factor within the Petroleum Refinery Industry.

Human Factor

procedure

Definitions
↓
Incidents at Petroleum Refineries
↓
Checklists

A4.1.1. Human Factor

Human Factors may be defined as "influences on human behavior that may increase or decrease the likelihood of human error in a task."

For example if a task is to fill in your name and address on a form, then a human factor might be that the room you are in is dark, hence making it very difficult to do the task without error. It should be recognized that human factors might be "positive" in that they reduce the likelihood of a human error. That is, in this same example, a computer completes the form and all that is required is for the person to point and click his mouse on the screen image of the form.

A4.1.2. Human Performance

Human Performance refers to the study of personnel (management personnel, operations, maintenance, etc.) in their work environment. It specifically refers to the factors or behaviors that affect the interrelation between the employee and the company and can increase or decrease the probability of human errors.

A4.1.3. Human Error

Human Error is defined as "departure from acceptable or desirable practice on the part of an individual that can result in unacceptable or undesirable results." It is obvious that human errors are not only restricted to wrong operations, but can occur in any part of the hierarchy (management, operations, work permit issuing, etc.).

In literature two characteristic types of human errors usually appear:

a) Intentional human errors are errors deliberately commited or performed omitting prescribed actions. Usually these are committed by operators or maintenance personnel. They are characteristic of the specific situation and have a direct effect. Usually they are further separated into errors that take place after an activity, or errors that take place because of the lack of an activity.

b) Unintentional human errors are actions committed or omitted with no prior thought. Unintentional errors can be a result of decisions at any level of a company (management, sales, etc.) but also of decisions taken outside the company (external decisions). They can also easily lead to an incident.

A4.2. Incidents at Petroleum Refineries

Incidents at petroleum refineries are of special concern due to the potential catastrophic effects of unintentional releases of combustible and toxic substances used or generated in the refining industry. An increased awareness of serious incidents at U.S. refineries prompted the U.S. refining industry to place increased emphasis on safety over the last two decades. In spite of the attention paid to serious incidents, until recently there has been no systematic effort to collect and analyze data. One of the most interesting studies on the subject is the one performed by the Battelle Memorial Institute in 1999 [Chadwell *et al.* 1999] which will be discussed in the following pages.

A4.2.1. Data Collection

The Battelle Memorial Institute survey [Chadwell *et al.* 1999] focuses on incidents in petroleum refineries in the U.S.A. in the 1990s. Sources of information included the Internet, local newspapers, scientific journals, announcements of safety committees, etc. A total of 136 incidents were identified for inclusion in the database.

A4.2.2. Analysis Method

Incidents were categorized under the following five headings (see Table A4.1):

- Equipment failure (no human error).
- Random human error.
- Incidents where there was contribution of human factors, separated into the following three categories:
    - facility design (Environment, Controls, Equipment),
    - procedural (Response to Upset, Operation/Maintenance Safe Work),
    - management systems (Training, Communication, Scheduling).

Any number of different human factors could contribute to each incident. To ensure that any single incident did not unfairly bias the analysis by being counted more than once, a weighing system was introduced.

The results of this study showed that 47% of the causes identified involved elements of human error (53% equipment failure). From these, 81% are characterized by human factor contributions, while only 19% from random human error. All results are shown in Figure A4.1. The high percentage attributed to procedural (52%) is quite interesting, and so are some other subcategories (lighting, overtime, etc.). These results clearly indicate the importance of the human factor contribution to hazard identification.

Table A4.1. Cause-Contribution Factor Characterization Scheme [Chadwell *et al.* 1999].

Contribution Factor			Example Situation
FACILITY DESIGN	Equipment	Labeling	*Mislabeled, or not labeled at all*
		Access	*Hard to reach or access*
		Operability	*Difficult to operate/change position*
		Layout	*Confusing/inconsistent arrangement*
		Uniqueness	*Several components look alike*
	Controls	Labeling	*Mislabeled, or not labeled at all*
		Mode	*Manual operation; many manual steps*
		Involvement	*Operator detached from process*
		Displays	*Unclear/complex/non-representational*
		Feedback	*None, or potentially misleading*
	Environment	Noise level	*Area where hearing protection required*
		Climate	*Extremes in temperature, humidity, wind, etc.*
		Visibility	*Often foggy or other visibility limitations*
		Lighting	*Inadequate lighting for task*
PROCEDURAL	Ops/Maint Safe Work	Content	*Incomplete/too general/out of date*
		Identification	*Ambiguous device/action identification*
		Format	*Confusing/inconsistent; difficult to read*
		Aids	*Task sequence done by memory*
	Response to Upset	Alarms	*Many simultaneous or false alarms*
		Coverage	*Operator not always present*
		Time	*Inadequate time to respond*
		Preparedness	*No drills/simulation of scenarios*
		Last-Resort	*Shutdown discouraged or unsafe*
MANAGEMENT SYSTEMS	Scheduling	Overtime	*Extreme enough to affect performance*
		Consistency	*Inconsistent shift rotation schedules*
		No. of Tasks	*Tasks required exceed available time*
		Task Frequency	*Very infrequent; lack of experience*
		Intensity	*Differing tasks in rapid succession*
	Commu-nication	Shift Changes	*Inadequate communication between shifts*
		Field/Control	*No/poor communication of control & operators*
		Supervision	*Little or no supervisory checks*
		Emergency	*No distinction between alarms in areas or types*
	Training	Initial	*Little or no job specific training*
		Refresher	*Overdue or non-existent*
		Safety Awareness	*Little or no training on procedural changes*
RANDOM HUMAN ERROR			*Human error with no contributing human factors*
EQUIPMENT FAILURE (No Human Error)			*Equipment failure with no contributing human factors*

HUMAN FACTOR INVOLVEMENT 81%

(RANDOM HUMAN ERROR 19%)

- FACILITY DESIGN 9%
  - Equipment 28%
    - Labeling 75%
    - Access
    - Operability 25%
    - Layout
    - Uniqueness
  - Controls 45%
    - Labeling
    - Mode
    - Involvement
    - Displays 20%
    - Feedback 80%
  - Environment 27%
    - Noise Level
    - Climate
    - Visibility
    - Lighting 100%
- PROCEDURAL 52%
  - Operations/ Maintenance Safe Work 80%
    - Content 86%
    - Identification 9%
    - Format
    - Aids 5%
  - Response to Upset 14%
    - Alarms 23%
    - Coverage
    - Time
    - Preparedness 45%
    - Last resort 32%
- MANAGEMENT SYSTEMS 39%
  - Scheduling 5%
    - Overtime 100%
    - Consistency
    - Number of Tasks
    - Task Frequency
    - Intensity
  - Communi-cations 10%
    - Shift Changes
    - Field/control
    - Supervision
    - Emergency 100%
  - Training 85%
    - Initial 4%
    - Refresher 3%
    - Safety Awareness 93%

Figure A4.1. Categorization of human factor in refinery incidents.

A4.3. Human Factors Checklists

Based on the aforementioned petroleum refinery survey, a checklist that could be used to identify the effect of human factor, before or during a HAZOP analysis, is given in Table A4.2.

As an example we can examine the following case [Chadwell *et al.* 1999]. The HAZOP study team is considering the Guide Word "Other Than" relative to a procedure step that directs the operator to open a specific valve to add a chemical to a reactor. This operation takes place once per shift (about 1000 times per year). The valve is operated from the control room by selecting the valve on the monitor with the mouse, and clicking on the "open" icon. The team considers the possibility that the operator opens the wrong valve. The consequence of this error could be a runaway reaction, challenging the temperature rate of change shutdown logic. The human factors are a key to evaluating the likelihood of this error. The team identifies the following human factors:

- The procedures specify the valve by number and function, and the number is displayed clearly on the screen.
- There are four other valves in parallel to the correct valve, all displayed on the screen above and below the correct valve.
- On the night shift, there is one less operator, so the board operator must attend to additional duties not in the control room, such as taking and recording plant parameters. This increases the time pressure on the operator to complete his tasks.

The team decides that the human factors make the valve selection error more likely than an "average" situation, and hence to expect the error to occur "frequently." Based on the high consequence and high likelihood, the team identifies that human factors must be improved to reduce the likelihood of human error.

For a typical HAZOP study there may be one to several hundred such scenarios, and hence numerous evaluations of human factors. In order to make the identification of human factors both comprehensive and consistent, a human factor checklist was developed for the team to use as part of the HAZOP study. This checklist, shown in Table A4.2, can be used in several ways. For example, the team could use this to help identify human factors for each scenario during the team discussion. Another way to use the checklist is to review it with the team before the HAZOP study, and have the team consider the plant as a whole with respect to the human factors. Developing and using a checklist such as this helps and ensures that human factors are identified and considered in the hazard and risk assessments.

Human Factor

procedure

Definitions
↓
Incidents at Petroleum Refineries
↓
Checklists

Table A4.2. Human Factors Checklist.

EXAMPLE OF HUMAN FACTORS IN PROCESS OPERATIONS
(Management Elements such as Training and Management of Change, not included)
"+" : Factors that may tend to reduce the likelihood of human error
"-" : Factors that may tend to increase the likelihood of human error

CATEGORY		+	-	FACILITY-WIDE PRACTICE
EQUIPMENT	Labeling	*Clearly labeled, uniform coding*	*Mislabeled or not labeled*	
	Access	*Immediately at hand, special tools available when needed*	*Hard to reach or access*	
	Operability	*Power-assisted operation*	*Difficult to operate, change position*	
	Layout	*Well planned. logical arrangement*	*Confusing/ inconsistent arrangement*	
	Uniqueness	*Only component of its kind in area*	*Several components look similar*	
CONTROLS	Labeling	*Clearly labeled, uniform coding*	*Mislabeled or not labeled*	
	Mode	*Fully automatic, well tuned*	*Manual operation, many manual steps*	
	Involvement	*Operator continually involved*	*Operator detached from process*	
	Displays	*Clear, simple, logical arrangements*	*Unclear, complex, illogical*	
	Feedback	*Immediate, unambiguous*	*None or potentially misleading*	
DEVIATIONS	Alarms	*Safety-critical*	*Many simultaneous*	
	Coverage	*2 continuous operators always*	*Operator not always present*	
	Time	*No time pressure for response*	*Inadequate time to respond*	
	Preparedness	*Periodic simulation exercise*	*No drills/simulation of scenarios*	
	Last-Resort	*Shutdown not discouraged; fast response*	*Shutdown discouraged or unsafe*	
TRANSIENTS	Procedures	*Complete, accurate, current, verified*	*Incomplete, general, out of date*	
	Identifying	*ID, location of devices/ actions given*	*Ambiguous devices/ action identification*	
	Format	*Graphical identification aids, details*	*Confusing, inconsistent difficult to read*	
	Aids	*Checklists, checks*	*Tasks done by memory*	

- Continued

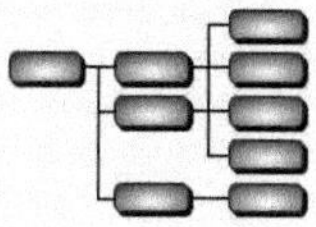

Human Factor

procedure

Definitions
↓
Incidents at Petroleum Refineries
↓
Checklists

Table A4.2 (cont.). Human Factors Checklist.

EXAMPLE OF HUMAN FACTORS IN PROCESS OPERATIONS
(Management Elements such as Training and Management of Change, not included)
"+" : Factors that may tend to reduce the likelihood of human error
"-" : Factors that may tend to increase the likelihood of human error

	CATEGORY	+	-	FACILITY-WIDE PRACTICE
SCHEDULING	Overtime	*Reasonable*	*Extreme enough to affect performance*	
	Consistency	*Permanent shift assignments*	*Inconsistent shift rotation/schedules*	
	No. of Tasks	*Tasks, workforce, and skills matched*	*Tasks required exceed time available*	
	Task Frequency	*Routine task*	*Very infrequent, no experience based*	
	Intensity	*Regular task at normal pace*	*Differing tasks in rapid succession*	
COMMUNICATION	Shift Changes	*Status communicated verbally, plus turnover sheet used*	*Inadequate communication between shifts of plant status*	
	Field/Control	*Constant communication with field*	*No communication with field operators*	
	Supervision	*Frequent supervisory communication*	*Little or no supervisory checks*	
	Emergency	*Rapid unambiguous plant alarm system*	*No distinction between area, type.*	
ENVIRONMENT	Noise Level	*Office environment noise level*	*Area where hearing protection required*	
	Climate	*Indoors, climate controlled*	*Temperature, humidity, precipitation, wind extremes*	
	Visibility	*Visibility enhancement of some kind*	*Often foggy or other visibility limitation*	
	Lighting	*Adequate lighting for tasks*	*Inadequate lighting for tasks*	
	Housekeeping	*Environment conducive to efficient performance*	*Area not kept clean / organized*	

B outflow

B1
Introduction

The subject of this chapter is the release, or better the incidental release of hazardous materials. It is obvious that this topic is much broader than just a chapter. Nevertheless, an attempt is made to present the basic concepts, mechanisms and algorithms for the calculation of some of the most characteristic cases of incidental releases which will help to understand overall issues of outflow.

At this point it should be noted that any mention of incidental release (gas or liquid) in this chapter will be treated under the general term of "outflow." Furthermore, the outflow of compressed gases or pressurized liquefied gases can lead to the creation of a cloud. It should be noted that depending on the prevailing conditions in each outflow case (continuous or instantaneous release), the cloud would be a plume or a puff (these will be described in Section C5 on Toxic Gas Dispersion). In the present chapter, the term "cloud" will be used to describe both cases, as here we examine the outflow incident and not its consequences.

The chapter is separated into three sections:

1) The first section refers to the outflow of compressed gases. Two cases are examined. In the case of the outflow of a gas out of a vessel, the outflow itself results in reduction of density and temperature, which themselves affect the outflow rate. Hence, the procedure is basically an iterative numerical algorithm in which the gas outflow from the vessel is described in small time steps, small enough for conditions in the vessel to be considered constant during each time step.

    The same algorithm is followed in the case of outflow from a hole in a pipe connected to a vessel, although in this case the calculations are a little more complicated because the outflow rate is dependent on the pressure in the vessel and on the pressure drop in the pipe.

    In the case of total pipe rupture, the calculation procedure adopted is based on the model of Bell [Hanna & Drivas 1987].

2) The second section refers to the outflow of pure liquid under pressure. Two cases are examined. In the case of liquid outflow from a vessel, the calculation depends on the difference between the pressure in the vessel and the atmospheric pressure. Thus we must account for the fact that the pressure in the vessel also decreases because of the liquid outflow. Hence, similarly to the gas case, the procedure is basically an iterative numerical procedure in which the outflow of liquid from the vessel is described in small time steps, small enough for conditions in the vessel to be considered constant during each time step.

   Moreover, in the case of outflow from a hole in a pipe connected to a vessel, a similar procedure is applied although, as already stated, the calculations are a little more complicated because of the dependence of the outflow rate on the pressure in the vessel and the pressure drop in the pipe.

3) The case of a two-phase mixture outflow constitutes the hardest case to simulate as it depends on many variables. In this section, a simplified presentation of the special case of the total rupture of a vessel containing pressurized liquefied gas in a temperature above its boiling point, is shown. The rupture of the vessel results in the sudden flash of a liquid content which will expand in all directions, forming a two-phase cloud of vapor and liquid droplets (as "aerosol"), until it is cooled to a temperature below its boiling point.

   The calculation algorithm is separated into three stages:

   - In the first stage, the liquid flashes and expands in all directions without being mixed with the atmospheric air, forming a mixture of vapors and liquid droplets, some of which will precipitate to the ground.
   - In the second stage, there is entrainment of the atmospheric air in the two-phase cloud of vapor and liquid droplets, resulting in additional mixing and further evaporation of the liquid droplets.
   - In the third stage, the liquid droplets evaporate, and the homogeneous gas cloud disperses in the atmosphere.

   This particular section concentrates on the first two stages, while for the third stage a dispersion model, as those described in Section C5, can be employed.

After each case, simple examples are given, in order to demonstrate the use of the algorithms and the procedures employed.

B2
Outflow of Compressed Gases

This section presents an initial study of the outflow of gas under pressure. We will first examine two cases of small releases (hole in vessel wall and hole in a pipe connected to a vessel, respectively), and then discuss the outflow resulting from a total pipe rupture.

B2.1. Gas Density

In the case of an outflow or release of gases under pressure, the knowledge of the density as a function of the pressure and the temperature is very important, because of their compressibility. For low pressures, the ideal-gas equation is valid. Hence

$$P = \rho_n R\,T \qquad \text{or} \qquad P = (\rho / W_{\text{g}})\, R\,T\,, \tag{B2.1}$$

where, P (Pa) is the pressure of a gas with molar density ρ_n (mol/m^3) in this pressure and temperature T (K). In the above equation, R denotes the universal gas constant (= 8.314 Jmol^{-1}K^{-1}). Usually, instead of the molar density it is customary to employ the mass density ρ (kg/m^3) and the molecular weight W_{g} (kg/mol) of the gas. The above two expressions are widely employed because of their accuracy and ease of use.

For higher pressures, the ideal-gas equation can also be used, but with the addition of the compressibility factor Z (-), as

$$P = Z\,\rho_n R\,T \qquad \text{or} \qquad P = Z\,(\rho / W_{\text{g}})\, R\,T\,. \tag{B2.2}$$

The compressibility factor can be calculated as a function of the critical temperature and pressure [Assael, Trusler & Tsolakis 1996].

When the pressure is considerably increased, then the Virial Equation, is recommended:

$$P = \rho_n R\,T\left(1 + B\rho_n + C\rho_n^2 + \ldots\right). \tag{B2.3}$$

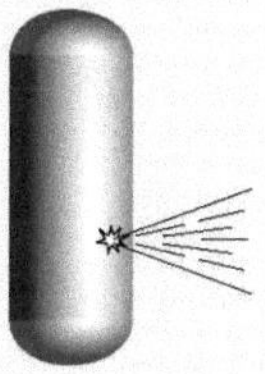

Density

Much of the importance of the virial equation of state lies in its rigorous theoretical foundation by which the virial coefficients appear not merely as empirical constants but with a precise relation to the intermolecular potential energy. Specifically, the second virial coefficient, B (m^3/mol), arises from the interaction between a pair of molecules, the third virial coefficient, C (m^6/mol^2), depends upon interactions in a cluster of three molecules, and so on. Most calculations employ either the coefficients B and C, or only B.

Although there are many experimental data for the second virial coefficient, it is often necessary to estimate its value for gases that have not been studied before. Hence many different correlations have been proposed [Assael, Trusler & Tsolakis 1996]. One of the most widely used correlations for non-polar gases is the extended scheme of corresponding states of Pitzer-Curl [Pitzer & Curl 1958], given by Eq. (B2.4) in Table B2.1. The parameter ω shown in the equation, is the acentric factor while T^c (K) and P^c (Pa) are the critical temperature and critical pressure, respectively. Pitzer and Curl initially proposed their own expressions for the dimensionless coefficients B_0 and B_1. Here, the most recent correlations (Eq. (B2.5)-(B2.6) in Table B2.1) as a function of the reduced temperature, $T_r = T/T^c$, proposed by Tsonopoulos [Tsonopoulos 1974] are shown. For the third virial coefficient for non-polar gases, in Table B2.1, the correlations of Orbey and Vera [Orbey & Vera 1983] are shown. In the case of a mixture, mixing rules of critical points are most commonly employed [Assael, Trusler & Tsolakis 1996]. In much higher pressures, the use of equations of state like Peng-Robinson or Benedict-Webb-Rubin is unavoidable [Assael, Trusler & Tsolakis 1996].

Table B2.1. Correlations for the Second and Third Virial Coefficients.

$$B = \left(RT^c / P^c\right)\left(B_0 + \omega B_1\right) \tag{B2.4}$$

$$B_0 = 0.1445 - 0.33/T_r - 0.1385/T_r^2 - 0.0121/T_r^3 - 0.000607/T_r^8 \tag{B2.5}$$

$$B_1 = 0.0637 + 0.331/T_r^2 - 0.423/T_r^3 - 0.008/T_r^8 \tag{B2.6}$$

$$C = \left(RT^c / P^c\right)^2 \left(C_0 + \omega C_1\right) \tag{B2.7}$$

$$C_0 = 0.01407 + 0.02432/T_r^{2.8} - 0.00313/T_r^{10.5} \tag{B2.8}$$

$$C_1 = -0.02676 + 0.0177/T_r^{2.8} + 0.040/T_r^3 - 0.003/T_r^6 - 0.00228/T_r^{10.5} \tag{B2.9}$$

Outflow of Gases from Vessel

calculation procedure

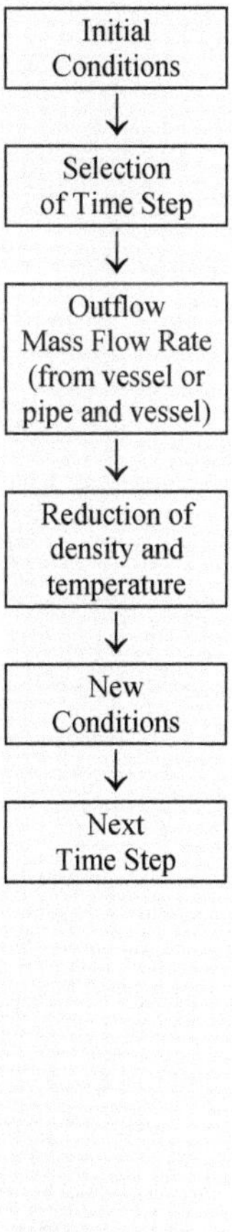

B2.2. Outflow from Vessels

B2.2.1. Small Outflow

The simulation of the dynamic behavior of a compressed gas in a vessel aims to estimate the reduction in its pressure and temperature due to gas outflow. As a part of the gas is released or leaks, the remaining gas will expand. This expansion generates cooling and decompression.

Usually this simulation takes place with an iterative numerical procedure, where the gas outflow is described in small steps in time. These steps should be small enough to consider the conditions in the vessel to be constant during one time-step. A possible algorithm will be as follows:

1) Data: vessel's volume V (m^3), initial vessel's pressure P_o (Pa) and initial temperature T_o (K).
   - From an equation of state, the density ρ_o (kg/m^3) is calculated.
   - Then the outflow mass flow rate $\dot{m}_o$ (kg/s) is obtained (see next sections).

2) First time step δt_1 (s) is selected (either by experience or trial and error).
   - The reduction in density (because of the mass that was released) is obtained from the equation

$$\delta\rho_1 = -\frac{\dot{m}_o}{V}\,\delta t_1 \; . \qquad \text{(B2.10)}$$

   - Then the reduction in temperature, because of the gas expansion, is calculated from the expression*

$$\delta T_1 = \frac{P_o}{\rho_o^2\, C_v}\,\delta\rho_1 \; , \qquad \text{(B2.11)}$$

   where C_v (Jkg^{-1}K^{-1}) is the specific heat capacity under constant volume.

3) New conditions are thus calculated from

$$\rho_1 = \rho_o + \delta\rho_1 \, , \quad T_1 = T_o + \delta T_1 \quad \text{and} \quad P_1 = R\,T_1\,(\rho_1/W_g). \qquad \text{(B2.12)}$$

   These conditions will be the initial conditions of the next time step. The algorithm continues until the vessel's pressure is equal to the ambient.

The algorithm is clearly demonstrated in the examples which follow immediately after the calculation of the outflow mass flow rate.

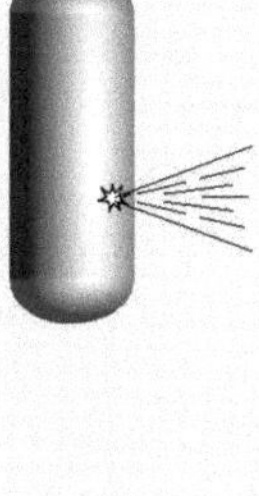

* The expression was derived from the change in internal energy δU (J/kg), assuming
- reversible adiabatic outflow and thus $\delta U = -P\,\delta V$,
- neglecting internal pressure (ideal gas) and thus $\delta U = C_v\,\delta T$,
- from the definition of density per unit mass in the vessel $\rho = 1/V$ and $\delta V = -\delta\rho/\rho^2$.

Outflow of Gases from Vessel

calculation procedure

Initial Conditions
↓
Selection of Time Step
↓
Outflow Mass Flow Rate (from vessel or pipe and vessel)
↓
Reduction of density and temperature
↓
New Conditions
↓
Next Time Step

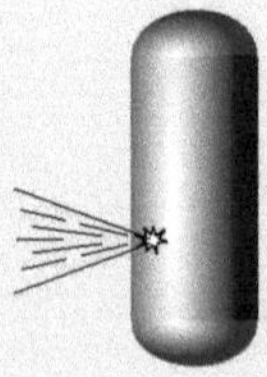

a) Outflow through Hole in Vessel's Wall

The calculation of the outflow mass flow rate, $\dot{m}$ (kg/s), of a gas under pressure through a hole in a vessel, Figure B2.1, can be accomplished with the following expression

$$\dot{m} = C_d A_h P_o K \sqrt{\frac{W_g}{\gamma R T}} \, . \qquad \text{(B2.13)}$$

In this relation, C_d (-) is the discharge coefficient, A_h (m^2) the hole's cross-sectional area, P_o (Pa) the initial gas pressure in the vessel (for each time step), and W_g (kg/mol) the molecular weight of the gas. Also, γ (-) denotes the Poisson ratio, i.e., the ratio of specific heat capacity at constant pressure, C_p ($Jkg^{-1}K^{-1}$), over the specific heat capacity at constant volume C_v ($Jkg^{-1}K^{-1}$), while R is the universal gas constant (= $(C_p\text{-}C_v)\, W_g$ = 8.314 $Jmol^{-1}K^{-1}$), and T (K) the temperature of the gas.

The coefficient K (-) depends upon whether the outflow is

a) subsonic (unchoked) : the gas exit velocity is smaller than the speed of sound. That is, the Mach number, $M_j < 1$.

b) sonic or supersonic (chocked): the gas exit velocity is equal or larger than the speed of sound. That is, the Mach number, $M_j \geq 1$.

The following expression can be considered as a criterion, as it declares that a flow can be considered as supersonic (or sonic) flow when

$$\frac{P_o}{P_a} \geq \left(\frac{\gamma+1}{2}\right)^{\frac{\gamma}{\gamma-1}} , \qquad \text{(B2.14)}$$

where P_a (Pa) is the ambient pressure.

The coefficient K (-) is calculated from the following expressions

for subsonic flow

$$K = \sqrt{\frac{2\gamma^2}{\gamma-1}\left(\frac{P_a}{P_o}\right)^{\frac{2}{\gamma}}\left[1-\left(\frac{P_a}{P_o}\right)^{\frac{\gamma-1}{\gamma}}\right]} , \qquad \text{(B2.15)}$$

for supersonic (or sonic) flow

$$K = \gamma \left(\frac{2}{\gamma+1}\right)^{\frac{\gamma+1}{2(\gamma-1)}} . \qquad \text{(B2.16)}$$

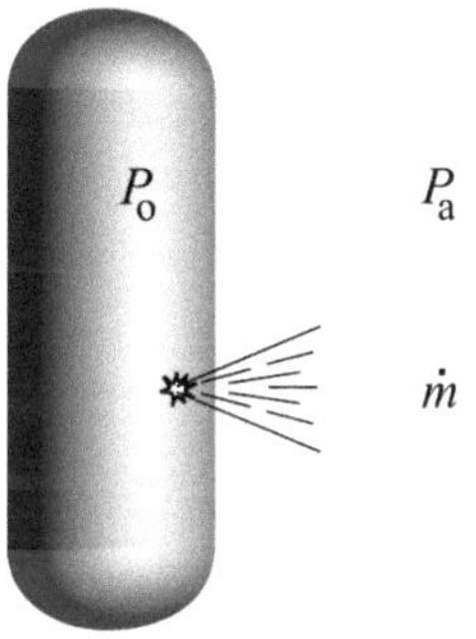

Figure B2.1. Outflow of compressed gas from a hole in a vessel.

The discharge coefficient, C_d (-) is a function of the type of the hole/orifice in the vessel and the gas velocity at the hole. For sharp orifices the discharge coefficient usually is taken equal to 0.62, while for rounded orifices it usually takes a value between 0.95 and 0.99 [Beek & Mutzall 1975].

The outflow mass flow rate calculated by Eq. (B2.13), as well as the equivalent mass flow rate from a pipe (that will be discussed in the next section), refer to initial pressure, P_o (Pa), in the vessel which remains stable during a time step. Hence, in essence it is valid for the time step in which the vessel's pressure P_o can be considered stable.

The above discussion refers solely to cases where there is no condensation of the exit gas, that is the gas pressure will not, under any means, be higher than the saturation pressure at the temperature of the process.

Propane fire erupted as a result of a ruptured one-ton chlorine container on February 16, 2007, at the Valero McKee Refinery in Sunray, TX, U.S.A. (Reproduced by kind permission of the U.S. Chemical Safety Board.)

Outflow of Gases from Vessel

calculation procedure

Initial Conditions

↓

Selection of Time Step

↓

Outflow Mass Flow Rate (from vessel or pipe and vessel)

↓

Reduction of density and temperature

↓

New Conditions

↓

Next Time Step

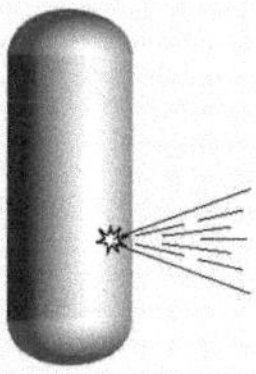

Outflow of Gases from Vessel

example

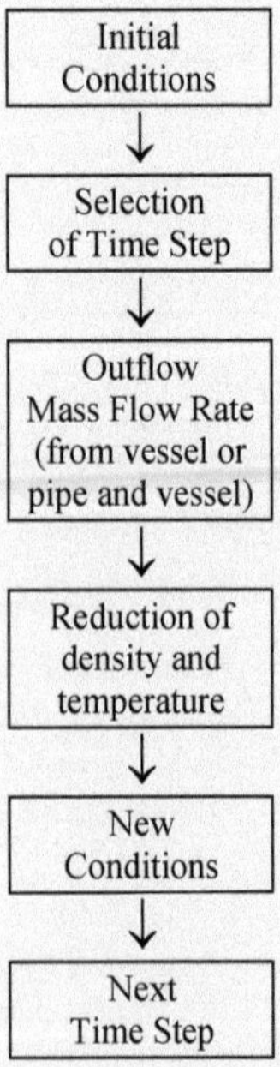

EXAMPLE B2.1. *Hole in Vessel's Wall*

Calculate the outflow mass flow rate of compressed hydrogen release from a hole in a vessel wall. The following data are available:

-	Vessel's volume, V	:	50	m^3
	Initial vessel's pressure, P_o	:	5	MPa
	Initial vessel's temperature, T_o	:	288.15	K
-	Hole's diameter, d_h	:	0.1	m
	Discharge coefficient, C_d	:	0.62	-
-	Molecular weight of Hydrogen, W_g	:	0.002	kg/mol
	Specific heat at constant volume, C_v	:	10.24	$kJkg^{-1}K^{-1}$
	Poisson ratio, γ	:	1.4	-

We follow the steps described in Section B2.2.1.

1) Data: vessel's volume V (m^3), initial pressure P_o (Pa), temperature T_o (K).
   - The density is calculated from the ideal-gas equation,

$$\rho_o = PW_g/(RT) = 5\times10^6 \times 0.002/(8.314\times288.15) = 4.17\ \text{kg/m}^3.$$

   - From Eq. (B2.13) the initial outflow mass flow rate $\dot{m}_o$ (kg/s) is obtained. The flow is supersonic ($P_o/P_a = 50 > 1.9 = ((\gamma+1)/2)^{\gamma/(\gamma-1)}$) and $K = 0.81$, independent of the pressure. From Eq. (B2.13), $\dot{m}_o = 15.23$ kg/s.

2) The first time step δt_1 (s) = 1 s, is selected.
   - The reduction in density and temperature, because of the released mass, is calculated from Eqs. (B2.10) and (B2.11), respectively

$$\delta\rho_1 = -\frac{\dot{m}_o}{V}\delta t_1 = -0.30\ \text{kg/s}$$

$$\delta T_1 = \frac{P_o}{\rho_o^2 C_v}\delta\rho_1 = -8.54\ \text{K}$$

3) New conditions are calculated from Eq. (B2.12), as

$\rho_1 = \rho_o + \delta\rho_1 = 3.87$ kg/m^3,

$T_1 = T_o + \delta T_1 = 279.6$ K,

and $P_1 = RT_1(\rho_1/W_g) = 4.50$ MPa.

These conditions will be the new initial conditions of the immediate next time step.

The results for the next 30 time steps (30 s) are shown in Table B2.2, while Figure B2.2 shows the reduction in temperature, pressure and outflow mass flow rate, respectively. In Table B2.2 the change in the mass M (kg) remaining in the vessel is also shown. Iterations stop when the vessel's pressure becomes equal to the ambient pressure, or the outflow mass flow rate drops below a specified value.

Finally, the reader can check that even if the selection of the time step was essentially arbitrary, the results are not significantly influenced by it.

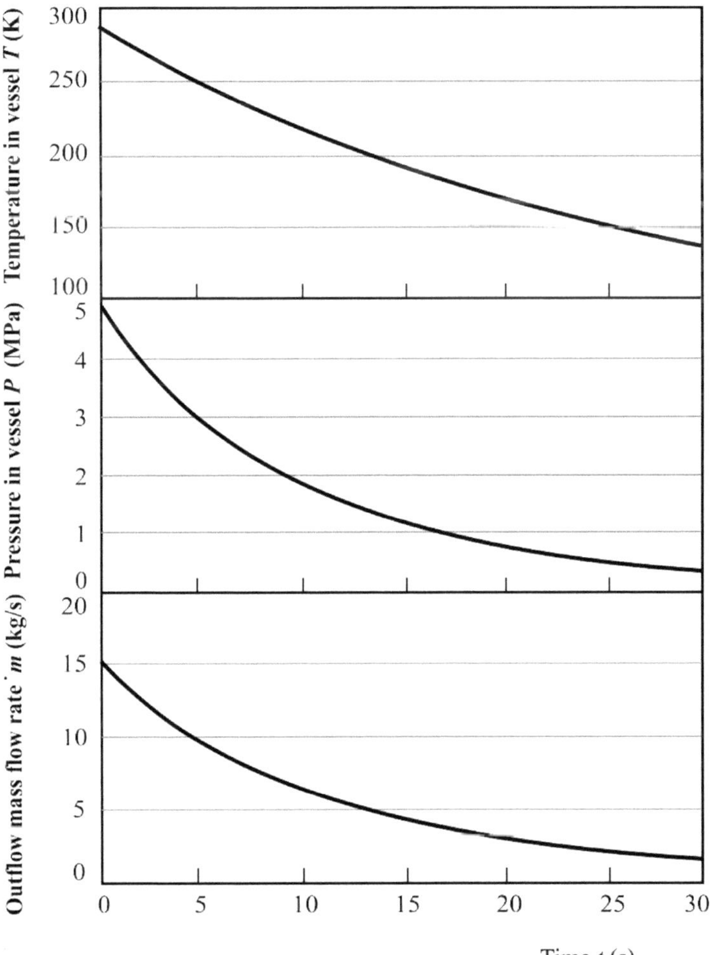

Figure B2.2. Reduction in temperature, pressure and outflow mass flow rate as a function of time.

Outflow of Gases from Vessel

example

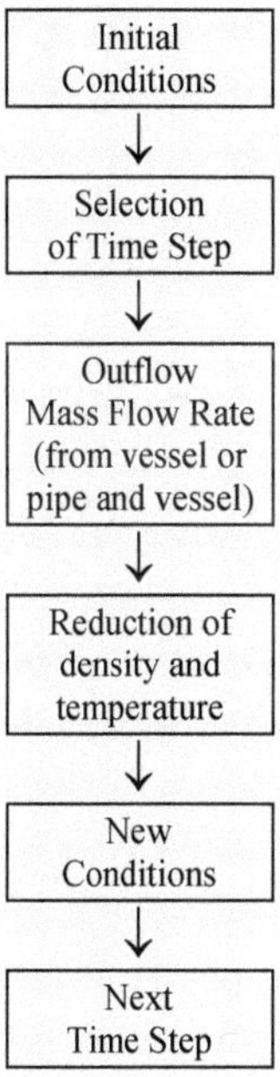

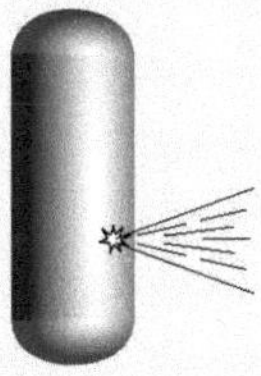

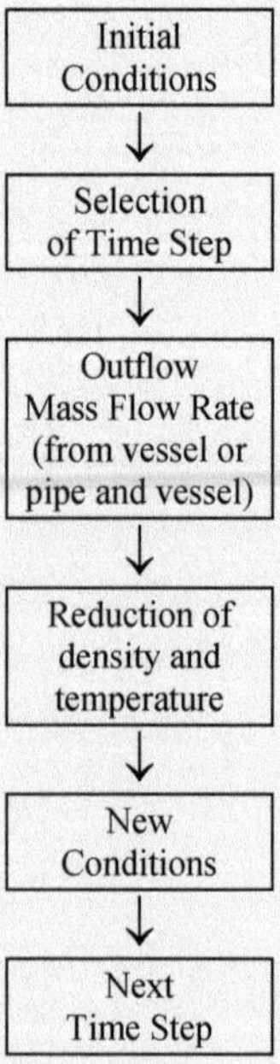

Table B2.2. Calculation of the Reduction of Mass.

M (kg)	t (s)	T (K)	ρ (kg/m^3)	P (MPa)	P_0/P_a (-)	$\dot{m}$ (kg/s)	$\delta\rho$ (kg/m^3)	δT (K)
209	0	288.2	4.17	5.00	50.0	15.23	-0.30	-8.54
193	1	279.6	3.87	4.50	45.0	13.91	-0.28	-8.16
180	2	271.5	3.59	4.05	40.5	12.72	-0.25	-7.80
167	3	263.6	3.34	3.66	36.6	11.65	-0.23	-7.47
155	4	256.2	3.10	3.31	33.1	10.68	-0.21	-7.16
145	5	249.0	2.89	2.99	29.9	9.80	-0.20	-6.86
135	6	242.2	2.69	2.71	27.1	9.01	-0.18	-6.58
126	7	235.6	2.51	2.46	24.6	8.29	-0.17	-6.31
117	8	229.3	2.35	2.24	22.4	7.64	-0.15	-6.06
110	9	223.2	2.20	2.04	20.4	7.05	-0.14	-5.82
103	10	217.4	2.05	1.86	18.6	6.51	-0.13	-5.59
96	11	211.8	1.92	1.69	16.9	6.02	-0.12	-5.38
90	12	206.4	1.80	1.55	15.5	5.57	-0.11	-5.18
85	13	201.3	1.69	1.42	14.2	5.16	-0.10	-4.98
79	14	196.3	1.59	1.30	13.0	4.79	-0.10	-4.80
75	15	191.5	1.49	1.19	11.9	4.44	-0.09	-4.62
70	16	186.8	1.40	1.09	10.9	4.13	-0.08	-4.46
66	17	182.4	1.32	1.00	10.0	3.84	-0.08	-4.30
62	18	178.1	1.25	0.92	9.2	3.57	-0.07	-4.15
59	19	173.9	1.17	0.85	8.5	3.33	-0.07	-4.00
55	20	169.9	1.11	0.78	7.8	3.10	-0.06	-3.87
52	21	166.1	1.05	0.72	7.2	2.90	-0.06	-3.73
49	22	162.3	0.99	0.67	6.7	2.70	-0.05	-3.61
47	23	158.7	0.93	0.62	6.2	2.53	-0.05	-3.49
44	24	155.2	0.88	0.57	5.7	2.36	-0.05	-3.38
42	25	151.9	0.84	0.53	5.3	2.21	-0.04	-3.27
40	26	148.6	0.79	0.49	4.9	2.07	-0.04	-3.16
37	27	145.4	0.75	0.45	4.5	1.94	-0.04	-3.06
36	28	142.4	0.71	0.42	4.2	1.82	-0.04	-2.96
34	29	139.4	0.67	0.39	3.9	1.71	-0.03	-2.87
32	30	136.5	0.64	0.36	3.6	1.61	-0.03	-2.78

❐

Initial Conditions

↓

Selection of Time Step

↓

Outflow Mass Flow Rate (from vessel or pipe and vessel)

↓

Reduction of density and temperature

↓

New Conditions

↓

Next Time Step

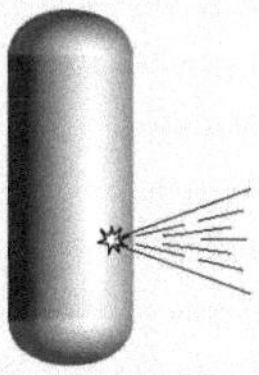

b) Outflow through Hole in Pipe Wall Connected to Vessel

In this section we refer to the specific outflow from a hole in a pipe connected to a vessel (see Figure B2.3). The outflow mass rate, $\dot{m}$ (kg/s) of the gas will directly depend upon the total difference between the vessel's pressure and the atmospheric pressure. More analytically, if P_o (Pa) denotes the pressure in the vessel and in the entrance of the pipe, P_e (Pa) the pressure before the hole and P_a (Pa) the ambient pressure, then the total pressure drop ΔP (Pa) will be equal to

$$\begin{aligned} \Delta P &= (P_o - P_e) + (P_e - P_a) \\ &= \Delta P_{pipe} + \Delta P_{hole} \end{aligned} \qquad \text{(B2.17)}$$

It should be noted that the pressure P_e (Pa) just before the hole, is not known.

The mass flow rate in the pipe, $\dot{m}_{pipe}$ (kg/s) directly depends on the pressure drop in the pipe, while the outflow mass flow rate, $\dot{m}$ (kg/s), depends on the pressure drop in the hole. Assuming that the mass through the pipe is equal to the mass that is released, we obtain that

$$\dot{m}_{pipe}(\Delta P_{pipe}) = \dot{m}(\Delta P_{hole}). \qquad \text{(B2.18)}$$

It is assumed that due to the lower pressure drop in the hole, the gas will be completely released. The above equation, in connection with Eq. (B2.13) - where, as the pressure before the hole, the pressure P_e, will be assumed - will be employed for the calculation of the unknown pressure, P_e (Pa), and consequently of the outflow mass flow rate, $\dot{m}$ (kg/ s), from the hole.

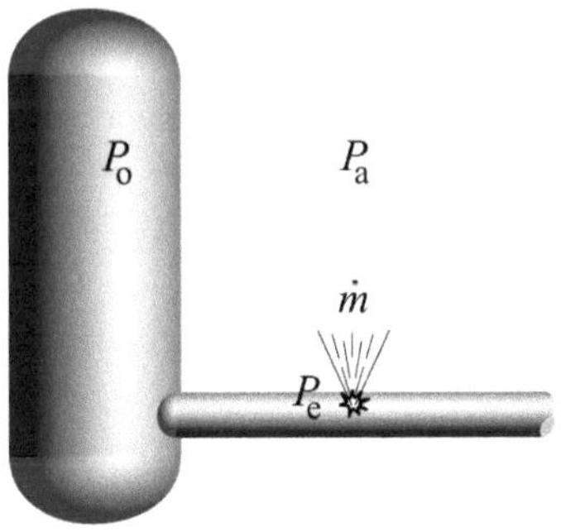

Figure B2.3. Outflow of compressed gas from pipe connected to vessel.

Initial Conditions
↓
Selection of Time Step
↓
Outflow Mass Flow Rate (from vessel or pipe and vessel)
↓
Reduction of density and temperature
↓
New Conditions
↓
Next Time Step

The mass flow rate, $\dot{m}_{\text{pipe}}$ (kg/s) inside the pipe can be obtained from the Fanning relation that expresses the pressure drop in a unit length pipe, $\delta P/\delta l_{\text{p}}$ (Pa/m), as a function of the Fanning friction factor, f_{F} (-), the gas density, ρ (kg/m^3), its velocity u (m/s), and the pipe's diameter d_{p} (m), as

$$\frac{\delta P}{\delta l_{\text{p}}} = 4 f_{\text{F}} \frac{\rho u^2}{2 d_{\text{p}}}. \tag{B2.19}$$

The velocity of the gas can also be expressed as

$$u = \frac{\dot{m}_{\text{pipe}}}{\rho A_{\text{p}}}, \tag{B2.20}$$

where A_{p} (m^2) is the pipe's cross-sectional area.

Integrating over the pipe's length, from Eqs. (B2.19) and (B2.20) one obtains

$$\dot{m}_{\text{pipe}} = A_{\text{h}} \sqrt{\frac{2 \int_{P_{\text{e}}}^{P_{\text{o}}} \rho(P)\, dP}{4 f_{\text{F}} (l_{\text{p}} / d_{\text{p}})}}. \tag{B2.21}$$

In the case of ideal gases ($P=C\rho^{\gamma}$, C=constant), there is an analytic solution of the above integral, and thus Eq. (B2.21) becomes

$$\dot{m}_{\sigma\omega\lambda} = A_{\text{h}} \sqrt{\frac{\rho P_{\text{o}}}{2 f_{\text{F}} (l_{\text{p}} / d_{\text{p}})} \frac{\gamma}{1+\gamma} \left(\left(\frac{P_{\text{o}}}{P_{\text{e}}} \right)^{(1+\gamma)/\gamma} - 1 \right)}. \tag{B2.22}$$

The solution of the above equation together with Eq. (B2.13) and Eq. (B2.18) will produce the value for the unknown pressure, P_{e} (Pa), and consequently the outflow mass flow rate, $\dot{m}$ (kg/s), from the hole.

The Fanning friction factor, f_{F} (-), can be calculated [Fanning 1877, Pope 2000] as a function of the Reynolds number, $Re=\rho u d_{\text{p}}/\eta$, where η (Pa s) is the fluid kinematic viscosity, as

for $Re < 2{,}000$ $$f_{\text{F}} = 16 / Re, \tag{B2.23}$$

for $4{,}000 < Re < 10^5$ $$f_{\text{F}} = 0.0791\, Re^{-0.25}. \tag{B2.24}$$

It is worth reminding that the Fanning friction factor is a fourth of the Darcy-Weisbach friction factor.

EXAMPLE B2.2. *Hole in Pipe Connected to Vessel*

Outflow of Gases from Pipe

example

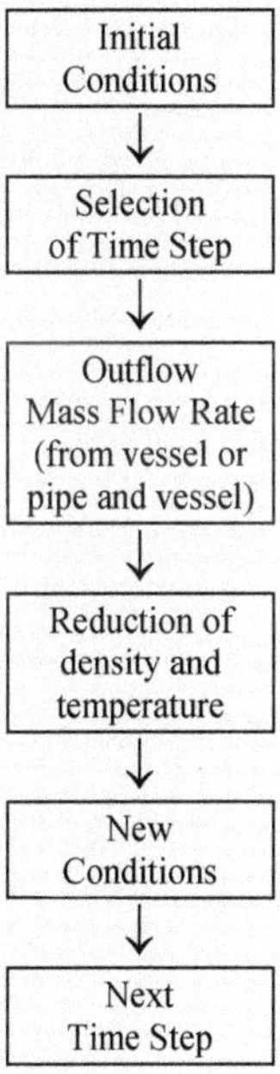

Calculate the outflow mass flow rate of compressed carbon monoxide, from a hole in a pipe connected to a vessel. The following data are available:

-	Vessel's Volume, V	:	50	m^3
	Initial vessel's pressure, P_o	:	1.5	MPa
	Initial vessel's temperature, T_o	:	288.15	K
-	Diameter of pipe, d_p	:	0.15	m
	Length of pipe (until the hole), l_p	:	100	m
-	Diameter of hole, d_h	:	0.1	m
	Discharge coefficient, C_d	:	0.62	-
-	Molecular weight of carbon monoxide, W_g	:	0.028	kg/mol
	Specific heat at constant volume, C_v	:	745	$J kg^{-1} K^{-1}$
	Poisson ratio, γ	:	1.4	-
	Kinematic viscosity, η	:	17.3	μPa s

The steps described in Section B2.2.1, (b), are followed

1) Data: volume of vessel+pipe $V\,(m^3) = 50 + (\pi/4)\times 0.15^2 \times 100 = 51.77\ m^3$, initial vessel's pressure P_o (Pa), and initial temperature T_o (K).
   - From the ideal-gas equation, the density is calculated,

   $$\rho_0 = PW_g/(RT) = 1.5\times10^6 \times 0.028/(8.314\times288.15) = 17.53\ \text{kg/m}^3.$$

   a) A value for the intermediate pressure P_e = 1.4 MPa, is assumed.

   b) The outflow mass flow rate, $\dot{m}$ (kg/s) is calculated from Eq. (B2.13). The flow is supersonic (P_e/P_a =14 > 1.9 = $((\gamma+1)/2)^{\gamma/(\gamma-1)}$) and K = 0.81, independent of pressure. From Eq. (B2.13), $\dot{m}$ = 15.96 kg/s.

   c) The mass flow rate, $\dot{m}_{pipe}$ (kg/s) in the pipe is calculated from Eq. (B2.22). The Fanning friction factor is obtained from Eq. (B2.24) equal to 0.0015. To calculate the velocity that is employed in the Reynolds number Re, the outflow mass flow rate obtained in step (b) is used. From Eq. (B2.22), $\dot{m}_{\sigma\omega\lambda}$ = 17.37 kg/s.

   d) If the intermediate pressure assumed in step (a) is correct, then the outflow mass flow rate in step (b) should be equal to the mass flow rate obtained in step (c).

   The steps are repeated with the new assumption for the intermediate pressure. Final results, P_e = 1.413 MPa and $\dot{m}$ = 16.10 kg/s.

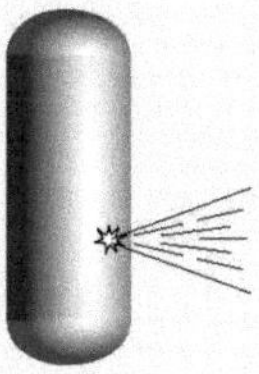

Outflow of Gases from Pipe

example

Initial Conditions
↓
Selection of Time Step
↓
Outflow Mass Flow Rate (from vessel or pipe and vessel)
↓
Reduction of density and temperature
↓
New Conditions
↓
Next Time Step

2) The first time step, δt_1 (s) = 2 s, is arbitrarily selected.
   - The reduction in density and temperature because of the released mass, is calculated from Eqs. (B2.10) and (B2.11), respectively

$$\delta\rho_1 = -\frac{\dot{m}_o}{V}\delta t_1 = -0.62.\ \text{kg/m}^3,$$

$$\delta T_1 = \frac{P_o}{\rho_o^2 C_v}\delta\rho_1 = -4.08\ \text{K}.$$

3) The new conditions are thus calculated from Eq. (B2.12), as

$\rho_1 = \rho_o + \delta\rho_1 = 16.91$ kg/m^3,

$T_1 = T_o + \delta T_1 = 284.07$ K

and $P_1 = R\,T_1\,(\rho_1/W_g) = 1.43$ MPa.

These conditions constitute the initial conditions of the immediate next time iteration.

The results for the next seven time steps (20 s), as well as, the change in mass M (kg) that remains in the vessel, are shown in Table B2.3.

Table B2.3. The Reduction of Mass with Time.

M (kg)	t (s)	T (K)	ρ (kg/m^3)	P (MPa)	P_e (MPa)	f_F x1000	$\dot{m}$ (kg/s)	$\delta\rho$ (kg/m^3)	δT (K)
908	0	288.15	17.53	1.50	1.4131	1.49	16.10	-0.62	-4.08
875	2	284.07	16.91	1.43	1.3464	1.51	15.45	-0.60	-4.00
844	4	280.08	16.31	1.36	1.2761	1.50	15.68	-0.61	-4.15
813	6	275.93	15.71	1.29	1.2134	1.52	14.99	-0.58	-4.05
783	8	271.88	15.13	1.22	1.1541	1.54	14.33	-0.55	-3.97
754	10	267.91	14.57	1.16	1.0980	1.55	13.70	-1.32	-9.70
686	15	258.21	13.25	1.02	0.9671	1.60	12.23	-1.18	-9.18
	20								

❐

B2.2.2. Total Vessel Rupture

Total
Vessel
Rupture

Introduction

The total rupture of a vessel with compressed gas, irrespective of the cause, usually has the following three consequences:

- release of the contained gas,
- rupture of the vessel with possible ejection of fragments,
- creation of a shock wave due to expansion of the compressed gas.

The release of the contained gas can result in secondary effects such as the creation of a fire ball (Section C2.2), flash fire (Section C2.4), vapor cloud explosion (Section C3) or dispersion of toxic cloud (Section C5). The appearance (or not) of these secondary effects depends entirely upon the flammability limits and the toxicity of the contained gas. All the above mentioned effects will be discussed in the corresponding sections.

Part of the internal energy of the compressed gas contained in the vessel transforms into kinetic energy of fragments, which can become high velocity projectiles that travel great distances and hit whatever they find in their way.

Another part of this internal energy is transformed into dynamic energy (expansion of the vessel's content). This type of mechanical energy is carried in the surrounding atmosphere as a shock wave. The shock wave is detected as a sudden change in the pressure, density and velocity of the gas. The resulting overpressure can destroy or eject objects and facilities (Section C3).

Usually there are two reasons for total rupture of a vessel either the internal pressure exceeds the vessel's design pressure, or the strength of the vessel is reduced due to its length of operation. Increase of the internal pressure of a vessel can be a consequence of overfilling, overheating from internal or external sources, failure of a pressure regulator, an inside run-away reaction, internal explosion, etc. Reduction of the strength of the vessel's wall can be the consequence of oxidation, rusting, overheating, metal fatigue, the impact of another object on it, etc.

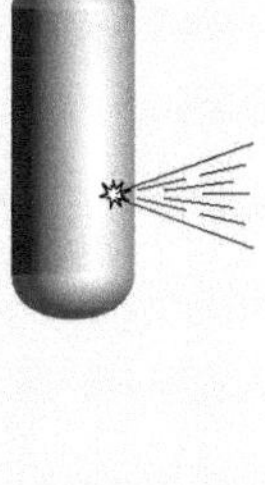

Total Vessel Rupture

calculation procedure

Characteristic Time
↓
Transient Outflow Mass Flow Rate
↓
Final Time

B2.3. Outflow Due to Total Pipe Rupture

The instantaneous rupture of a pipe with compressed gas is another case which can lead to the outflow of a large volume of gas from a pipe. The instantaneous rupture of one side of the pipe produces an instantaneous shock wave that moves with the speed of sound from the rupture point to the other end of the pipe (that is, in direction opposite to the flow of gas).

There are many models in literature that describe this situation [Bell 1978, Wilson 1979, Picard & Bishnoi 1989, Chen, Richardson & Saville 1992]. Here, the empirical model of Bell, as modified by Hanna & Drivas [Hanna & Drivas 1987, Lees 2003], is described. This model agrees well with the rest, but additionally is easy in its use. Its validity is restricted to the time taken for the shock wave to travel from the rupture point to the other side of the pipe. It also estimates the outflow mass flow rate based on the initial conditions.

According to the model of Bell [Hanna & Drivas 1987], the transient mass flow rate $\dot{m}(t)$ (kg/s) from the pipe is calculated as a function of the time t (s), from the expression

$$\dot{m}(t) = \frac{\dot{m}}{1+S}\left\{ S \exp\left(-\frac{t}{t_\mathrm{B}} \right) + \exp\left(-\frac{t}{t_\mathrm{B} S^2} \right) \right\}, \qquad \text{(B2.25)}$$

where
$$S = \frac{M}{\dot{m}\, t_\mathrm{B}} . \qquad \text{(B2.26)}$$

In relation to Eqs. (B2.25) and (B2.26) we note the following:

- The outflow mass flow rate $\dot{m}$ (kg/s), in the above two equations, refers to the outflow mass flow rate from a hole, obtained from Eq. (B2.13). In the case of the total pipe rupture examined here, the discharge coefficient, C_d (-), has a value equal to 1.
- The initial mass, M (kg), of the gas in the pipe is calculated from the pipe's dimensions (length, l_p (m), and diameter, d_p (m)), the density, ρ (kg/m^3), of the gas, and the pipe cross-sectional area A_p (m^2), as

$$M = \rho\, l_\mathrm{p}\, A_\mathrm{p} = \rho\, l_\mathrm{p}\, (\pi/4)\, d_\mathrm{p}^2 . \qquad \text{(B2.27)}$$

- The characteristic time, t_B (s), in Eq. (B2.26), is given by the empirical expression

$$t_B = \frac{4}{3}\frac{l_P}{u_{sound}}\sqrt{\frac{\gamma\, f_F\, l_P}{d_p}}\,, \tag{B2.28}$$

where f_F (-) is the Fanning friction factor, Eqs. (B2.23)-(B2.24) and u_{sound} (m/s) is the speed of sound, equal to

$$u_{sound} = \sqrt{\frac{\gamma\, R\, T}{W_g}}\,. \tag{B2.29}$$

In the above expression, γ (-) is the Poisson ratio, R the universal gas constant (= 8.314 $Jmol^{-1}K^{-1}$), T (K) the gas temperature and W_g (kg/mol) its molecular weight.

As mentioned before, the model of Bell [Hanna & Drivas 1987] is valid for the time period until the shock wave that moves with the speed of sound, u_{sound} (m/s), travels the pipe's length, l_p (m). Hence it is valid for time t_E (s), equal to

$$t_E = \frac{l_p}{u_{sound}}\,. \tag{B2.30}$$

Fire and explosion on February 18, 2008, in Big Spring Refinery, TX, U.S.A. (Reproduced by kind permission of Texas Forest Service U.S.A.)

Characteristic Time
↓
Transient Outflow Mass Flow Rate
↓
Final Time

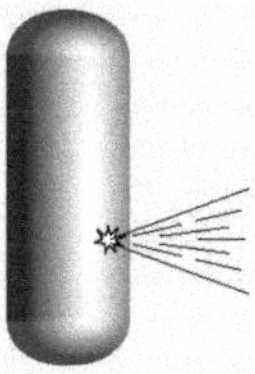

Total Vessel Rupture

example

Characteristic Time
↓
Transient Outflow Mass Flow Rate
↓
Final Time

EXAMPLE B2.3. *Sudden Total Pipe Rupture*

Calculate the transient outflow mass flow rate of propane from a pipe that suddenly ruptures. The following data are available:

-	Initial pressure of pipe, P_o	:	0.5	MPa
	Initial temperature of pipe, T_o	:	288.15	K
-	Pipe diameter, d_p	:	1	m
	Pipe length (until rupture point), l_p	:	10,000	m
-	Discharge coefficient, C_d	:	1	-
-	Molecular weight of propane, W_g	:	0.0441	kg/mol
	Poisson ratio, γ	:	1.19	-
	Viscosity of propane, η	:	82	μPa s

The transient outflow mass flow rate will be calculated from Eq. (B2.25)

$$\dot{m}(t) = \frac{\dot{m}}{1+S}\left\{S\exp\left(-\frac{t}{t_B}\right) + \exp\left(-\frac{t}{t_B S^2}\right)\right\} \qquad \text{where} \quad S = \frac{M}{\dot{m}\, t_B}.$$

Subsequently, we will calculate in turn
a) the outflow mass flow rate $\dot{m}$ (kg/s),
b) the characteristic time t_B (s) and
c) the parameter S (-).

a) Initially we need to calculate the outflow mass flow rate $\dot{m}$ (kg/s), from the hole, calculated from Eq. (B2.13).

$$\dot{m} = C_d A_h P_o K \sqrt{\frac{W_g}{\gamma R T}}.$$

The flow is supersonic as

$$50 = \frac{P_o}{P_a} \geq \left(\frac{\gamma+1}{2}\right)^{\frac{\gamma}{\gamma-1}} = 1.7,$$

where P_a (Pa) is the ambient pressure.

Hence the coefficient K (-) is calculated from Eq. (B2.16) as

$$K = \gamma \left(\frac{2}{\gamma + 1} \right)^{\frac{\gamma+1}{2(\gamma-1)}} = 0.71.$$

Thus the outflow mass flow rate $\dot{m}$ (kg/s) obtained is equal to 1,089.2 kg/s. In the calculations we employed $C_d = 1$ and $A_h = (\pi/4) \times 1^2$ m^2.

Characteristic Time → Transient Outflow Mass Flow Rate → Final Time

b) The characteristic time, t_B (s), is obtained from the empirical equation, Eq. (B2.28), as

$$t_B = \frac{4}{3} \frac{l_P}{u_{sound}} \sqrt{\frac{\gamma \, f_F \, l_P}{d_p}} \; .$$

The speed of sound, u_{sound} (m/s), is obtained from Eq. (B2.29) as

$$u_{sound} = \sqrt{\frac{\gamma R T}{W_g}} = 254.3 \text{ m/s},$$

while the Fanning friction factor, f_F (-), is calculated from Eq. (B2.24),

$$f_F = 0.0791 \, Re^{-0.25} = 1.23 \times 10^{-3}$$

where $Re = \rho u d_p / \eta$ and

$\rho = P_o W_g / (R T_o) = 9.2$ kg/m^3,

$u = \dot{m} / (\rho A_h) = 150.7$ m/s and finally

$Re = 1.69 \times 10^7$.

Hence according to the above, the characteristic time $t_B = 201$ s.

c) The initial mass, M (kg), of the gas in the pipe is obtained from the dimensions of the pipe (length, l_p (m), and diameter, d_p (m)) and the density, ρ (kg/m^3). Thus the initial mass of the gas is equal to 72,275 kg.

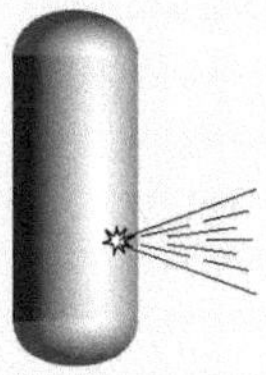

Therefore $$S = \frac{M}{\dot{m} \, t_B} = 0.33.$$

Table B2.4. Temporal Change of Transient Outflow Mass Flowrate of Propane.

t, s	$\dot{m}(t)$, kg/s
0	1,089
25	500
50	294
100	173

Characteristic Time → Transient Outflow Mass Flow Rate → Final Time

In Table B2.4 the temporal change of the transient outflow mass flow rate of propane is shown.

As already mentioned, the model of Bell [Hanna & Drivas 1987] is valid for the time period until the shock wave that moves with the speed of sound, u_{sound} (m/s) covers the entire length, l_p (m) of the pipe.

Hence the model is valid for time

$$t_E\,(s) = l_p / u_{sound} = 39\ s$$

For this reason, in Table B2.4 no higher times are given.

❐

B3
Outflow of Liquids

This section presents a preliminary study of liquid outflow. With the term 'liquid' we refer to a fluid that remains in its liquid state in all conditions during outflow. The study will concentrate on the outflow from a vessel or a pipe.

B3.1. Outflow from Vessel

B3.1.1. Small Outflow

In the case of a liquid leak from a vessel, the study aims to predict the liquid mass reduction inside the vessel during the outflow. The outflow mass flow rate is directly related to the hydrostatic pressure inside the vessel, which in turn depends upon the difference in height between the liquid level and the point of outflow. During the outflow, the mass inside the vessel will continuously decrease, and hence the pressure will also correspondingly decrease. The calculations, as in the case of the compressed gas outflow (Section B2.2.1), will take place in an iterative numerical procedure, where the liquid outflow (leak) will refer to small time steps, inside which the conditions in the vessel can be considered as stable. More analytically, a proposed calculation algorithm is the following:

1) Data: vessel's volume V_d (m^3), initial vessel's pressure P_o (Pa) and intial temperature T (K).
   - The initial liquid mass, M_o (kg), inside the vessel can easily be calculated from the volume of the vessel and the density, ρ (kg/m^3), of the liquid in the initial conditions of the vessel, as

$$M_o = \phi V_d \rho \ . \tag{B3.1}$$

   In the above equation, ϕ (-), is the fill factor of the vessel ($\phi = 0$ the vessel is empty, $\phi = 1$ the vessel is full).

Outflow of Liquids from Vessel

calculation procedure

Initial Conditions
↓
Selection of Time Step
↓
Outflow Mass Flow Rate (from vessel or pipe and vessel)
↓
Reduction of Mass
↓
New Conditions
↓
Next Time Step

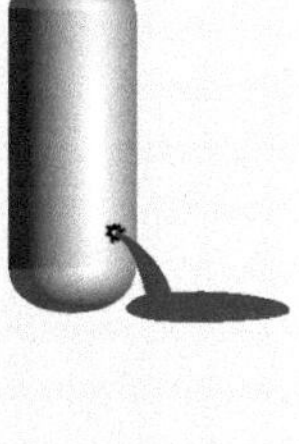

Outflow
of Liquids
from Vessel

calculation
procedure

Initial
Conditions
↓
Selection
of Time Step
↓
Outflow
Mass Flow Rate
(from vessel or
pipe and vessel)
↓
Reduction
of Mass
↓
New
Conditions
↓
Next
Time Step

- The initial volume, V_o (m^3) of the liquid is thus calculated from

$$V_o = M_o / \rho \; . \tag{B3.2}$$

The initial liquid level, h_o (m), in the vessel can be obtained from the initial volume, i.e., for the simple case of a cylindrical vessel of base area A_d (m^2)

$$h_o = V_o / A_d \; . \tag{B3.3}$$

- The outflow mass flow rate $\dot{m}_o$ (kg/s) is calculated (see following sections).

2) The first time step δt_1 (s) is selected.
The duration of the time steps is arbitrary and is directly dependent upon their number so that the whole time of outflow is covered. Empirical observations [TNO 2005] have shown that usually 50 time steps are sufficient for any such process.
   - The reduction of the mass in the vessel, because of the leaked mass, is calculated from the expression

$$\delta M_1 = \dot{m}_o \, \delta t_1 \; . \tag{B3.4}$$

   - The reduction in the volume of the liquid in the vessel, because of the leaked mass, is then calculated from the expression

$$\delta V = \delta M_1 / \rho \; , \tag{B3.5}$$

   and hence the new liquid level in the vessel.
   - From the new liquid level, the new hydrostatic pressure is obtained and thus the outflow mass flow rate corresponding to this time step can be calculated.

3) The next time step is selected and the two previous steps are repeated.

The procedure is better understood with the example which is presented immediately after the calculation of the outflow mass flow rate.

a) Outflow through Hole in Vessel's Wall

The outflow mass flow rate, $\dot{m}$ (kg/s), of liquid through a hole, is defined by the Bernoulli equation as

$$\dot{m} = C_\mathrm{d}\, A_\mathrm{h}\, \sqrt{2(P - P_\mathrm{a})\rho}\ , \tag{B3.6}$$

where C_d (-) denotes the discharge coefficient, A_h (m^2) the hole cross-sectional area, P (Pa) the total pressure in the hole, P_a (Pa) the ambient pressure and ρ (kg/m^3) the density of the liquid.

The total pressure P (Pa) is calculated as the sum of the hydrostatic pressure, P_h (Pa), due to the liquid level in the vessel, plus the absolute pressure P_T (Pa), exerted on the liquid's surface inside the vessel. That is,

$$P = P_\mathrm{h} + P_\mathrm{T} = \rho g h + P_\mathrm{T}\ , \tag{B3.7}$$

where g (m/s^2) denotes the gravitational acceleration (=9.81 m/s^2).

The discharge coefficient, C_d (-), is a function of the type of hole (or orifice) and the velocity of the fluid through the hole. Usual average values [Beek & Mutzall 1975] are:

- hole with sharp and steep edges : 0.62,
- hole with straight edges : 0.86,
- hole with rounded edges : 0.96,
- pipe rupture : 1.

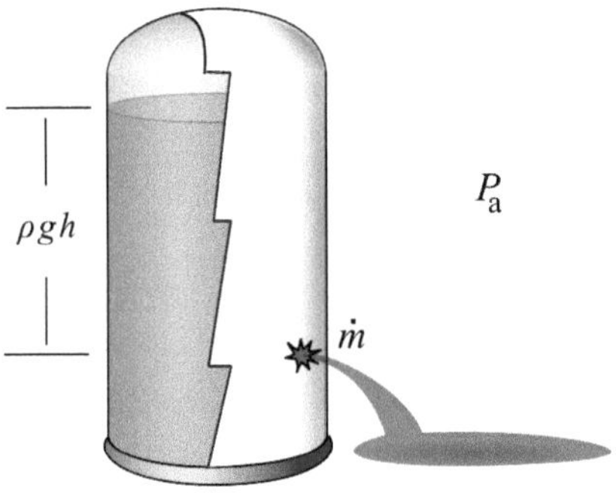

Figure B3.1. Outflow of liquid from hole in vessel's wall.

Outflow of Liquids from Vessel

calculation procedure

Initial Conditions
↓
Selection of Time Step
↓
Outflow Mass Flow Rate (from vessel or pipe and vessel)
↓
Reduction of Mass
↓
New Conditions
↓
Next Time Step

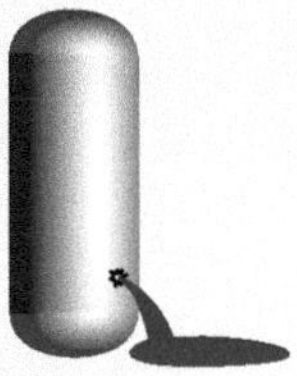

Outflow of Liquids from Vessel

example

Initial Conditions
↓
Selection of Time Step
↓
Outflow Mass Flow Rate (from vessel or pipe and vessel)
↓
Reduction of Mass
↓
New Conditions
↓
Next Time Step

EXAMPLE B3.1. *Hole in Vessel's Wall*

Calculate the outflow mass flow rate of liquid cyclohexane from a hole in a cylindrical vessel's wall (very near its bottom) under atmospheric pressure. The following data are available:

-	Initial vessel's temperature, T_o	: 298.15	K
-	Vessel's volume, V_d	: 6,000	m^3
	Vessel's height, h_v	: 15	m
	Fill factor, ϕ	: 0.75	m^3/m^3
-	Hole diameter, d_h	: 0.1	m
	Discharge coefficient, C_d	: 0.62	-
-	Density of cyclohexane, ρ	: 773.1	kg/m^3

We follow the steps described in Section B3.1.1.

1) Data: vessel's volume V_d (m^3).

- The initial mass, M (kg), of the liquid in the vessel can be easily calculated from the volume of the vessel and the density, ρ (kg/m^3), of the liquid in the initial conditions of the vessel, from Eq. (B3.1), as

$$M_o = \phi\, V_d\, \rho = 0.75 \times 6{,}000 \times 773.1 = 3{,}478{,}950 \text{ kg}$$

- Since the height of the vessel and its fill factor are given, then by analogy, the liquid level, h (m), in the vessel will be equal to

$$h = \phi\, h_v = 0.75 \times 15 = 11.25 \text{ m.}$$

- The outflow mass flow rate $\dot{m}$ (kg/s), can be calculated from Eq. (B3.6),

$$\dot{m} = C_d\, A_h \sqrt{2(P - P_a)\rho}\ ,$$

where $C_d = 0.62$, $A_h = (\pi/4) \times 0.1^2\ m^2$, and

$$P = \rho\, g\, h + P_T = 773.1 \times 9.81 \times 11.25 + 101{,}325 = 186{,}646 \text{ Pa.}$$

Substituting in the above equation for the outflow mass flow rate, we obtain $\dot{m} = 55.92$ kg/s.

2) The first time step, δt_1 = 3,000 s, is selected arbitrarily (one can always return after the end of the procedure and optimize the choice of time step, so as to obtain the best solution with the smallest possible number of time steps).
   - We calculate the reduction of the mass in the vessel, due to the leaked mass, from Eq. (B3.4), as

$$\delta M_1 = \dot{m}_0 \, \delta t_1 = 55.92 \times 3{,}000 = 167{,}758 \text{ kg.}$$

   - We calculate the new liquid level in the vessel, either through the reduction in the volume due to the leaked mass from Eq. (B3.5) or directly, as the new level will be proportional to the new mass,

$$h_1 = h\,(M_1 - \delta M_1)/M_1 = 10.71 \text{ m.}$$

   From the new liquid level the new hydrostatic pressure is obtained and thus the outflow mass flow rate that corresponds to this time step is calculated.

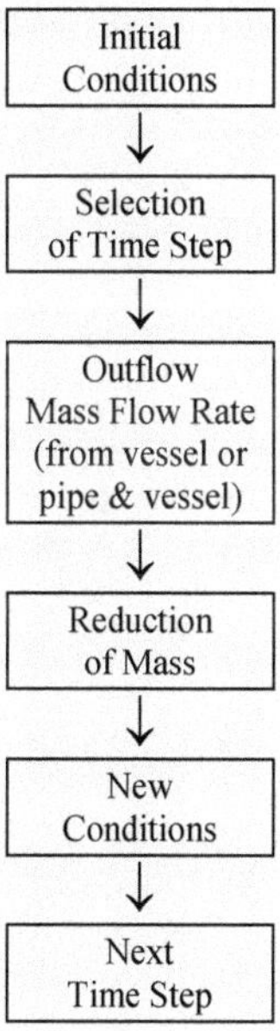

Table B3.1 shows typical results. The iterative procedure is continued until there is no more mass of cyclohexane in the vessel.

Table B3.1. Reduction of Mass with Time.

M (kg)	t (s)	ϕ (-)	h (m)	P (Pa)	$\dot{m}$ (kg/s)	δM (kg)	$M_{leaking}$ (kg)
3,478,950	0	0.750	11.25	186,646	55.92	167,758	
3,311,192	3,000	0.714	10.71	182,532	54.55	163,663	331,421
3,147,529	6,000	0.679	10.18	178,518	53.19	159,567	490,988
2,987,962	9,000	0.644	9.66	174,605	51.82	155,470	646,457
2,832,493	12,000	0.611	9.16	170,792	50.46	151,371	797,828
2,681,122	15,000	0.578	8.67	167,080	49.09	147,271	945,099
...	...	...	...	...	...	...	...
117,727	99,000	0.025	0.38	104,212	10.29	30,860	3,392,083
86,867	102,000	0.019	0.28	103,455	8.84	26,509	3,418,591
60,359	105,000	0.013	0.20	102,805	7.37	22,097	3,440,688
38,262	108,000	0.008	0.12	102,263	5.86	17,593	3,458,281
20,669	111,000	0.004	0.07	101,832	4.31	12,931	3,471,212
7,738	114,000	0.002	0.03	101,515	2.64	7,738	3,478,950
0	116,934	0.000	0.00	101,325	0.02		

❐

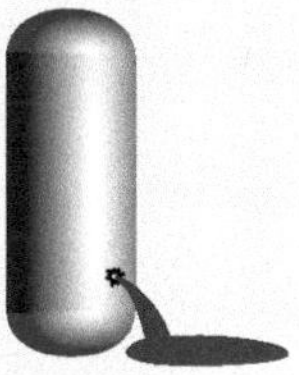

Outflow of Liquids from Pipe

calculation procedure

Initial Conditions
↓
Selection of Time Step
↓
Outflow Mass Flow Rate (from vessel or pipe & vessel)
↓
Reduction of Mass
↓
New Conditions
↓
Next Time Step

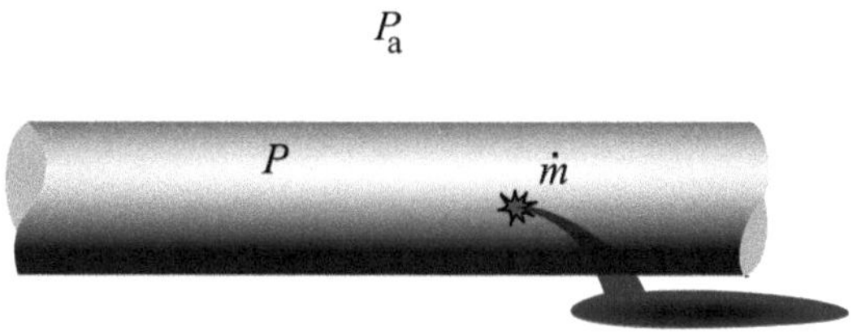

Figure B3.2. Outflow of liquid from hole in pipe.

b) Outflow through Hole in Pipe Wall

In the case of liquid flow in a pipe, we can use the Fanning expression, Eq.(B2.19), in order to calculate the pressure drop ΔP (Pa) in a pipe length l_p (m), as

$$\Delta P = 2 f_F \, \rho \, u^2 \, \frac{l_p}{d_p} \,. \tag{B3.8}$$

In the above expression, f_F (-) denotes the Fanning friction factor, u (m/s) the liquid velocity and d_p (m) the pipe's diameter. (We note that in this case no integration over the liquid density is necessary, as in the case of gases, as the density of the liquid can be assumed to be constant.)

Solving the above expression for the velocity, the outflow mass flow rate is calculated as,

$$\dot{m} = A_h \, u \, \rho \; = A_h \sqrt{\frac{\Delta P \, d_p \, \rho}{2 f_F \, l_p}} \,. \tag{B3.9}$$

The Fanning friction factor, f_F (-), is obtained [Fanning 1877, Pope 2000], Eqs. (B2.23)-(B2.24), as a function of the Reynolds number, $Re = \rho u d_p / \eta$, where η (Pas) is the liquid kinematic viscosity, as

for $Re < 2{,}000$ $\qquad f_F = 16 / Re \,,$ (B3.10)

for $4{,}000 < Re < 10^5$ $\qquad f_F = 0.0791 \, Re^{-0.25} \,.$ (B3.11)

The case of liquid outflow from a pipe connected to a vessel is solved exactly as the corresponding case with gases (see Section B2.2.1, subsection b.)

B4
Outflow of Pressurized Liquefied Gases

In the previous two chapters the outflow of compressed gas (Chapter B2) and the outflow of liquid (Chapter B3) was studied. The outflow of pressurized liquefied gases, examined here, is a much more difficult case as it depends upon many variables. In this chapter a simplified presentation of the special case of the total rupture of a vessel containing a pressurized liquefied gas in a temperature over its boiling point will be presented. The total rupture of the vessel results in instantaneous flashing of its liquid content in all directions, forming a vapor cloud with liquid droplets (also known as aerosol) until this cloud is cooled to a temperature lower than the liquid's boiling point.

This phenomenon mostly appears in high-pressure storage vessels of low volatility liquids, and it is especially dangerous when the liquid is toxic. The resulting toxic cloud can be very harmful to persons depending upon the toxicity of the liquid and the size of the droplets.

- Droplets of diameters 1-10 μm stay in the air for a long period of time as thick fog and enter the human body through inhalation, while
- Droplets of diameters 70-100 μm affect personnel through the skin or indirectly through deposition on the ground.

If the liquid is flammable, then the flammable cloud can result in a flash fire (Section C2.4) or vapor cloud explosion (Chapter C3).

B4.1. Description of Event

As already mentioned, the total rupture of a pressurized vessel results in the instantaneous flashing of the liquid content in all directions, forming a vapor cloud with liquid droplets (also known as aerosol) until this cloud is cooled to a temperature lower than the liquid's boiling point. Part of these droplets can precipitate (rain out) on the ground.

The flashing process is accompanied by simultaneous entrainment of air in the cloud. The incoming air results in further evaporation of the liquid droplets. In order to simulate the explosion, it is necessary to determine the change of the cloud's size, and of the concentration of the liquid inside it, before atmospheric dispersion begins. For this reason different models are required in order to describe the flashing of the liquid phase, the dimensions of the cloud, the expansion velocity, the liquid concentration inside the cloud, the fraction deposited on the ground, and the evaporation of the droplets inside the cloud due to the entering air.

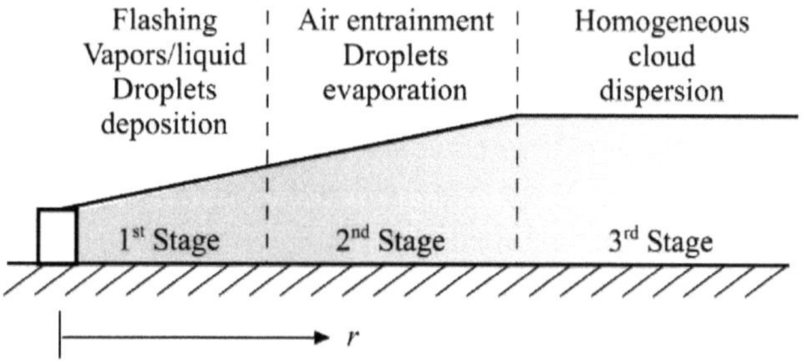

Figure B4.1. Three dispersion stages after the vessel's explosion.

The whole behavior of the liquid substance after the rupture of the vessel can be separated into three stages [Hardee & Lee 1975, TNO 2005]:

1. In the first stage the liquid flashes instantaneously and expands without any premixing with atmospheric air. A cloud composed of air and liquid droplets is formed. Part of the droplets precipitates to the ground.
2. In the next stage, air entrains inside the cloud of droplets and vapor, causing mixing and further evaporation of the liquid phase.
3. In the third stage, the droplets evaporate and the homogeneous cloud disperses in the atmosphere.

In this chapter special emphasis will be given to the first two stages and the procedure adopted will follow the recent literature [TNO 2005]. For a more detailed description of the air entrainment in the cloud as well as the estimation of its quantity, the reader should refer to the corresponding literature.

B4.2. Initial Flashing

In this section we will first examine the conditions immediately preceding and directly following, the vessel's explosion. Then the cloud's radius and expansion velocity, and finally the fraction of the droplets deposited on the ground will be calculated.

Initial conditions before the vessel's explosion are calculated from: initial temperature T_o (K) and initial pressure P_o (Pa), total mass M (kg) of the liquid content and the initial mass fraction, w_o (-) of the vapor phase. The initial pressure that will be exerted is the vessel's rupture pressure, easily calculated from its design characteristics. If these data are not available, one can instead assume the vessel's critical pressure and temperature. Depending on the case, the initial mass fraction of the vapor phase can be taken as equal to zero.

Final conditions after the explosion will be: the final temperature T_f (K), the final pressure P_f (Pa) and the final mass fraction, w_f (-) of the vapor phase. The final pressure is the ambient pressure P_a (Pa), and therefore $P_f = P_a$. Also after the expansion of the liquid phase from initial pressure P_o to the atmospheric P_a, the temperature decreases to the normal boiling point, T_b (K), that is $T_f = T_b$. The final mass fraction, w_f (-), of the vapor phase can be calculated assuming isentropic change, through the mass balance

$$w_o S_{v,o} + (1-w_o) S_{l,o} = w_f S_{v,f} + (1-w_f) S_{l,f}, \tag{B4.1}$$

or

$$w_f = \frac{w_o(\lambda_o / T_o) + S_{l,o} - S_{l,f}}{(\lambda_f / T_f)}, \tag{B4.2}$$

where S_v ($J \cdot kg^{-1} \cdot K^{-1}$) and S_l ($J \cdot kg^{-1} \cdot K^{-1}$) denote the entropy of the vapor and liquid phase in "o" initial conditions or "f" final conditions, and λ (J/kg) is the heat of vaporization in the corresponding conditions. If there are no entropy data, the entropy difference ($S_{l,o}$ - $S_{l,f}$) can be calculated as a function of the heat capacity under constant pressure, C_{PL} ($J \cdot kg^{-1} \cdot K^{-1}$)*, of the liquid phase, and assuming that $(\lambda_o/\lambda_f) \sim 1$, Eq. (B4.2) becomes

$$w_f = w_o \frac{T_f}{T_o} + \frac{T_f}{\lambda_f} C_{PL} \ln \frac{T_o}{T_f}. \tag{B4.3}$$

Eqs. (B4.2) and (B4.3) are simplified in the case when initially in the vessel there was only liquid present ($w_o = 0$), while usually it is assumed that, $T_f = T_b$.

---

* $C_p = T\, dS/dT$, and therefore $(S_{l,o} - S_{l,f}) = C_{PL} \ln(T_o/T_f)$

B4.2.1. Cloud's Radius

The volume, V_{f} (m^3), of the cloud after its initial sudden expansion, can be calculated as a function of the total mass, M (kg), that existed, as

$$V_{\mathrm{f}} = \frac{M}{\rho_{\mathrm{f}}}, \tag{B4.4}$$

where ρ_{f} (kg/m^3) is the average density at the final conditions of temperature and pressure (T_{f}, P_{f}), calculated from the expression

$$\frac{1}{\rho_{\mathrm{f}}} = \frac{(1-w_{\mathrm{f}})}{\rho_{\mathrm{l,f}}} + \frac{w_{\mathrm{f}}}{\rho_{\mathrm{v,f}}}. \tag{B4.5}$$

In the above expression, $\rho_{\mathrm{l,f}}$ (kg/m^3) and $\rho_{\mathrm{v,f}}$ (kg/m^3) denote the density of the substance in the liquid and vapor phase, respectively, in the final conditions of temperature and pressure. The density, $\rho_{\mathrm{l,f}}$ (kg/m^3), of the liquid phase can be calculated from an equation of state, while the density, $\rho_{\mathrm{v,f}}$ (kg/m^3), of the vapor phase is calculated from the ideal-gas law.

The cloud resulting from the flashing (due to the explosion), will be hemispherically shaped, as the vessel is originally on the ground, and therefore the radius, R_{f} (m), of the cloud can be obtained from the expression

$$R_{\mathrm{f}} = \left(\frac{3\,V_{\mathrm{f}}}{2\pi}\right)^{1/3}. \tag{B4.6}$$

B4.2.2. Cloud's Expansion Velocity

In its initial state inside the vessel, the liquid is considered at rest. During the vessel's explosion, the resulting cloud expands with a velocity u_{f} (m/s), calculated from the principle of energy conservation, taking into account the transfer of energy to the atmospheric air. If $H_{\mathrm{v,o}}$ and $H_{\mathrm{v,f}}$ (J/kg) denote the enthalpy of the vapors at initial "o" and final "f" conditions, respectively, and W (J/kg) is the work carried out to the atmospheric air, then the energy balance per unit mass can be written as

$$\left\{H_{\mathrm{v,o}} + (1-w_{\mathrm{o}})\lambda_{\mathrm{o}}\right\} + \frac{1}{2}u_{\mathrm{o}}^2 = \left\{H_{\mathrm{v,f}} + (1-w_{\mathrm{f}})\lambda_{\mathrm{f}}\right\} + \frac{1}{2}u_{\mathrm{f}}^2 + W. \tag{B4.7}$$

In the above equation, since the liquid inside the vessel is at rest, $u_o = 0$. Moreover

$$W_f = \frac{P_f - P_o}{\rho_o} = \frac{P_a - P_o}{\rho_o}, \tag{B4.8}$$

where P_a (Pa) is the ambient pressure and ρ_o (kg/m^3) the average density of the substance in the vessel (at initial conditions) equal to

$$\frac{1}{\rho_o} = \frac{(1 - w_o)}{\rho_{l,o}} + \frac{w_o}{\rho_{v,o}} . \tag{B4.9}$$

Experimental observations have shown that the actual cloud's expansion velocity is smaller than the calculated one, as the process is not entirely isentropic and part of the kinetic energy is transformed into heat. The actual value of the cloud's expansion velocity is estimated [Melhem & Groce 1993] to be about 20% lower than the theoretically calculated one. Thus the cloud's expansion velocity, u_f (m/s), can be calculated from Eq. (B4.7) as

$$u_f = 0.8\sqrt{2\left[H_{v,o} - H_{v,f} - (1 - w_f)\lambda_f + (1 - w_o)\lambda_o + \frac{P_o - P_a}{\rho_o}\right]} . \tag{B4.10}$$

B4.2.3. Droplet Rain-out on the Ground

There are no theoretical models of the fraction of the droplet rain-out on the ground in the case of a sudden rupture of a vessel. The proposed calculation procedure [TNO 2005] is based upon experimental observations.

Estimation of droplet rain-out directly depends upon the final mass fraction, w_f (-), of the vapor phase after the flashing. The larger the final mass fraction of the vapor phase is, the less liquid will be present in the mixture and therefore the droplet rain-out fraction will be smaller.

Two cases can be distinguished:

- $w_f > 0.5$: It is assumed that no droplet rain-out will occur.
- $w_f < 0.5$: It is assumed that double the quantity of originally flashed material will remain in the air. That is

$$M_{rem} = 2\, w_f\, M . \tag{B4.11}$$

Outflow of Pressurized Liquefied Gases

calculation procedure

Initial Flashing
↓
Cloud's Radius
↓
Cloud's Expansion Velocity
↓
Liquid Droplet Rain-Out
↓
Droplets' Diameter

Air Entrainment and Atmospheric Dispersion
↓
Cloud's Radius and Velocity

Outflow of Pressurized Liquefied Gases

calculation procedure

Initial Flashing
↓
Cloud's Radius
↓
Cloud's Expansion Velocity
↓
Liquid Droplet Rain-Out
↓
Droplets' Diameter

Air Entrainment and Atmospheric Dispersion
↓
Cloud's Radius and Velocity

In the previous equation, M_{rem} (kg) denotes the remaining mass in the cloud (vapors and droplets), while M (kg) is the initial mass in the vessel. It can also be noted that the new mass fraction, $w_{f,rem}$ (-), of the remaining vapor mass will always be equal to 0.5, as a result of the rain-out droplets.

To illustrate the aforementioned discussion, let us assume that initially in a vessel there were 1,000 kg of propane and that the mass fraction, w_f (-) of the vapor phase after the flashing was found to be equal to 0.3. Thus 300 kg of the gas mass are vapors and 700 kg are liquid droplets. Since $w_f < 0.5$ (according to the aforementioned empirical rule), the remaining mass, M_{rem} (kg), in the cloud will be equal to $2 \times 0.3 \times 1{,}000 = 600$ kg. This result indicates that during this first stage of expansion, liquid droplets of a quantity equal to 400 kg will be deposited on the ground. Therefore 300 kg of vapor and 300 kg of liquid droplets remain in the cloud. Thus, the new mass fraction, $w_{f,rem}$ (-), of the remaining mass in the cloud will be equal to 0.5.

Closing up this subsection, it should be noted that in the case where liquid droplet rain-out is observed, using the value $w_{f,rem} = 0.5$, one must recalculate at the final conditions

- the density of the gas mixture ρ_f (kg/m^3) - Eq. (B4.5),
- the volume of the vapor cloud V_f (m^3) - Eq. (B4.4),
- the radius of the vapor cloud R_f (m) - Eq. (B4.6),

and these new values should be employed further in the calculations.

B4.2.4. Diameter of Deposited Droplets

In practice all cloud droplets are not of equal size. For calculation and simulation purposes it is usual to define a "representative" average diameter, d_d (m), of droplets, which can be estimated [TNO 2005] from the empirical expression

$$d_d = C_{ds} \frac{\sigma_s}{\rho_a\, u_f^2}\,, \tag{B4.12}$$

where, σ_s (Pa m) is the surface tension (its value for organic substances is between 0.02 and 0.03 Pa m), u_f (m/s) the expansion velocity and ρ_a (kg/m^3) the density of atmospheric air. The parameter, C_{ds} (-), is obtained experimentally and uses values between 10 and 20; thus usually $C_{ds} = 15$.

The estimation of the droplets' diameter is necessary in the determination of the effects of toxic substances as it defines the way they enter the human body (through inhalation or skin deposition).

B4.3. Air Entrainment and Atmospheric Dispersion

The previous section referred to the variables associated with the sudden flashing of the liquid in the vessel. This section refers to the phenomena that characterize the entrance of air to the cloud. An approximate approach will be discussed for the calculation of cloud's expansion velocity and the temporal change of the cloud's radius.

B4.3.1. Cloud's Radius and Velocity

As already mentioned the cloud resulting from the explosion will be of hemispherical shape with an initial radius, R_f. This radius will increase with time, as $R(t)$, because of the air entraining, which also evaporates the remaining droplets and makes the cloud homogeneous. The increase of the cloud's radius will continue until its final value R_{fin}, where the cloud's expansion velocity will become equal to the atmospheric air velocity at the particular area.

Unfortunately there is no well-established model for the simulation of the cloud which now resembles aerosol, i.e., vapors with droplets and air. The simplest approach is based on the ideas of Hardee & Lee [Hardee & Lee 1975], who assume that during flashing, the expansion velocity u_f (m/s) is constant, and can be calculated from Eq. (B4.10), with a small correction for non-isentropic process and the presence of turbulence. Following this, they assume a cloud which expands with homogeneous concentration. From mass conservation (and neglecting the difference in density between cloud and air) they proposed a normal radial velocity equal to the cloud's expansion velocity, $u(t)$ (m/s), as

$$u(t) = u_f \left(\frac{R_f}{R(t)} \right)^3 . \qquad \text{(B4.13)}$$

In the above equation, R_f (m) denotes the cloud's radius at the moment the expansion is finalized, while $R(t)$ (m), is the cloud's radius at any time after.

The cloud's expansion rate calculated from the above equation agrees quite well with the experimental observations of Giesbrecht [Giesbrecht *et al.* 1981], assuming the expansion of a hemispherical cloud from the ground.

Applying $u = dR/dt$, in Eq.(B4.13), we can write the corresponding expression for the radius increase as

$$R(t) = \left[4\, u_f\, R_f^3 (t - t_f) + R_f^4 \right]^{1/4} , \qquad \text{(B4.14)}$$

Outflow of Pressurized Liquefied Gases

calculation procedure

Initial Flashing
↓
Cloud's Radius
↓
Cloud's Expansion Velocity
↓
Liquid Droplet Rain-Out
↓
Droplets' Diameter

Air Entrainment and Atmospheric Dispersion
↓
Cloud's Radius and Velocity

where t_f (s) is the time required for the conclusion of the flashing after the explosion (usually considered negligible). The increase in the cloud's radius will continue until its final value R_{fin}, during which the cloud's expansion velocity, $u(t)$ (m/s), will become equal to the velocity, u_s (m/s), of the atmospheric air at that height.

The cloud's volume and mass are calculated from its radius. Since the mass of the chemical substance is known, the mass of the air entrained can be obtained. This calculation is approximate as it does not take into consideration the droplets evaporated because of the air.

EXAMPLE B4.1. *Dimensions of Propane Cloud*

Calculate the dimensions of a cloud resulting from the instantaneous rupture of a vessel containing propane mass $M = 5$ tn at a temperature $T_o = 291$ K, pressure $P_o = 0.78$ MPa, and initial mass fraction of the vapor phase $w_o = 0.05$. The following data are available:

-	Boiling temperature at P_a, T_b	:	230.9	K
	Ambient pressure, P_a	:	0.1	MPa
-	Liquid density at (T_o, P_o), $\rho_{l,o}$	:	505	kg/m^3
	Vapor density at (T_o, P_o), $\rho_{v,o}$	:	14.3	kg/m^3
	Liquid density at (T_b, P_a), $\rho_{l,f}$	:	584	kg/m^3
	Vapor density at (T_b, P_a), $\rho_{v,f}$	:	2.33	kg/m^3
-	Vapor enthalpy at (T_o, P_a), $H_{v,o}$	:	483	kJ/kg
	Vapor enthalpy at (T_b, P_a), $H_{v,f}$	:	383	kJ/kg
-	Heat of vaporization, λ_o	:	342	kJ/kg
	Heat of vaporization, λ_f	:	426	kJ/kg
-	Liquid heat capacity, C_{pL}	:	2.41	kJ/kg K
-	Wind velocity, u_s	:	3.5	m/s

The variables associated with the initial flashing will be calculated first, and subsequently we will examine the cloud's expansion.

a) Initial Flashing

The final mass fraction, w_f (-), of the vapor phase can be calculated assuming isentropic change, from Eq. (B4.3). All the variables are known. Hence, assuming that $T_f = T_b$,

$$w_f = w_o \frac{T_f}{T_o} + \frac{T_f}{\lambda_f} C_{PL} \ln \frac{T_o}{T_f} = 0.342 \,.$$

To calculate the volume and the radius of the cloud, its density is needed. Using Eq. (B4.5), we obtain,

$$\frac{1}{\rho_f} = \frac{(1 - w_f)}{\rho_{l,f}} + \frac{w_f}{\rho_{v,f}} \Rightarrow \rho_f = 6.76 \text{ kg/m}^3,$$

and from Eq. (B4.4) $$V_f = \frac{M}{\rho_f} = 738.6 \text{ m}^3 \,.$$

Having calculated the volume of the cloud (and knowing its shape is hemispherical as the vessel is on the ground) we find its radius, R_f (m), using Eq. (B4.6),

$$R_f = \left(\frac{3 V_f}{2\pi} \right)^{1/3} = 7.07 \text{ m}.$$

Finally the cloud's expansion velocity, u_f (m/s), is calculated from Eq. (B4.10), as

$$u_f = 0.8 \sqrt{2 \left[H_{v,o} - H_{v,f} - (1 - w_f)\lambda_f + (1 - w_o)\lambda_o + \frac{P_o - P_a}{\rho_o} \right]} = 424.9 \text{ m/s}$$

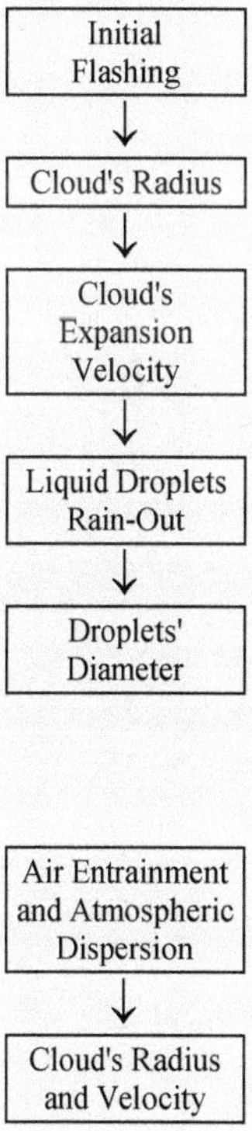

The droplet rain-out directly depends upon the final vapor mass fraction, w_f (-), after the flashing. Since w_f = 0.342, it is assumed that an amount equal to approximately twice the amount originally flashed will remain in the atmosphere. That is from Eq. (B4.11),

$$M_{rem} = 2\, w_f\, M = 2 \times 0.342 \times 5{,}000 = 3{,}422 \text{ kg},$$

and from now on we assume $w_f = 0.5$. Because of this new value of w_f, and of the new mass M_{rem}, a re-evaluation of the density ρ_f, and consequently of the volume V_f and the radius R_f, of the cloud is necessary. The results are

$$\rho_f = 4.64 \text{ kg/m}^3, \quad V_f = 737.5 \text{ m}^3 \quad \text{and} \quad R_f = 7.06 \text{ m}.$$

Outflow of Pressurized Liquefied Gases

example

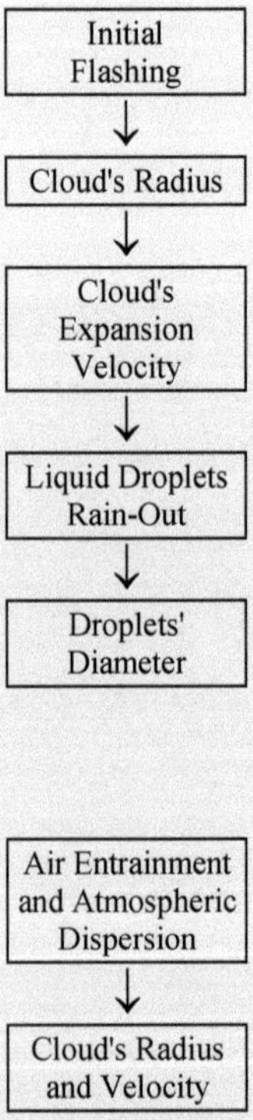

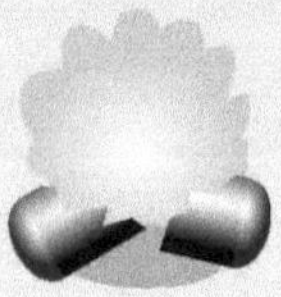

b) Air Entrainment

Based on the ideas of Hardee & Lee [Hardee & Lee 1975], the cloud's expansion velocity, $u(t)$ (m/s), and its radius $R(t)$ (m), are calculated from Eqs. (B4.13) - (B4.14), as

$$u(t) = u_{\mathrm{f}} \left(\frac{R_{\mathrm{f}}}{R(t)} \right)^3 \qquad \text{and} \qquad R(t) = \left[4\, u_{\mathrm{f}}\, R_{\mathrm{f}}^3 (t - t_{\mathrm{f}}) + R_{\mathrm{f}}^4 \right]^{1/4}.$$

The results are shown in Table B4.1 and in Figure B4.2. The increase of the cloud's radius will continue until the cloud's expansion velocity becomes equal to the wind velocity, u_s=3.7 m/s. At that point the final radius will be 29.2 m.

Table B4.1. Temporal Increase of Cloud's Radius.

t (s)	$R(t)$ (m)	$u(t)$ (m/s)
0.00	7.06	426.4
0.10	15.8	38.1
0.50	23.4	11.7
1.00	27.8	7.0
1.50	30.8	5.1
2.00	33.1	4.1
2.30	34.3	3.7
2.50	35.0	3.5

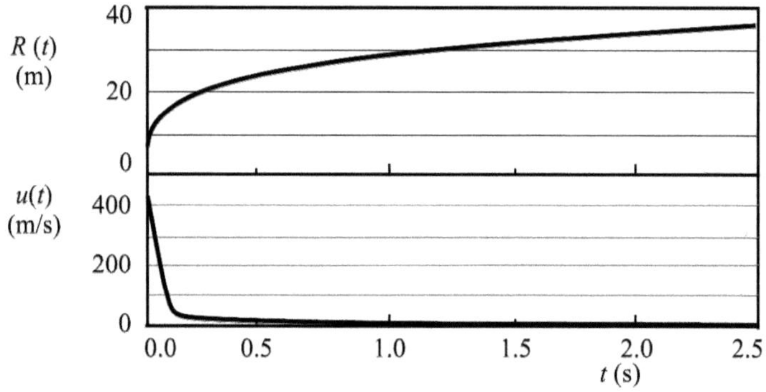

Figure B4.2. Temporal increase of cloud's radius.

❐

C effects and consequences analysis

C1 Introduction

In this chapter, an attempt will be made to analyze and calculate the major hazards facing the industrial sector, with the aim of quantifying their consequences. The chapter consists of three large sections: fires, explosions and toxic gas dispersion. In Figure C1.1, the probable consequences of a gas or liquid fuel leak are shown.

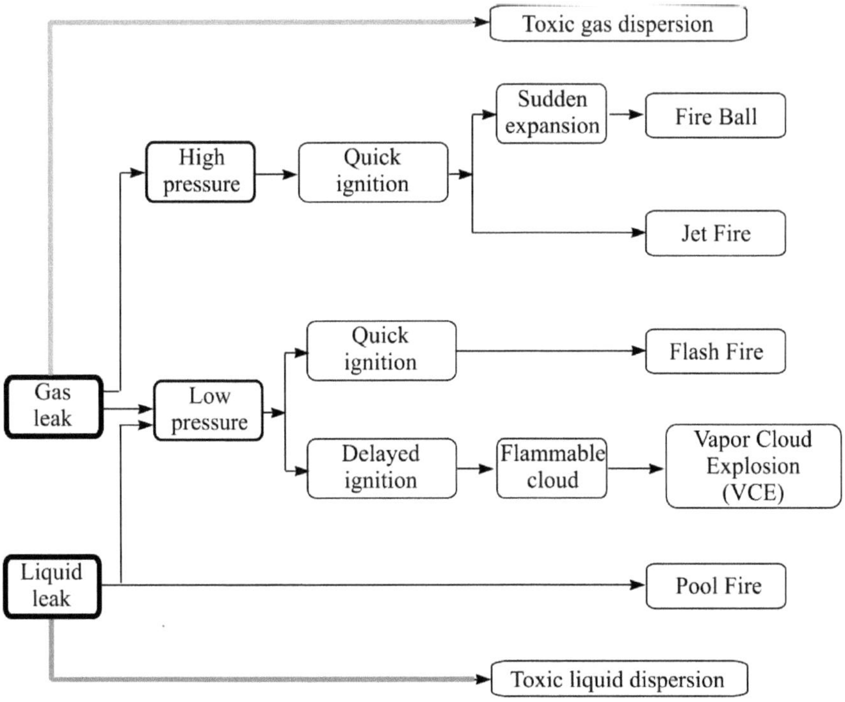

Figure C1.1 Probable consequences resulting from a gas or liquid fuel leak.

The cases shown in Figure C1.1 are
a) fire ball, jet fire, flash fire and pool fire,
b) vapor cloud explosion (VCE), and
c) toxic gas and/or liquid dispersion.
The only case not included in Figure C1.1 is that of BLEVE (Boiling Liquid Expanding Vapor Explosion), which refers to the failure of a pressure vessel as a result of its external heating from a fire.

Table C1.1 The 15 Major Industrial Disasters in Relation to Material Damages, During the Years 1976-2006 [MinA 2006].

Date	Location	Unit	Initial cause	Cost ($ millions)
6.7.1988	UK - North Sea	Oil platform	VCE	1,503
23.10.1989	USA - Pasadena	Petrochemicals	VCE	1,030
21.9.2001	France - Toulouse	Petrochemicals	VCE	888
20.1.2004	Algeria - Skilkda	Refinery	VCE	845
15.3.2001	Brazil - C.Basin	Oil platform	VCE	610
24.4.1988	Brazil - Enchova	Oil platform	Fire	546
25.6.2000	Kuwait - Al Ahmadi	Refinery	VCE	512
5.5.1988	USA - Norco	Refinery	VCE	398
9.11.1992	France - La Mede	Refinery	VCE	376
24.2.1986	Greece - Thessaloniki	Oil tank	Fire	368
25.12.1997	Malaysia - Sarawak	Petrochemicals	Explosion	348
14.11.1987	USA - Pampa	Petrochemicals	VCE	340
23.7.1984	USA - Romeoville	Refinery	VCE	325
16.4.2001	UK - Grimsby	Refinery	VCE	308
16.10.1992	Japan - Sodegaura	Refinery	VCE	232

- In the table, cost of industries destroyed during the Kuwait war is not included. This cost is estimated to be up to $2,546,000,000.
- Cost in 2005 prices.

Table C1.1 shows the 15 major industrial disasters during the years 1976-2006 [MinA 2006]. The most common cause for large material damages is the explosions (and often VCE or Vapor Cloud Explosion). This is expected as explosions result in very large material damages and usually are followed by fires. Typically we note that the total cost of material damages shown in the table is equal to U.S. $8,629,000,000 (in 2005 prices).

Tables C1.2 and C1.3 present the 15 major industrial disasters over the same period for numbers of deaths and casualties, respectively. It is worth noticing the inclusion of non-industrial countries (along with the presence of industrialized nations) where safety procedures are often not properly followed.

Table C1.2 The 15 Major Industrial Disasters in Relation to Number of Deaths, During the Years 1976-2006 [MinA 2006].

Date	Location	Unit	Initial cause	Number of deaths
3.12.1984	India - Bhopal	Petrochemicals	MIC[1]	2,500
3.11.1982	Afghanistan - Salang	Oil transport	Oil	2,000
4.6.1989	USSR - Siberia	Pipeline	LPG[2]	645
19.11.1984	Mexico - Mexico City	Depot	LPG	550
25.2.1984	Brazil - Sao Paolo	Oil pipeline	Oil	508
4.8.1993	Columbia - Remeios	Petrochemicals	Oil	430
2.11.1994	Egypt - Donca	Depot	Oil	410
1.5.1983	Egypt - Nile River	Transport	LPG	317
14.2.1998	Cameroon - Yaoundi	Transport	Oil	220
11.6.1978	Spain - San Carlos	Transport	Propylene	211
6.7.1988	UK - North Sea	Oil platform	Natural gas	165
19.2.1981	Venezuela - Caracas	Storage	Oil	153
19.12.1982	Venezuela - Tacos	Storage	Oil	150
10.4.1991	Italy - Livorno	Transport	Naphtha	141
11.4.1996	USA - Alberton	Transport	Chlorine	140

[1] MIC : Methyl IsoCyanate.
[2] LPG : Liquefied Propane Gas.

Table C1.3 The 15 Major Industrial Disasters in Relation to Number of Casualties, During the Years 1976-2006 [MinA 2006].

Date	Location	Unit	Initial cause	Number of casualties
3.12.1984	India - Bhopal	Petrochemicals	MIC[1]	25,000[3]
19.11.1984	Mexico - Mexico City	Depot	Propane	6,400
26.7.1993	USA - Richmond	Petrochemicals	Acid	6,250
21.9.2001	France - Toulouse	Petrochemicals	Ammonium Nitrate	3,000
1.12.1995	India - Maharashtra	Transport	Ammonium Nitrate	2,000
9.9.2000	USA - Manhattan	Transport	Anvil	2,000
5.3.1982	Australia - Melbourne	Transport	Butadiene	1,000
20.3.1995	Japan - Tokyo	Metro	Sarin	980
10.1.1997	Pakistan - Lahore	Transport	Ammonium Nitrate	900
4.6.1989	USSR - Siberia	Pipeline	LPG[2]	706
1.9.1991	China - Shaxi	Petrochemicals	Pesticides	650
19.12.1981	Venezuela - Caracas	Storage	Oil	500
19.12.1982	Venezuela - Tacos	Storage	Oil	500
21.1.1997	India - Bhopal	Transport	Ammonium Nitrate	400
8.7.1986	USA - Miamisburg	Transport	Acid	400

[1] MIC : Methyl IsoCyanate.
[2] LPG : Liquefied Propane Gas.
[3] Initially, many more later on.

In 1988 Garrison [Garrison 1988] published a very interesting survey examining the causes of industrial damages during the years 1957 through 1986. In this survey, the relation between explosion and fire as a cause of disaster was examined. The conclusions are shown in Figure C1.2. At a first glance, these results do not seem to agree with those of Table C1.1. This is attributed to the fact that the table had the large material cost as a criterion, while the survey showed average values.

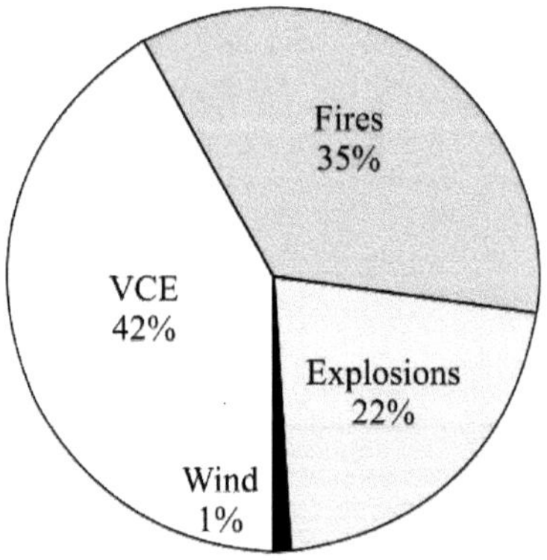

Figure C1.2 Causes of disasters [Garrison 1988].

In conclusion, material damage because of destruction resulting from explosions and fires is very high. Toxic gas or liquid dispersion, although it does not incur material damage and cost, usually results in a large number of human deaths. In the following sections, an attempt will be made to estimate, calculate and predict the consequences of these hazards.

C1.1. Definitions

In this section, some useful definitions will be presented.

Atmospheric Stability
Describes the extent to which vertical temperature (=density) gradients promote or suppress turbulence in the atmosphere. We distinguish three types of stability of the surface atmospheric layer:
- stable (temperature increases with height),
- unstable (temperature decreases with height), and
- neutral (no significant change of temperature with height).

Boiling Liquid Expanding Vapor Explosion (BLEVE)
The BLEVE is an explosion due to flashing of liquids when a vessel with a high vapor pressure substance fails. The failure of the vessel is often caused by an external fire. If the substance released is a fuel, the BLEVE can result in very large fire balls. Rocketing vessels are also hazards related to BLEVE.

Continuous & Instantaneous Source
Every leak-dispersion is characterized as a continuous or instantaneous source depending upon its time duration. There is no specific time limit that exactly distinguishes the two cases, rather, it is up to the model employed. There are also cases where the source is initially treated as continuous and then as instantaneous.

Explosion
We define an explosion as an event leading to a rapid increase of pressure. This pressure increase can be caused by: nuclear reactions, loss of containment in high pressure vessels, high explosives, vapor explosions, runaway reactions, combustion of dust, mist or gas (including vapors) in air or in other oxidizers.

Explosion - Deflagration
A deflagration is defined as a combustion wave propagating at subsonic velocity relative to the unburned gas immediately ahead of the flame. The velocity of the unburned gas ahead of the flame is a consequence of the expansion of the combustion products. In an accidental gas explosion the deflagration is the common mode of flame propagation. In this mode the flame speed ranges from order of 1 m/s up to 500-1000 m/s corresponding to explosion pressures between a few mbar and several bar.

Effects Analysis

definitions

Explosion - Detonation

A detonation is defined as a combustion wave propagating at supersonic velocity relative to the unburned gas immediately ahead of the flame. In simple terms, a detonation wave can be described as a shock wave immediately followed by a flame. For fuel-air mixtures at ambient pressure the detonation velocity can be up to 2000 m/s and the maximum pressures produced are close to 20 bar.

Explosion - Gas Explosion

A gas explosion is defined as a process where combustion of a premixed gas cloud (i.e., fuel-air or fuel-oxidizer) is causing rapid increase of pressure. Gas explosions are classified according to the environment where the explosion takes place as

Confined Gas Explosion

An explosion within vessels, pipes, channels or tunnels.

Partly Confined Gas Explosion

An explosion in a compartment, buildings or off-shore modules.

Unconfined Gas Explosion

An explosion in plants and other unconfined areas.

Explosion - Shock Wave

The term shock wave is used to describe the overpressure created during an explosion from the rapid expansion of gases produced by the explosion. It propagates at supersonic velocity and is the main cause of damages.

Fire

Fire is an exothermic oxidation reaction occurring in the gas phase, which results from the mixing of flammable gases with air or other oxidative means. The following types of fire can be distinguished:

Pool Fire

A fire of turbulent dispersion taking place over a pool of vaporizing flammable liquid under conditions of negligible initial fuel momentum.

Fire Ball

A fire ball can occur during a sudden leak and ignition of pressurized flammable gases.

Jet Fire

A fire of turbulent dispersion occurring from the burning of a flammable fluid continuously released with considerable momentum towards a specific direction.

Flash Fire

A flash fire can appear during the sudden ignition of a cloud of flammable gases, when the flame is not accelerated by the presence of obstacles or the influence of turbulent dispersion.

Ignition - Flammability Limits (Low and Upper)
A well-mixed combination of fuel and air can only ignite when the concentration of the fuel is between a low flammability limit (LFL) and an upper flammability limit (UFL).

Light and Dense Gas
A gas is characterized as light or dense according to its density in relation to that of the surrounding ambient air. This distinction is necessary as it determines the type of dispersion model to be employed.

Safety Zone
A Safety Zone or "Risk Zone" is a specific circular area where the consequences of a toxic release in its center can reach levels that can seriously affect public health (death, serious injuries, possible damage).

Temperature - Adiabatic Flame Temperature
It is the maximum temperature recorded during the combustion of a fuel-air mixture under constant pressure and negligible heat losses to the surroundings (walls, facilities, etc.).

Temperature - Autoignition Temperature
When a flammable mixture is heated up to a specific temperature, its combustion reaction can start by itself. The autoignition temperature is the lowest wall temperature adjacent to a flammable mixture that can initiate its ignition.

Temperature Inversion
The temperature in the atmosphere follows a specific pattern (i.e., decreases with height in the troposphere). Temperature inversion is noted when in a specific atmospheric layer the temperature does not follow this pattern.

Temperature - Flash Temperature
The flash temperature of a fuel is the lowest temperature in which the fuel can still produce enough vapors to create a flammable mixture with air. Operating below the flash temperature of a mixture will result in a situation where no flammable mixture with air is produced.

Turbulence
In fluid dynamics flow is divided into laminar and turbulent regimes. Laminar flow means that the fluid flows in laminars or layers, while turbulent flow is characterized by an irregular random fluctuation imposed on mean (time-averaged) flow velocity (see Figure C1.3).

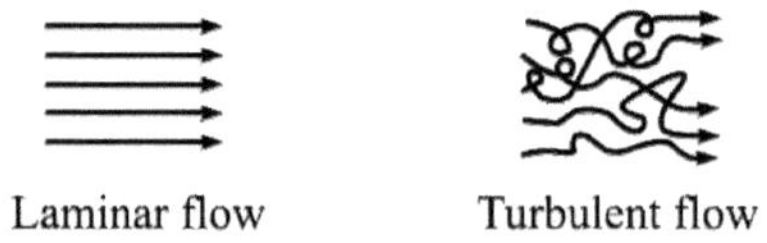

Figure C1.3 Laminar and turbulent flow.

Whether the flow is laminar or turbulent is determined by the Reynolds number (see Figure C1.4), defined as the product of the fluid velocity and a characteristic length, divided by the kinematic viscosity.

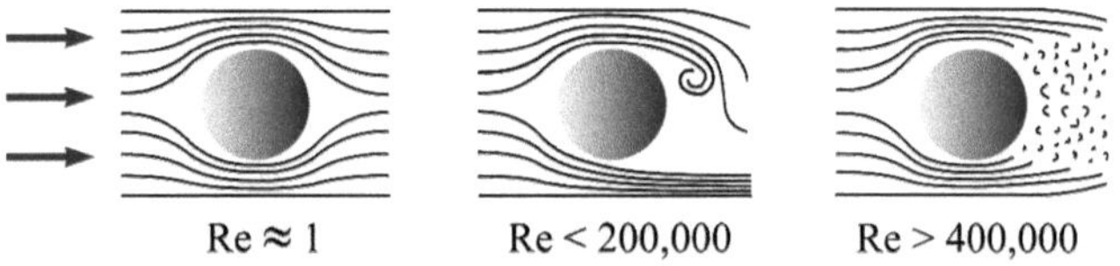

Figure C1.4 Cylinder in a crossflow at different Reynolds numbers.

The turbulence is very important in determining how fast the flame can propagate in a premixed gas cloud. Turbulence wrinkles the flame front and increases diffusion of heat and mass and thus causes a higher burning rate.

Vapor Cloud Explosion (VCE)
A vapor cloud explosion (VCE) is an unconfined or partly confined gas explosion.

C2
Fires

Fire is an exothermic oxidation reaction occurring in the gas phase, which results from the mixing of flammable gases with air or other oxidative means. If the concentration of the flammable substance reaches its critical mass for ignition, and a proper ignition source capable of supplying the required power is present, then there will be a fire.

Fires and explosions are the most significant and most common causes of damage to equipment and of injuries and deaths in industry. This is especially true in offshore oil drilling, where there is a high concentration of equipment in very close spaces.

Explosion and fire at the Barton Solvents distribution facility in Valley Center, KS, U.S.A. (Reproduced by kind permission of the U.S. Chemical Safety Board.)

Damages are a direct consequence of the generated heat flux. To a first approximation for the calculation of the heat flux, q' (W/m^2), in the flame surface, the Stefan-Boltzmann equation can be employed:

$$q' = \varepsilon \sigma \left(T_{\mathrm{f}}^4 - T_{\mathrm{a}}^4 \right), \tag{C2.1}$$

where ε denotes the grey-body emissivity (-) and σ the Stefan-Boltzmann constant (= 5.6703×10^{-8} W$\cdot$m$^{-2}\cdot$K^{-4}). The temperatures, T_{f} and T_{a} (K), refer to the temperature at the flame surface and the ambient temperature, respectively.

In fact, however, this equation cannot be employed, since the temperature differs all over the flame and hence it is not a unique temperature that can be determined. Furthermore the flame does not radiate from its whole surface, since a part of it is covered by soot, and a large part of the heat flux is absorbed by the carbon dioxide and the humidity in the atmosphere. As a consequence of these, the heat flux calculated by the Stefan-Boltzmann equation is significantly larger than the actual heat flux.

In an effort to estimate the heat flux, and its effects, many models appear in the literature. The most important groups of such models are the following:

a) Point-Source Models

The point-source models do not consider the shape of the flame, but assume that the heat-flux originates from a point source. The heat flux, q' (W/m^2), in a distance, X (m), from the center of the fire, can be expressed as:

$$q' = \frac{1}{4\pi X^2} \eta \, m_{\mathrm{k}} \Delta H_{\mathrm{c}}, \tag{C2.2}$$

where, m_{k} (kg/s) denotes the burning rate with which the flammable material is burnt, ΔH_{c} (J/kg) the heat of combustion, and η the combustion efficiency. While the above relation has the advantage of simplicity, the analogy of the heat flux with the square of the distance has not been experimentally observed. These models usually overestimate the heat flux, but produce good results in a distance of about 10 radii from the center of the fire.

b) Solid-Flame Models

The solid-flame models assume that the flame is of a solid shape that radiates heat only from its surface. Models take into consideration the shape of the flame and calculate the heat flux as a function of the Surface Emitting Power, the Shape Factor and the Atmospheric Transmissivity. These models are simple in their application, easy to program and produce relatively good results.

c) Field Models

The field models, or computational fluid dynamics models (CFDs), are based upon the numerical solution of the partial differential Navier-Stokes equations. These models require careful validation against real or experimental data. Their main disadvantage is the large requirements in computing time, the difficulty in programming and the inflexibility in compatibility with many applications.

d) Integral Models

The integral models constitute a compromise between the semi-empirical models and the CFD models. They are based upon the solution of differential equations for the conservation of mass, momentum and energy, but their mathematical approach is more simplified and refers to the specific case to be examined. In this way, a significant reduction in computing time is achieved.

e) Zone Models

According to the zone models, space is separated into homogeneous space zones of unified approach, which are connected through empirical equations and mass and energy balances. These models are employed in structural areas, but not in open spaces.

Of the aforementioned methods, **solid-flame models** are the most widely used today, as they can, with care, easily be applied in most cases. Additionally, integral models are increasingly in use in industry as they take into consideration the exact geometry of the unit.

In this work, and specifically in the section of fires, we will deal mostly with the application of the solid-flame models. As noted, in the **"solid-flame" approach the flame is treated as a solid object that radiates heat from its visible surface.** According to this model, the heat flux q' (W/m^2), in a certain distance, is expressed as:

$$q' = SEP_{\mathrm{act}}\, F_{\mathrm{view}}\, \tau_{\alpha}, \qquad \text{(C2.3)}$$

where, SEP_{act} (W/m^2) denotes the actual surface emitting power, F_{view} (-), the view factor, and τ_{α} (-) the atmospheric transmissivity. In the above relation, the surface emitting power is calculated empirically as a function of the burning rate and takes into consideration the fraction of the flame covered by soot, while the view factor depends upon the shape of the flame, the presence of wind and the distance of the receptor from the external flame surface. Finally, the atmospheric transmissivity takes into consideration the part of the heat flux absorbed by the air which is between the flame and the receptor of the radiation. In the following sections of

Fires models

Chapter C2, this approach will be adopted for the calculation of the heat flux for many types of fire. Once the heat flux is calculated, the following effects are examined:

- On humans (probable number of fatalities because of fires, as well as the probable number of injuries with 1st and 2nd degree burns).
- On materials (1st and 2nd degree damages).

Pool fire

Fire ball

Jet fire

Flash fire

Four different models of fires have been developed: pool fire, fire ball, jet fire and flash fire. Depending on the outflow, the following cases can be distinguished:

- If the outflow is in the liquid state, then a pool of flammable substance is formed, probably followed by a pool fire.
- If the outflow is a high-pressure flammable gas and ignites immediately, a jet fire is formed. If the outflow is very sudden then the result can be a flash fire.
- If the outflow is a low-pressure flammable gas and it does not ignite immediately, then a cloud of flammable gases can be created.

In the following sections of this chapter, a first attempt on the calculation of the heat flux of different types of fire, as well as its effects on humans and materials, is presented.

Table C2.1 Magnitude of Typical Heat Fluxes.

	Heat flux
Cigarette	5 W
Filament lamp	60 W
Wax	80 W
Burning dustbin	100,000 W
Pool fire 1 m^2 oil	2,000,000 W
Wooden palettes of 7 m height	7,000,000 W
Nuclear power production	3,200,000,000 W

Reproduced by kind permission of M. Al-Thani.

C2.1. Pool Fire

A pool fire is usually defined as a turbulent diffusion fire that burns over horizontal pool of vaporizing flammable material, under conditions where the flammable material has zero or very low initial momentum.

In a pool fire, the following can be observed:

- The fire characteristics depend, to a large degree, upon the meteorological conditions, and specifically the wind velocity.
- The duration of the pool fire is not instantaneous, but depends upon the quantity of the flammable material that evaporates.
- The burning rate of the flammable material is equal to its vaporization rate from the pool.
- There is a degree of feedback between the fire and the flammable material. Up to a limit, heat is transferred from the fire to the pool of the flammable material, and thus influences its vaporization rate, and consequently the size and other characteristics of the fire.

Pool fires usually appear in jet fuels and diesel oils, in hydrocarbons (heavier than hexane), glycols, oils and hydraulic fluids. For the calculation of the heat flux, the shape of the flame is considered cylindrical.

Pool fires are usually separated into three categories:

- Confined pool fires on land - fires of constant diameter, a pool of flammable material is created, ignites and starts burning.
- Unconfined pool fires on land - fires of continuously changing diameter, while the pool is already burning.
- Fires on water.

In this section, a thorough presentation of confined pool fires on land, the most common case, will be attempted. Analysis of the other two types of fires is based on the full understanding of this case.

The algorithm adopted for the calculation of the heat flux in the case of pool fires follows the general methodology, i.e., first the burning rate is calculated, and from it the maximum surface emitting power. Following that, the actual surface emitting power is obtained, and from that and the view factor, the heat flux is calculated.

Pool Fire

calculation procedure

Burning Rate
↓
Maximum Surface Emitting Power
↓
Actual Surface Emitting Power
↓
View Factor
↓
Heat Flux

Pool Fire

calculation procedure

Burning Rate
↓
Maximum Surface Emitting Power
↓
Actual Surface Emitting Power
↓
View Factor
↓
Heat Flux

C2.1.1. Burning Rate

The burning rate, m' (kg/m^2s), expresses the rate with which the flammable material that forms the pool burns. In the case of a pool fire, the temperature of the surface of the pool is very near the boiling point of the liquid fuel. The liquid is heated up to its boiling temperature and then it evaporates and burns in its vapor phase. Hence, since the burning rate is considered equal to the liquid mass lost (burnt) over a unit surface, a simple energy balance on the pool liquid will result in [Hottel 1958]

$$m'[\Delta H_{\mathrm{v}} + C_{\mathrm{p}}(T_{\mathrm{b}} - T_{\mathrm{a}})] = q_{\mathrm{r}} + q_{\mathrm{c}} - q_{\mathrm{rr}} - q_{\mathrm{misc}}\,. \qquad \text{(C2.4)}$$

In the above expression, ΔH_v (J/kg) is the heat of vaporization, while C_p (J/kg K) denotes the specific heat capacity of the fuel. The temperatures, T_b (K) and T_a (K), represent the boiling temperature and the ambient temperature of the fuel. q_r (W/m^2) denotes the radiative heat flux that is absorbed by the pool, q_c (W/m^2) is the heat transferred to the pool by convection and q_{rr} (W/m^2) is the heat flux that radiates back from the pool, as its temperature is relatively high. Finally the term q_{misc} (W/m^2) includes the heat flux lost to enclosure walls as well as other transient terms. Usually the last two terms are very small and are thus ignored.

The analysis performed by Hottel [Hottel 1958] on the data of Blinov and Khudiakov and according to Eq. (C2.4) showed two distinct areas of interest: pool fires in pools of large diameter, D (m), where heat transferred by radiation prevails, and pool fires in small-diameter pools where heat transferred by convection prevails. More analytically four regions were proposed.

D (m)	Heat transferred
< 0.05	convection, laminar flow
0.05 έως 0.2	convection, turbulent flow
0.2 έως 1.0	radiation, optically transparent flame
> 1.0	radiation, optically non-transparent flame

The burning rate can be calculated according to the following methods:
a) Zabetakis-Burgess Method [Zabetakis & Burgess 1961],
b) Burgess-Strasser-Grumer Method [Burgess *et al.* 1961],
c) Mudan Method [Mudan 1985].

a) Zabetakis-Burgess Method

Although there are no established analytic expressions for the region where heat transferred by convection prevails, for the region where radiation prevails ($D > 0.2$ m), Zabetakis and Burgess [Zabetakis & Burgess 1961] proposed the following modification of Eq. (C2.4) - in combination with Eq. (C2.1) - for the calculation of burning rate

$$m' = \frac{\sigma T_{\mathrm{f}}^4 \left(1 - e^{-k\beta D}\right)}{[\Delta H_{\mathrm{v}} + C_{\mathrm{p}}(T_{\mathrm{b}} - T_{\mathrm{a}})]} \, . \qquad \text{(C2.5)}$$

In the above expression, the terms in parentheses in the numerator represent the emissivity ε of Eq. (C2.1). Furthermore, β (-) is the mean-beam-length corrector and k (1/m), the absorption extinction coefficient of the flame. However, in the above expression, the flame temperature, T_{f} (K), which is not known, is employed. Thus, Eq. (C2.5) was substituted by the empirical expression that produces the burning rate as a function of the diameter, D (m), of the pool fire and the equivalent burning rate, m'_{∞} (kg/m^{2}s), of an infinite-diameter pool:

$$m' = m'_{\infty}\left(1 - e^{-k\beta D}\right). \qquad \text{(C2.6)}$$

Typical values for the parameters k and β, as well as the equivalent burning rate of an infinite-diameter pool, are given in Table C2.2 [Babrauskas 1983].

Table C2.2 Values of Various Parameters.

Flammable substance	m'_{∞} (kg/m^{2}s)	$k\beta$ (1/m)	k (1/m)	T_{f} (K)
Liquid H_2	0.169	6.1	-	1600
Liquefied Natural Gas (LNG)	0.078	1.1	0.5	1500
Liquefied Propane Gas (LPG)	0.099	1.4	0.4	-
Methanol	0.017	-	-	1300
Ethanol	0.015	-	0.4	1490
Butane	0.078	2.7	-	-
Hexane	0.074	1.9	4	-
Benzene	0.085	2.7	2	1490
Xylene	0.090	1.4	-	-
Gasoline	0.055	2.1	2	1450
Kerosine	0.039	3.5	2.6	1480
Jet fuel JP-5	0.054	1.6	0.5	1250

Note: Values for enthalpy and other physical properties are given in Appendix A.

Pool Fire

calculation procedure

Burning Rate
↓
Maximum Surface Emitting Power
↓
Actual Surface Emitting Power
↓
View Factor
↓
Heat Flux

Pool Fire

calculation procedure

Burning Rate
↓
Maximum Surface Emitting Power
↓
Actual Surface Emitting Power
↓
View Factor
↓
Heat Flux

b) Burgess-Strasser-Grumer Method

Burgess, Strasser and Grumer examined experimental values of the burning rate vs. the ratio ($\Delta H_c/\Delta H_G$) where $\Delta H_G = \Delta H_v + C_p(T_b - T_a)$. Here, ΔH_c (J/kg) and ΔH_v (J/kg) represent the heat of combustion and the heat of vaporization, respectively, C_p (J/kgK) the specific heat capacity of the fuel, and ρ_L (kg/m^3), the density of the liquid fuel at its boiling temperature. The temperatures T_b (K), and T_a (K), represent the boiling temperature of the fuel and the ambient temperature, respectively.

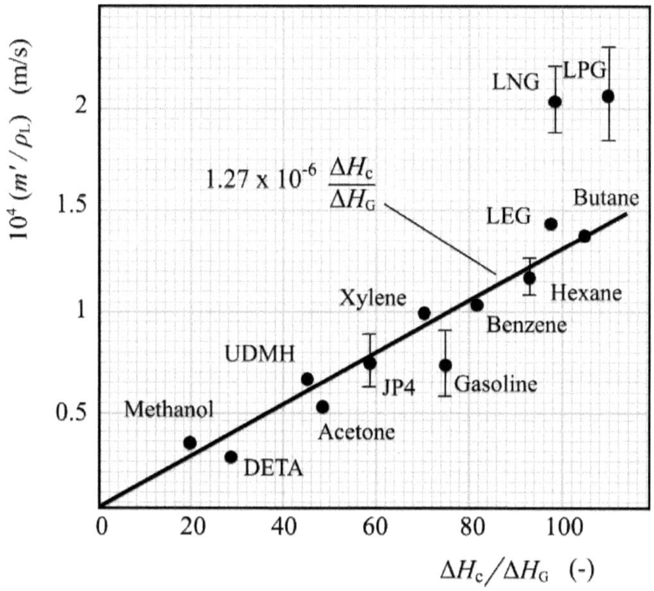

Figure C2.1 Expression of Burgess, Strasser and Grumer.

According to Figure C2.1, the burning rate, m' (kg/m^{2}s), can be calculated from the expression

$$m' = \rho_L c_1 \frac{\Delta H_c}{\Delta H_v + C_p(T_b - T_a)}, \qquad \text{(C2.7)}$$

where $c_1 = 1.27 \times 10^{-6}$ m/s.

This expression, as can be seen in the figure, produces good results for the calculation of the burning rate of hydrocarbons and liquid fuels in general, but it greatly underestimates the burning rate of liquefied gases such as LNG and LPG.

c) Mudan Method

In response to Burgess, Strasser and Grumer's inability to sufficiently describe the burning rate of liquefied gases, Mudan [Mudan 1984] proposed the following expression

$$m' = \frac{c_1\ \Delta H_c}{\Delta H_v + C_p(T_b - T_a)}, \qquad \text{(C2.8)}$$

where $c_1 = 0.001$ kg/m^{2}s.

The above expression, as can be seen in Figure C2.2, in comparison with the expression of Burgess, Strasser and Grumer, gives a better representation of the burning rate of liquefied gases, but predicts less accurately the burning rate of other fuels.

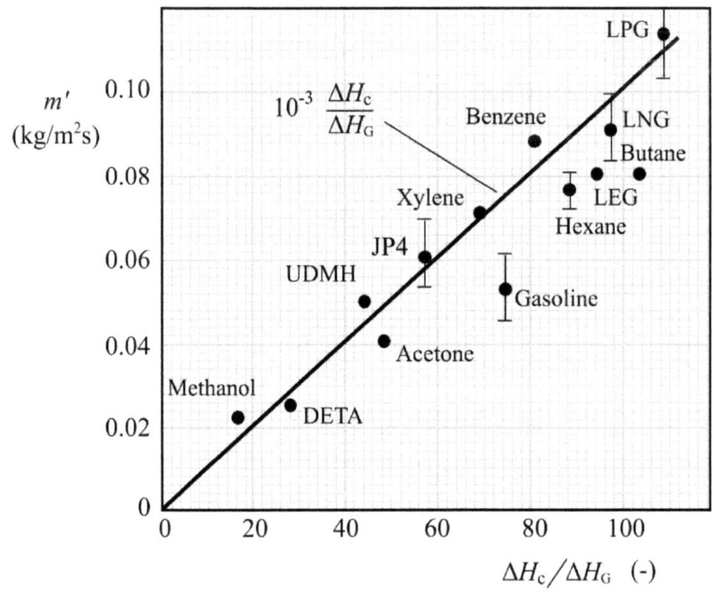

Figure C2.2 Expression of Mudan.

In conclusion the expressions of Zabetakis-Burgess and Burgess-Strasser-Grumer predict values of similar uncertainties. The Mudan expression is less accurate than the Burgess-Strasser-Grumer expression, but it also addresses the liquefied gases that are not covered by the first expression.

Pool Fire

example

Burning Rate

↓

Maximum Surface Emitting Power

↓

Actual Surface Emitting Power

↓

View Factor

↓

Heat Flux

EXAMPLE C2.1. *Burning Rate*

Calculate the burning rate of *n*-hexane in a pool fire of 10 m diameter. The following data are available:

-	Boiling temperature, T_b	:	341.9	K
	Ambient temperature, T_a	:	298	K
-	Heat of combustion, ΔH_c	:	44,700	kJ/kg
	Heat of vaporization, ΔH_v	:	450	kJ/kg
	Specific heat capacity, C_p	:	2.31	kJ/kg K
-	Liquid density, ρ_L (at T_b)	:	616	kg/m^3

a) Zabetakis-Burgess Method

The burning rate, m' (kg/m^{2}s), is calculated from Eq. (C2.6) and values obtained from Table C2.2, as

$$m' = m'_{\infty}\left(1 - e^{-k\beta D}\right) = 0.074\left(1 - e^{-1.9\times 10}\right) = 0.074 \text{ kg/m}^2\text{s}$$

b) Burgess-Strasser-Grumer Method

The burning rate, m' (kg/m^{2}s), is calculated from Eq. (C2.7), as

$$m' = \rho_L 1.27\times 10^{-6}\frac{\Delta H_c}{\Delta H_v + C_p(T_b - T_a)}$$

$$= 616\cdot 1.27\times 10^{-6}\frac{44{,}700}{450 + 2.31(341.9 - 298)} = 0.063 \text{ kg/m}^2\text{s}$$

c) Mudan Method

The burning rate, m' (kg/m^{2}s), is calculated from Eq. (C2.8) as

$$m' = \frac{c_1\,\Delta H_c}{\Delta H_v + C_p(T_b - T_a)} = \frac{0.001\cdot 44{,}700}{450 + 2.31(341.9 - 298)} = 0.081 \text{ kg/m}^2\text{s}$$

The experimental value for the burning rate of *n*-hexane is shown in Figure C2.2 as equal to 0.076 kg/m^{2}s. In this case, all three methods produce reasonable results, an expected fact as they all perform well in the case of hydrocarbons.

❐

Pool Fire

calculation procedure

Burning Rate
↓
Maximum Surface Emitting Power
↓
Actual Surface Emitting Power
↓
View Factor
↓
Heat Flux

C2.1.2. Maximum Surface Emitting Power

The maximum surface emitting power, SEP_{max} (kW/m^2), expresses the emitting power from the flame's surface if no soot is present. It is calculated as a function of the burning rate, m' (kg/m^{2}s) and the heat of combustion, ΔH_c (kJ/kg). However, as the burning rate refers to a pool of diameter D, while the emitting power refers (approximately) to the surface of a cylinder, a correction factor proportional to these two surfaces must be applied to the burning rate, that is $(\pi D^2/4)/(\pi DL+\pi D^2/4) = 1/(1+4(L/D))$. Therefore, in the final expression the mean flame length, L (m), and the pool diameter D (m), enter as

$$SEP_{max} = F_s \frac{1}{1+4(L/D)} m' \, \Delta H_c \qquad \text{(C2.9)}$$

In the above expression, the radiation fraction, F_s (-), represents the fraction of the combustion energy radiated from the flame temperature. A selection of values for the radiation fraction can be found in Table C2.3 [Burgess & Herzberg 1974]. Furthermore, in this expression it has been assumed that the radiated energy is approximately equal to the heat of combustion; the heat of vaporization as well as the enthalpy required to raise the temperature from ambient to the boiling point have both been ignored. This assumption is valid, as the value of the heat of combustion is much larger.

Table C2.3 Radiation Fraction.

Substance	D (m)	F_s (-)	Substance	D (m)	F_s (-)
Methanol	0.076	0.162	Gasoline	1.22	0.30-0.40
	0.152	0.165		1.53	0.16-0.27
	1.220	0.177		3.05	0.13-0.14
				>3.05	0.2
Methane	0.305	0.210			
	0.760	0.230	Benzene	0.076	0.350
	1.530	0.15-0.24		0.457	0.345
	3.050	0.24-0.34		0.760	0.350
	6.100	0.20-0.27		1.220	0.360
			Butane	0.305	0.199
				0.457	0.205
				0.700	0.269

Pool Fire

calculation procedure

Burning Rate
↓
Maximum Surface Emitting Power
↓
Actual Surface Emitting Power
↓
View Factor
↓
Heat Flux

In Eq. (C2.9), the mean flame length, L (m), can be calculated from the following two empirical methods:

a) Thomas Method [Thomas 1963],

b) Pritchard-Binding Method [Pritchard & Binding 1992].

a) Thomas Method

The expression proposed by Thomas [Thomas 1963] is empirical and is based upon laboratory measurements of fire on wood. It calculates the mean flame length, L (m), as

$$L/D = 42\left[\frac{m'}{\rho_{\text{air}}\sqrt{g\,D}}\right]^{0.61} . \qquad \text{(C2.10)}$$

In the above expression, ρ_{air} (kg/m^3), represents the density of air at ambient conditions, and g (m/s^2) the acceleration due to gravity (=9.814 m/s^2). The above equation is recommended for the calculation of the mean flame length in cases when the influence of wind is small, and there is no soot.

If a strong wind is present, a new dimensionless wind velocity, u^*, is introduced. This quantity is directly related to the wind velocity, u_w (m/s), at a 10 m height, as

$$u^* = u_w\left(\frac{g\,m'D}{\rho_{\text{air}}}\right)^{-1/3}, \qquad \text{if} \quad u^*<1 \quad \text{then} \quad u^*=1 \qquad \text{(C2.11)}$$

and Eq. (C2.10) becomes

$$L/D = 55\left[\frac{m'}{\rho_{\text{air}}\sqrt{g\,D}}\right]^{0.67}\left(u^*\right)^{-0.21} . \qquad \text{(C2.12)}$$

This equation usually underestimates the length of the flame.

b) Pritchard-Binding Method

As the expression of Thomas underestimates the mean flame length, Pritchard-Binding [Pritchard & Binding 1992] proposed an improved form as

$$L/D = 10.615\left[\frac{m'}{\rho_{\text{air}}\sqrt{g\,D}}\right]^{0.305}\left(u^*\right)^{-0.03} , \qquad \text{(C2.13)}$$

where the dimensionless wind velocity u^* was defined in Eq. (C2.11).

EXAMPLE C2.2. *Mean Flame Length*

Pool Fire

calculation procedure

Burning Rate
↓
Maximum Surface Emitting Power
↓
Actual Surface Emitting Power
↓
View Factor
↓
Heat Flux

Calculate the mean flame length of a 5-m diameter pool fire that resulted from a leak of JP-5 fuel.

Burning rate, m'	:	0.054	kg/m^2s (Table C2.2)
Wind velocity, u_w (10 m)	:	5	m/s
Density of air, ρ_{air}	:	1.21	kg/m^3

According to the given data, both the Thomas method and the Pritchard-Binding method can be employed.

a) Thomas Method

Eq. (C2.12) can be used,

$$L/D = 55\left[\frac{m'}{\rho_{air}\sqrt{g\,D}}\right]^{0.67}(u^*)^{-0.21}$$

where

$$u^* = u_w\left(\frac{g\,m'D}{\rho_{air}}\right)^{-1/3}.$$

By direct substitution, one obtains $u^* = 3.85$, and $L = 7$ m.

b) Pritchard-Binding Method

The expression of Pritchard-Binding [Pritchard & Binding 1992], Eq. (C2.13), is

$$L/D = 10.615\left[\frac{m'}{\rho_{air}\sqrt{g\,D}}\right]^{0.305}(u^*)^{-0.03}$$

Employing the value $u^* = 3.85$, one obtains $L = 10.8$ m.

Both methods produce different values (see also Figure C2.3). The expression of Thomas results in a lower mean flame length.

❐

Pool Fire

example

Burning Rate
↓
Maximum Surface Emitting Power
↓
Actual Surface Emitting Power
↓
View Factor
↓
Heat Flux

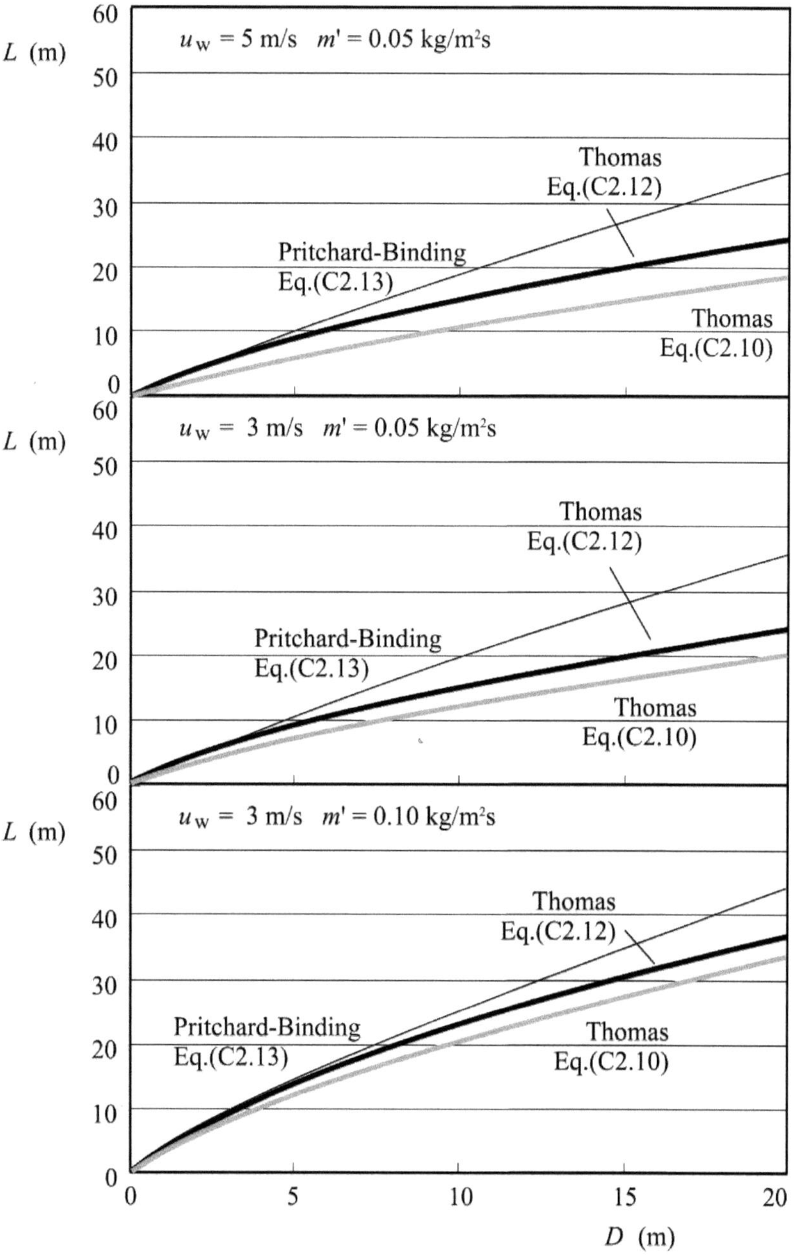

Figure C2.3 Influence of the wind velocity and the burning rate in the calculation of the mean flame length.

C2.1.3. Actual Surface Emitting Power

The maximum surface emitting power, SEP_{max} (kW/m^2), that was calculated in the previous section, refers to the radiation from the flame's surface in cases when there is no soot. Soot, however, results in a considerable reduction of this radiation. The actual surface emitting power, SEP_{act} (kW/m^2), takes into consideration the presence of soot and it is calculated from the expression

$$SEP_{act} = SEP_{max}(1-s) + SEP_{soot}\, s \, . \qquad \text{(C2.14)}$$

In the above expression the term SEP_{soot} (kW/m^2) represents the surface emitting power of soot, while s (-) denotes the surface fraction that is covered by soot. For oil products, s equals 80%, while SEP_{soot} = 20 kW/m^2 [Hagglund & Person, 1976]. In the case of fires of small diameter (1-2 m), where not much smoke is formed, it can be assumed that $SEP_{act} = SEP_{max}$.

The actual surface emitting power can alternatively be calculated with the following two empirical methods, according the material burnt:

a) Mudan-Croce Method for Hydrocarbon Fuels Fires,
b) Non-Hydrocarbon Fuels Fire Method.

Table C2.4 Actual Surface Emitting Power (SEP_{act}) as a Function of the Ratio Length/Diameter of Pool Dire (T_b> 20°C) [TNO 2005].

	D = 1 M		D = 10 M	
Substance	L/D (-)	SEP_{act} (kW/m^2)	L/D (-)	SEP_{act} (kW/m^2)
Acetone	3.06	42	1.52	79
Benzene	4.16	71	2.06	135
n-Hexane	4.53	87	2.24	166
Methanol	1.59	19	0.79	34
Methyl ester	2.59	26	1.28	48

Pool Fire

calculation procedure

Burning Rate
↓
Maximum Surface Emitting Power
↓
Actual Surface Emitting Power
↓
View Factor
↓
Heat Flux

a) Mudan-Croce for Hydrocarbon Fuels Fire Method

In the case of pool fires of hydrocarbons with a molecular weight of around 60 the actual surface emitting power, SEP_{act} (kW/m^2), can alternatively be calculated as a function of the pool diameter, D (m), from the empirical expression [Mudan & Croce 1988]

$$SEP_{act} = 140\, e^{-0.12D} + 20\,(1 - e^{-0.12D})\,. \qquad \text{(C2.15)}$$

The value of 20 kW/m^2 corresponds to the emitting power of soot [Hagglund & Person 1976] and is valid for large diameter pool fires caused by jet plane fuel JP-4, benzene, kerosene, etc. The expression produces good results for pool diameters well above 15 m.

b) Non-Hydrocarbon Fuels Fire Method

For non-hydrocarbon pool fires the following empirical expression can also be employed [TNO 2005]

$$SEP_{act} = \frac{c_3\ m'\,\Delta H_c}{1 + 72\,(m')^{0.61}}\,, \qquad \text{(C2.16)}$$

where parameter $c_3 = 0.35$ and ΔH_c (kJ/kg) is the heat of combustion.

Explosion and fire at the Barton Solvents distribution facility in Valley Center, KS, U.S.A. (Reproduced by kind permission of the U.S. Chemical Safety Board.)

EXAMPLE C2.3. *Actual Surface Emitting Power*

Calculate the actual surface emitting power of a 5-m diameter *n*-hexane pool fire. The following data are available:

-	Burning rate, m'	:	0.074	kg/m^2s (Example C2.1, Zabetakis-Burgess equation)
	Wind velocity, u_w (10 m)	:	5	m/s
	Density of air, ρ_{air}	:	1.21	kg/m^3
-	Heat of combustion, ΔH_c	:	44,700	kJ/kg
	Radiation fraction, F_s	:	0.2	-

Initially the maximum surface emitting power must be calculated. This is a function of the burning rate and the ratio (*L/D*). To calculate the ratio (*L/D*) there are two equations available, those of Thomas (C2.10) and of Pritchard-Binding (C2.13). Here, the Pritchard-Binding equation (C2.13) will be employed as it is more accurate.

$$L/D = 10.615 \left[\frac{m'}{\rho_{air}\sqrt{g\,D}} \right]^{0.305} (u^*)^{-0.03} \quad \text{where} \quad u^* = u_w \left(\frac{g\, m' D}{\rho_{air}} \right)^{-1/3}$$

By direct substitution, we obtain u^* = 3.47, L/D =2.41 and L = 12 m. The maximum surface emitting power, SEP_{max} (kW/m^2), is given by Eq. (C2.9) as

$$SEP_{max} = F_s \frac{1}{1+4(L/D)} m' \,\Delta H_c \,,$$

and as $F_s = 0.2$, one obtains $SEP_{max} = 62\ kW/m^2$.

The actual surface emitting power, SEP_{act} (kW/m^2), is obtained by Eq. (C2.14)

$$SEP_{act} = SEP_{max}(1-s) + SEP_{soot}\, s \,.$$

Assuming $s = 0.8$ and $SEP_{soot} = 20\ kW/m^2$, we obtain $SEP_{act} = 28.4\ kW/m^2$.

If Eq. (C2.15) was employed, we would have found that $SEP_{act} = 86\ kW/m^2$. This value is quite different from the previous one, because Eq. (C2.15) produces good results for a pool diameter $D >> 15$ m.

❐

Pool Fire

example

Burning Rate
↓
Maximum Surface Emitting Power
↓
Actual Surface Emitting Power
↓
View Factor
↓
Heat Flux

C2.1.4. View Factor

The view factor, F_{view} (-), is defined as the fraction of the emitted radiation that reaches the receptor per unit area (the receptor can be a person or some material). To describe the emitted radiation in pool fires, the solid-flame model is employed. In Figure C2.4, the flame is considered as a tilted cylinder to account for wind which is the most typical (for a better understanding of the meaning of the view factor, the reader is recommended to study Section C2.2.5, which refers to the view factor for fire balls where the equations are considerably simpler).

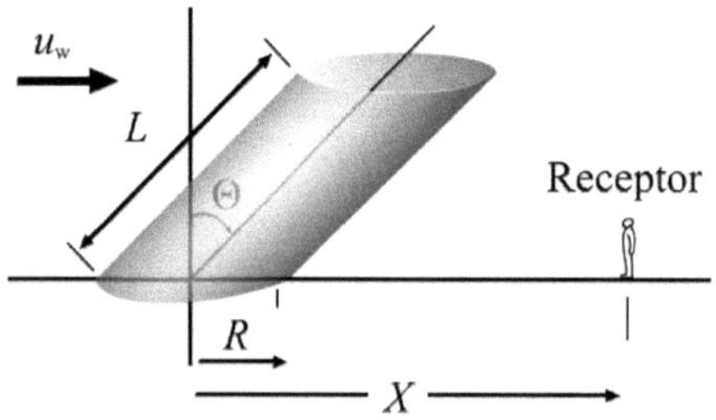

Figure C2.4 Flame as tilted cylinder.

The angle of tilt, Θ, represents the tilt of the cylinder because of the wind, and is measured with respect to the vertical.

The view factor, F_{view}, is calculated as a function of the perpendicular contribution F_{v}, and the horizontal contribution F_{h} as

$$F_{\text{view}} = \sqrt{F_{\text{v}}^2 + F_{\text{h}}^2}\,. \tag{C2.17}$$

The two contributions F_{v} and F_{h} have been derived [Mudan 1987] for the above geometry as

$$\begin{aligned}\pi F_{\text{v}} &= -E\tan^{-1} D + E\left[\frac{\alpha^2+(\beta+1)^2-2\beta(1+\alpha\sin\Theta)}{AB}\right]\tan^{-1}\left(\frac{AD}{B}\right)\\ &+\frac{\cos\Theta}{C}\left[\tan^{-1}\left(\frac{\alpha\beta-F^2\sin\Theta}{FC}\right)+\tan^{-1}\left(\frac{F\sin\Theta}{C}\right)\right],\end{aligned} \tag{C2.18}$$

and

$$\begin{aligned}\pi F_{\text{h}} &= \tan^{-1}\left(\frac{1}{D}\right)+\frac{\sin\Theta}{C}\left[\tan^{-1}\left(\frac{\alpha\beta-F^2\sin\Theta}{FC}\right)+\tan^{-1}\left(\frac{F\sin\Theta}{C}\right)\right]\\ &-\left[\frac{\alpha^2+(\beta+1)^2-2(\beta+1+\alpha\beta\sin\Theta)}{AB}\right]\tan^{-1}\left(\frac{AD}{B}\right),\end{aligned} \tag{C2.19}$$

where

$$\alpha = L/R \quad \text{and} \quad \beta = X/R, \tag{C2.20}$$

and

$$A = \sqrt{\alpha^2 + (\beta+1)^2 - 2\alpha(\beta+1)\sin\Theta}\,, \qquad D = \sqrt{(\beta-1)/(\beta+1)}\,, \qquad \text{(C2.21)}$$

$$B = \sqrt{\alpha^2 + (\beta-1)^2 - 2\alpha(\beta-1)\sin\Theta}\,, \qquad E = \alpha\cos\Theta/(\beta - \alpha\sin\Theta)\,, \qquad \text{(C2.22)}$$

$$C = \sqrt{1 + (\beta^2 - 1)\cos^2\Theta}\,, \qquad F = \sqrt{(\beta^2 - 1)}\,. \qquad \text{(C2.23)}$$

The angle of tilt, Θ, that enters the above equations is obtained as a function of the wind velocity from the expressions [Pritchard & Binding 1992]

$$\tan\Theta / \cos\Theta = 0.666\, Fr^{0.333} Re^{0.117}\,, \qquad \text{(C2.24)}$$

where the Froude number, *Fr*, and the Reynolds number, *Re*, are given as

$$Fr = \frac{u_w^2}{g\,D} \quad \text{and} \quad Re = \frac{u_w \rho_{air} D}{\eta_{air}}\,. \qquad \text{(C2.25)}$$

In Table C2.5, values for the view factor, F_{view} (-), for fires of radius R (m) and length L (m) are shown. These values have been calculated for wind velocity u_w = 5 m/s, in various distances X (m) from the center of the fire.

Table C2.5 Calculated Values of the View Factor F_{view} from Eq. (C2.17)-(C2.23).

		$\alpha = L/R$ (u_w= 5 m/s, Θ = 49.6°)				
$\beta = X/R$		0.1	0.5	1	5	10
1.1	F_v	0.458	0.394	0.380	0.375	0.375
	F_h	0.313	0.782	0.798	0.803	0.803
	$\boldsymbol{F_{view}}$	**0.555**	**0.876**	**0.884**	**0.886**	**0.886**
1.3	F_v	0.119	0.457	0.404	0.379	0.378
	F_h	0.016	0.554	0.690	0.722	0.723
	$\boldsymbol{F_{view}}$	**0.120**	**0.718**	**0.800**	**0.815**	**0.816**
1.5	F_v	0.057	0.397	0.415	0.365	0.364
	F_h	0.004	0.268	0.566	0.657	0.658
	$\boldsymbol{F_{view}}$	**0.057**	**0.479**	**0.702**	**0.752**	**0.752**
2	F_v	0.021	0.149	0.332	0.320	0.316
	F_h	0.001	0.036	0.225	0.530	0.535
	$\boldsymbol{F_{view}}$	**0.021**	**0.154**	**0.401**	**0.619**	**0.621**
5	F_v			0.026	0.168	0.158
	F_h			0.003	0.159	0.228
	$\boldsymbol{F_{view}}$			**0.026**	**0.235**	**0.277**
10	F_v				0.041	0.084
	F_h				0.012	0.071
	$\boldsymbol{F_{view}}$				**0.043**	**0.110**

Pool Fire

calculation procedure

Burning Rate
↓
Maximum Surface Emitting Power
↓
Actual Surface Emitting Power
↓
View Factor
↓
Heat Flux

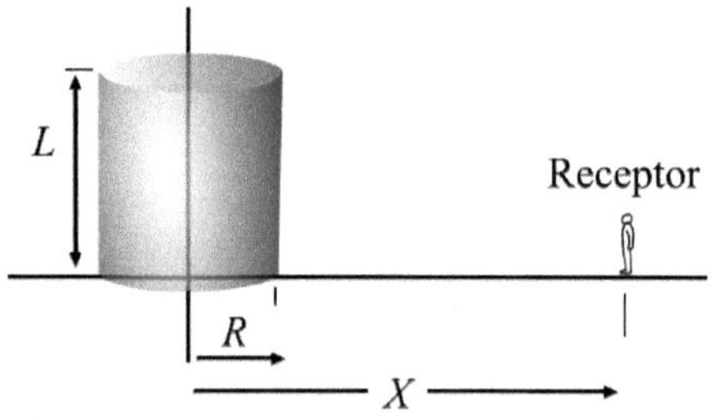

Figure C2.5 Flame as straight cylinder.

$\alpha = L/R$

and

$\beta = X/R$

In cases where no wind is present, then the previous equations simplify [Mudan 1987] as

$$F_{\text{view}} = \sqrt{F_{\text{v}}^2 + F_{\text{h}}^2}\,, \tag{C2.26}$$

$$\pi F_{\text{v}} = -\frac{\alpha}{\beta}\tan^{-1}(D) + \left[\frac{\alpha}{\beta}\frac{(\alpha^2+\beta^2+1)}{AB}\right]\tan^{-1}\left(\frac{AD}{B}\right) + \frac{1}{\beta}\tan^{-1}\left(\frac{\alpha}{\sqrt{\beta^2-1}}\right), \tag{C2.27}$$

$$\pi F_{\text{h}} = \tan^{-1}\left(\frac{1}{D}\right) - \left[\frac{\alpha^2+\beta^2-1}{AB}\right]\tan^{-1}\left(\frac{AD}{B}\right), \tag{C2.28}$$

όπου

$$A = \sqrt{\alpha^2 + (\beta+1)^2}\,,\quad B = \sqrt{\alpha^2 + (\beta-1)^2}\,,\quad D = \sqrt{(\beta-1)/(\beta+1)}\,. \tag{C2.29}$$

Table C2.6 presents values for the view factor, F_{view} (-), for pool fires of radius R (m) and flame length L (m), in various distance X (m) from the center of the fire, for the case of no wind present (wind velocity is zero).

In Section C2.1.4, equations for the view factor were given. These were for the cases when the flame can be considered as a cylinder, as well as for the special case when the wind blows towards the receptor "worst case scenario." For all other cases the reader is referred to the corresponding literature [Mudan 1987].

Table C2.6 Calculated Values of the View Factor F_{view} from Eq. (C2.26)-(C2.29).

		$\alpha = L/R$ (u_w= 0 m/s, Θ = 0°)				
$\beta = X/R$		0.1	0.5	1	5	10
1.1	F_v	0.331	0.450	0.454	0.455	0.455
	F_h	0.132	0.332	0.355	0.363	0.363
	$\boldsymbol{F_{\text{view}}}$	**0.356**	**0.560**	**0.576**	**0.582**	**0.582**
1.3	F_v	0.131	0.345	0.377	0.385	0.385
	F_h	0.020	0.178	0.243	0.278	0.279
	$\boldsymbol{F_{\text{view}}}$	**0.132**	**0.388**	**0.448**	**0.474**	**0.475**
1.5	F_v	0.072	0.254	0.312	0.333	0.333
	F_h	0.007	0.097	0.171	0.229	0.231
	$\boldsymbol{F_{\text{view}}}$	**0.072**	**0.272**	**0.356**	**0.404**	**0.406**
2	F_v	0.028	0.126	0.195	0.249	0.250
	F_h	0.001	0.028	0.074	0.158	0.165
	$\boldsymbol{F_{\text{view}}}$	**0.029**	**0.129**	**0.208**	**0.295**	**0.299**
5	F_v		0.015	0.029	0.087	0.097
	F_h		0.001	0.003	0.038	0.055
	$\boldsymbol{F_{\text{view}}}$		**0.015**	**0.030**	**0.095**	**0.112**
10	F_v				0.029	0.042
	F_h				0.007	0.017
	$\boldsymbol{F_{\text{view}}}$				**0.030**	**0.046**

C2.1.5. Heat Flux

The heat flux, q'(kW/m^2), in a certain distance from the center of the fire (distance enters equations via the view factor and the atmospheric transmissivity), is calculated as discussed in the introduction - see Eq. (C2.3) - from the product of the actual surface emitting power, SEP_{act} (kW/m^2), the view factor, F_{view} (-), and the atmospheric transmissivity, τ_α (-), as

$$q' = SEP_{\text{act}}\, F_{\text{view}}\, \tau_{\text{a}}\,. \tag{C2.30}$$

For the calculation of the atmospheric transmissivity, τ_α (-), the following empirical expression can be employed [Bagster & Pittblado 1989]

$$\tau_{\text{a}} = c_4 \left[P_{\text{w}}(X - R)\right]^{-0.09}. \tag{C2.31}$$

In the above expression, P_{w} (Pa) denotes the partial water vapor pressure in air while X (m) is the distance of the receptor from the center of the fire of radius R (m). The constant c_4 is equal to 2.02 Pa$^{0.09}$m$^{0.09}$. The partial water vapor pressure in air can easily be calculated from the saturation vapor pressure, P_{w}^{o} (Pa), in air and the relative humidity, RH (fraction 0-1), as

$$P_{\text{w}} = RH\, P_{\text{w}}^{\text{o}}\,. \tag{C2.32}$$

Pool Fire

example

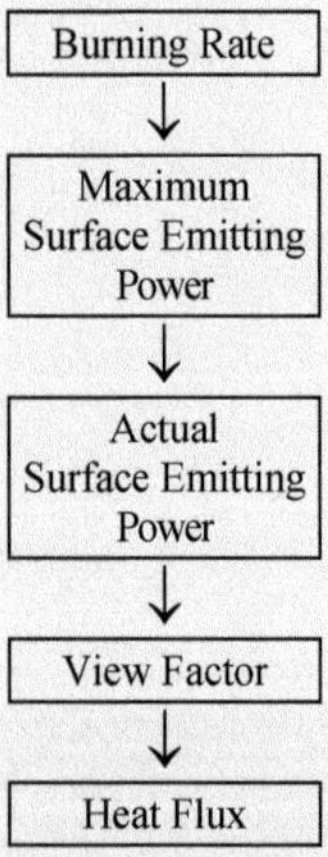

EXAMPLE C2.4. *Heat Flux*

Calculate the heat flux (a) on the flame's surface, and (b) 20 m from the flame's surface in the direction of the wind, from a 0.02-m thick pool formed after spillage of 28.3 m^3 of petrol. The following data are available:

-	Boiling temperature, T_b	:	423	K
	Heat of combustion, ΔH_c	:	45,000	kJ/kg
	Heat of vaporization, ΔH_v	:	370	kJ/kg
	Specific heat capacity, C_p	:	2.21	kJ/kg K
	Ambient temperature, T_a	:	298	K
-	Soot surface emitting power, SEP_{soot}	:	20	kW/m^2
-	Wind velocity, u_w	:	5	m/s
	Density of air, ρ_{air}	:	1.21	kg/m^3
	Viscosity of air, η_{air}	:	16.7	μPa s
	Saturation water vapor pressure, P_w^o	:	2,320	Pa
	Relative humidity, RH	:	0.7	-

a) Burning Rate

The Mudan expression, Eq. (C2.8), can be employed to calculate the burning rate. According to this expression the burning rate, m' (kg/m^{2}s), is given as

$$m' = \frac{c_1\,\Delta H_c}{\Delta H_v + C_p(T_b - T_a)} = \frac{0.001 \cdot 45{,}000}{370 + 2.21(423 - 298)} = 0.069\ \text{kg/m}^2\text{s}$$

b) Maximum Surface Emitting Power

- To start with, the pool diameter is required. Since 28.3 m^3 spilled and a 0.02 m thick pool was formed, the pool diameter, D (m), can be obtained from D^2 = 28.3x4/(0.02π) and thus D = 42.4 m.

- Using the Pritchard-Binding expression, Eq. (C2.13), which produces the best results,

$$L/D = 10.615\left[\frac{m'}{\rho_{air}\sqrt{g\,D}}\right]^{0.305}(u^*)^{-0.03} \quad \text{where} \quad u^* = u_w\left(\frac{g\,m'D}{\rho_{air}}\right)^{-1/3}$$

By direct substitution one obtains u^* = 1.74 and L = 74 m.

Pool Fire

example

Burning Rate → Maximum Surface Emitting Power → Actual Surface Emitting Power → View Factor → Heat Flux

Alternatively, if the expressions of Thomas, Eqs. (C2.11) and (C2.12) were used, the length of the flame would have been $L = 40.7$ m (for $D >> 5$ m, the two expressions produce different results as expected; see Figure C2.3).
The example will be continued with $L = 74$ m.

The maximum surface emitting power, SEP_{max} (kW/m^2), is calculated from Eq. (C2.9). Since all variables are known, employing $F_s = 0.2$ (Table C2.3), we obtain

$$SEP_{max} = \frac{F_s}{1 + 4(L/D)} m' \Delta H_c = 78.6 \text{ kW/m}^2$$

(For $L = 40.7$ m, we would have obtained $SEP_{max} = 129.6$ kW/m^2, which is much higher).

c) Actual Surface Emitting Power
The actual surface emitting power, SEP_{act} (kW/m^2), is calculated from Eq. (C2.14), as the soot surface emitting power, $SEP_{soot} = 20$ kW/m^2. Hence for $s = 0.8$,

$$SEP_{act} = SEP_{max}(1 - s) + SEP_{soot}\, s = 31.7 \text{ kW/m}^2$$

If the empirical expression of Mudan-Crose, Eq. (C2.15) was used, as it can be applied to benzene fires, then

$$SEP_{act} = 140\, e^{-0.12D} + 20\,(1 - e^{-0.12D}) = 20.7 \text{ kW/m}^2 .$$

This value is too low. Since the Mudan-Crose expression is empirical and independent of the type of fuel, the previous result is considered as more acceptable. Hence the example is continued with $SEP_{act} = 31.7$ kW/m^2.

d) View Factor
Solving Eqs. (C2.17)-(C2.25), one obtains $\Theta = 49.6°$ and

- on the flame's surface $F_{view} = 0.938$ ($F_v = 0.324$, $F_h = 0.881$)
- 20 m from the flame's surface $F_{view} = 0.629$ ($F_v = 0.331$, $F_h = 0.535$)

Pool Fire

example

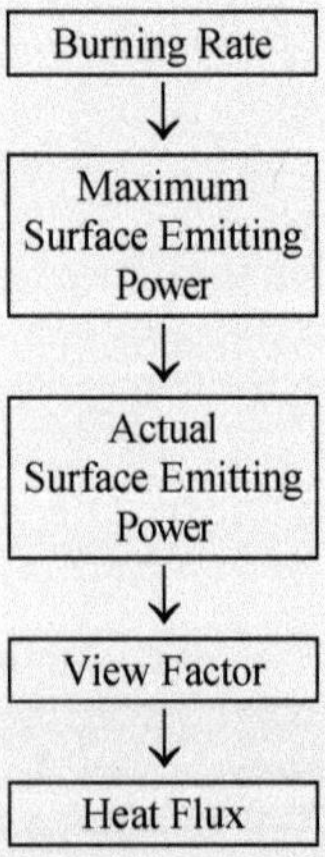

e) Heat Flux

The heat flux is calculated from Eq. (C2.30), as

$$q' = SEP_{act}\, F_{view}\, \tau_a$$

Therefore, on the flame's surface $q' = 31.7 \times 0.938 \times 1 = 29.7$ kW/m^2
and in 20 m from the flame's surface $q' = 31.7 \times 0.629 \times 0.79 = 15.8$ kW/m^2
For the second case, the atmospheric transmissivity was calculated from Eqs. (C2.31)-(C2.32), as $\tau_\alpha = 0.79$, while for the first case $\tau_\alpha = 1$, as it refers to the flame's surface.

Figure C2.6 shows the change of heat flux with distance (distance enters through the view factor and the atmospheric transmissivity).

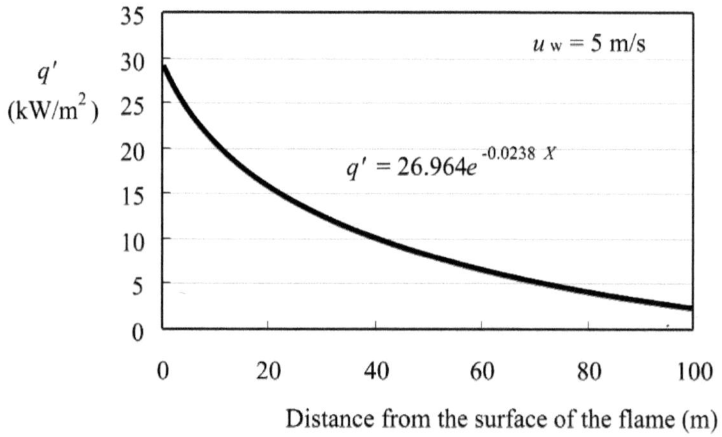

Figure C2.6 Change of heat flux with distance.

It can also be noted that if velocity drops to 0.1 m/s, then in a distance of 20 m from the flame,

$$\Theta = 4.8\,°, \quad F_{view} = 0.322 \quad \text{and} \quad q' = 7.7 \text{ kW/m}^2.$$

That is, the wind velocity strongly influences the heat flux (referring of course to wind velocity in a direction towards the receptor).

❐

Fire Ball

calculation procedure

Dimensions and Duration
↓
Burning Rate
↓
Maximum Surface Emitting Power
↓
Actual Surface Emitting Power
↓
View Factor
↓
Heat Flux

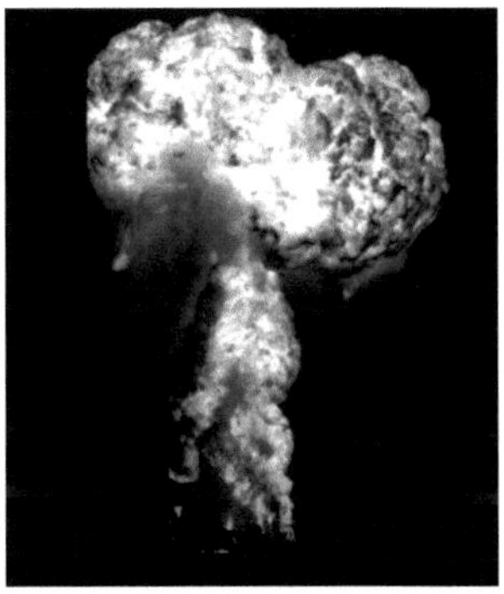
Reproduced by kind permission of D. Garcia.

C2.2. Fire Ball

A **Fire Ball** can be the result of a very rapid outflow and ignition of pressurized flammable gases. Even today, the conditions under which the rapid outflow of a pressurized flammable gas can result in a fire ball are not yet fully understood.

Usually a fire ball results from the rapid ignition that follows an event known as a BLEVE* (Boiling Liquid Expanding Vapor Explosion), but it can also appear during the ignition of a flammable gas mixture. Fire balls can radiate very large amounts of heat causing material damages, injuries or deaths in an area much larger than the fire radius.

In a fire ball, the following can be observed:

- The formation of a fire ball, to a large extent, does not depend upon the meteorological conditions because of the high pressure with which the gas is liberated.
- The duration of a fire ball is very small or instantaneous.
- The burning rate is equal to the total quantity of the flammable substance divided by the duration of the fire.

For the calculation of the heat flux, the shape of the fire ball is considered to be spherical.

The algorithm adopted for the calculation of the heat flux in the case of fire balls, follows the general methodology, i.e., firstly, the dimensions and the duration of the fire as well as the burning rate are calculated and from them, the maximum surface emitting power. Following that, the actual surface emitting power is obtained and from that, and the view factor, the heat flux is calculated.

---

* BLEVE (Boiling Liquid Expanding Vapor Explosion): In cases when a vessel with pressurized liquefied gas is heated externally by fire, the pressure inside the vessel increases and the wall weakens, and the vessel ruptures resulting in the instantaneous release of its contents to the atmosphere (see Section C4).

Fire Ball

calculation procedure

Dimensions and Duration
↓
Burning Rate
↓
Maximum Surface Emitting Power
↓
Actual Surface Emitting Power
↓
View Factor
↓
Heat Flux

C2.2.1. Dimensions and Duration of a Fire Ball

In this section, expressions (mostly empirical) for the calculation of the dimensions of a fire ball and its duration are presented. These expressions are based upon the knowledge of the mass, M (kg), of the flammable substance. This mass is usually obtained from the initial volume, V (m^3), of the vessel, the fraction, f (-), of this volume which is filled with pressurized gas in the liquid phase, and the density, ρ_{mat} (kg/m^3), of the flammable substance, as

$$M = f\,V\,\rho_{mat}\,. \tag{C2.33}$$

Actual observations of fire balls lead to the following typical guidelines. During the first third of the total time duration, t_{max} (s), of the fire ball, its diameter, D (m), increases until it reaches its maximum value, D_{max} (m). Following that, in the remaining time, the fire ball rises while its diameter remains constant. The maximum height that the center of this sphere rises from the ground is about equal to its diameter.

According to these observations, the increase of the diameter, D (m), of the fire ball sphere for the first third of the total time is given by the empirical expression [Hardee, Lee & Benedict 1978], as a function of the time, t (s),

$$D = c_1 M^{1/4} t^{1/3}\,. \tag{C2.34}$$

In the above expression $c_1 = 8.664$ mkg$^{-1/4}$s$^{-1/3}$.

In order to calculate the maximum diameter, D_{max} (m), and the total time duration, t_{max} (s), of the fire ball, two empirical algorithms have been proposed:

a) Roberts Method [Roberts 1982] and

b) TNO Method [TNO 2005]

a) Roberts Method

According to Roberts [Roberts 1982], the maximum diameter, D_{max} (m), and the total time duration, t_{max} (m), of the fire ball sphere are calculated from the following empirical expressions:

$$D_{max} = c_2\, M^{1/3}\,, \tag{C2.35}$$

$$t_{max} = c_3\, M^{1/3}\,, \tag{C2.36}$$

where, $c_2 = 5.8$ mkg$^{-1/3}$ and $c_3 = 0.45$ skg$^{-1/3}$.

b) TNO Method

TNO [TNO 2005] proposed that the maximum diameter, D_{max} (m), and the total time duration, t_{max} (s), of the fire ball must be calculated from the following empirical expressions:

$$D_{max} = c_4\, M^{0.325}, \tag{C2.37}$$

$$t_{max} = c_5\, M^{0.26}, \tag{C2.38}$$

where, $c_4 = 6.48$ mkg$^{-0.325}$ and $c_5 = 0.852$ skg$^{-0.26}$.

Figure C2.7 shows the comparison of the two methods for two cases: (a) mass of 10,000 kg and (b) mass of 20,000 kg. It can be noticed that the values for both the total duration time and the maximum diameter are very close. In the same figure, the increase in diameter, Eq. (C2.34) for the first third of the total duration time is also shown. In this time, the diameter has reached its maximum value.

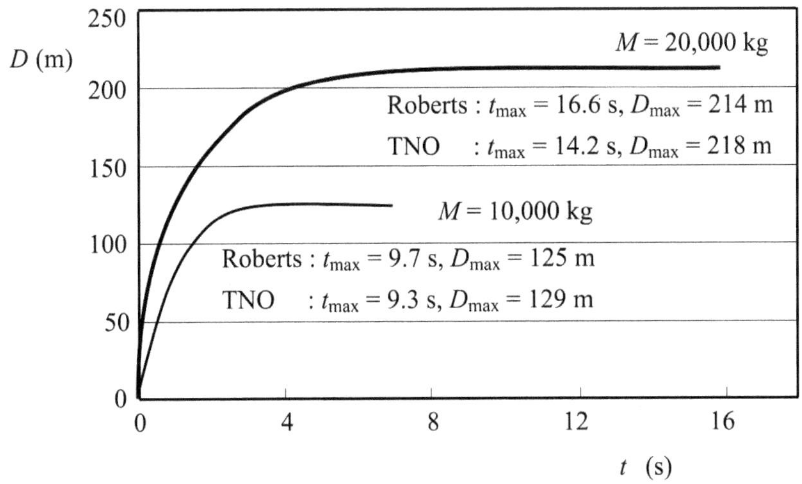

Figure C2.7 Comparison of the methods of Roberts and TNO.

Finally, the **height of the fire ball center from the ground**, H (m), is usually considered equal to the maximum diameter, D_{max} (m).

$$H = D_{max}\,. \tag{C2.39}$$

Fire Ball

calculation procedure

Dimensions and Duration
↓
Burning Rate
↓
Maximum Surface Emitting Power
↓
Actual Surface Emitting Power
↓
View Factor
↓
Heat Flux

Fire Ball

calculation procedure

Dimensions and Duration
↓
Burning Rate
↓
Maximum Surface Emitting Power
↓
Actual Surface Emitting Power
↓
View Factor
↓
Heat Flux

C2.2.2. Burning Rate

The burning rate, m' (kg/m^{2}s), can be calculated as a function of the mass, M (kg), of the flammable substance and the total fire ball duration, t_{max} (s), as

$$m' = \frac{M}{(0.888\pi\, D_{max}^2)\, t_{max}}, \qquad \text{(C2.40)}$$

where $(0.888\pi\, D_{max}^2)$ is the time-average surface of the fire ball sphere.

C2.2.3. Maximum Surface Emitting Power

The maximum surface emitting power, SEP_{max} (kW/m^2), as in the case of pool fires, Eq. (C2.9) can be calculated as a function of the burning rate, m' (kg/m^{2}s), the heat of combustion, ΔH_c (kJ/kg), and the radiation fraction, F_s (-), that represents the fraction of the combustion energy radiated from the flame surface, as

$$SEP_{max} = F_s\, m'\, \Delta H_c\,. \qquad \text{(C2.41)}$$

In this expression it has been assumed that the radiated energy is approximately equal to the heat of combustion; that is, the heat of vaporization as well as the enthalpy required to raise the temperature from ambient to the boiling point have both been ignored. This assumption is valid, as the value of the heat of combustion is much larger. Finally, in Eq. (C2.37) no surface correction proportional to that in Eq. (C2.9) is necessary, as the burning rate in this case already refers to the surface of the sphere.

For the calculation of the radiation fraction, F_s (-), the following expression has been proposed [TNO 2005, Roberts 1982]

$$F_s = c_6 P_{sv}^{0.32}, \qquad \text{(C2.42)}$$

where c_6 = 0.00325 Pa$^{-0.32}$, and P_{sv} (Pa), denotes the vapor pressure inside the vessel. Usually the radiation fraction takes values between 0.2 and 0.4 [Roberts 1982].

C2.2.4. Actual Surface Emitting Power

In the case of a fire ball which is a very rapid expansion of gases with a very small duration of fire, usually it is assumed that not enough soot is formed to be able to influence the radiative heat flux. Hence, it is assumed that

$$SEP_{act} = SEP_{max} \,. \qquad \text{(C2.43)}$$

C2.2.5. View Factor

The view factor, F_{view} (-), expresses the fraction of the emitted radiation that reaches the receptor per unit area (the receptor can be human or any object). In this case the shape of the fire is considered as a perfect sphere (Figure C2.8).

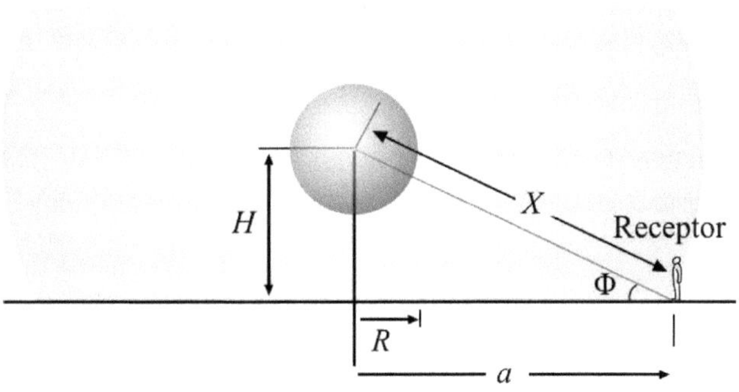

Figure C2.8. Fire ball.

The total emitted heat from a sphere of radius R is equal to $SEP_{act}(4\pi R^2)$ (kW). In the distance, X (m), of the receptor, the same emitted heat per unit area is equal to $SEP_{act}(4\pi R^2)/(4\pi X^2)$ (kW/m^2). Therefore, the view factor, F_{view}, that the receptor actually "faces" is

$$F_{view} = (R/X)^2 \,, \qquad \text{(C2.44)}$$

where

$$X = \sqrt{H^2 + a^2} \,. \qquad \text{(C2.45)}$$

Fire Ball

calculation procedure

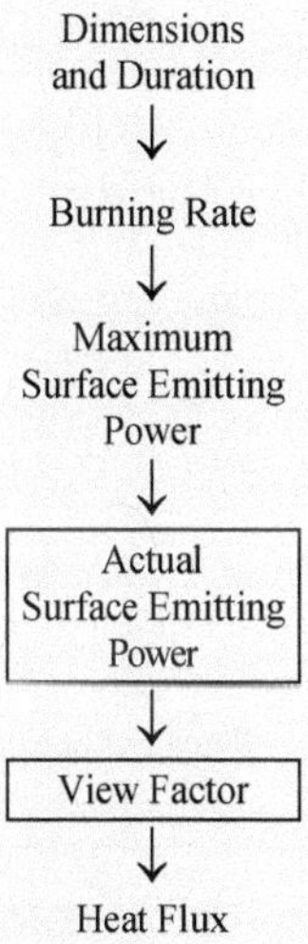

C2.2.6. Heat Flux

The heat flux, q' (kW/m^2), in a specific distance from the center of the fire (the distance enters the calculations through the view factor and the atmospheric transmissivity), is calculated as discussed in the introduction - see Eq. (C2.3) - from the product of the actual surface emitting power, SEP_{act} (kW/m^2), the view factor, F_{view} (-), and the atmospheric transmissivity, τ_α (-), as

$$q' = SEP_{act}\, F_{view}\, \tau_a \,. \tag{C2.46}$$

For the calculation of atmospheric transmissivity, τ_α (-), the following empirical expression [Bagster & Pittblado 1989] can be employed

$$\tau_a = c_7 \left[P_w (X - R)\right]^{-0.09} . \tag{C2.47}$$

In the above expression, P_w (Pa), denotes the partial water vapor pressure in air while X (m) is the distance of the receptor from the center of the fire of radius R (m). The constant c_7 is equal to 2.02 Pa$^{0.09}$m$^{0.09}$.

The partial water vapor pressure in air can easily be calculated from the saturation vapor pressure, P_w^o (Pa), in air, and the relative humidity, RH (fraction 0-1), as

$$P_w = RH\, P_w^o \,. \tag{C2.48}$$

Typical values of the actual surface emitting power, SEP_{act}, from a fire ball are usually between 150 and 300 kW/m^2, while measurements on butane fire balls have shown values up to 350 kW/m^2.

Fire Ball

example

EXAMPLE C2.5. *Heat Flux*

A road accident involving a tanker resulted in an increase of pressure inside the tanker and a consequent BLEVE. The tanker has a 50-m^3 capacity and was 85% full with LPG (liquefied propane gas). All the propane was released, and ignited immediately, forming a fire ball. Calculate the heat flux 150 m away from the tanker. The following data are available:

- Heat of combustion, ΔH_c	:	46,000	kJ/kg
Pressure in tanker, P_{sv}	:	1.6	MPa
Ambient temperature, T_a	:	298	K
- Density, ρ_{LPG}	:	517	kg/m^3
Saturation water vapor pressure, P_w^o	:	2,320	Pa
Relative humidity, RH	:	0.7	-

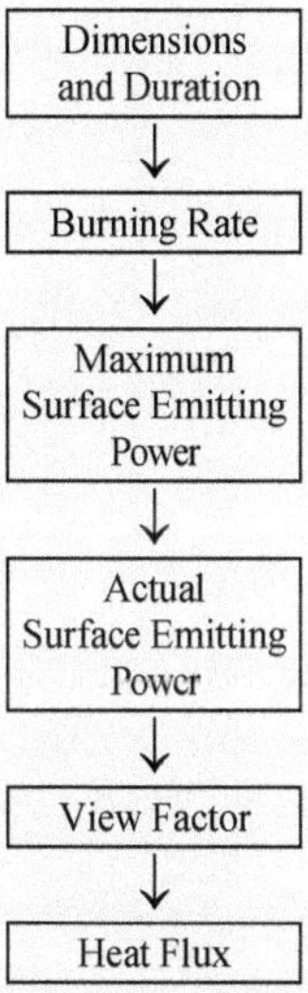

The mass of propane is $M = 0.85 V \rho_{LPG} = 0.85 \times 50 \times 517 = 21{,}973$ kg.
Employing the expressions of Roberts (TNO equations produce very similar results), one obtains

$$D_{max} = 5.8\, M^{1/3} = 162 \text{ m},$$

$$t_{max} = 0.45\, M^{1/3} = 12.6 \text{ s},$$

while the height of center of the fire ball is $H = D_{max} = 162$ m.

a) Burning Rate

The burning rate, m' (kg/m^2s), can be calculated from Eq. (C2.40), as

$$m' = \frac{M}{(0.888\pi\, D_{max}^2)\; t_{max}} = 0.024 \text{ kg/m}^2\text{s}.$$

b) The Maximum and the Actual Surface Emitting Power

To obtain the maximum surface emitting power, SEP_{max} (kW/m^2), we must first calculate the radiation fraction, F_s (-), Eq. (C2.42), as

$$F_s = 0.00325 P_{sv}^{0.32} = 0.314\,,$$

and thus from Eq. (C2.41),

$$SEP_{max} = F_s\, m' \Delta H_c = 342.3 \text{ kW/m}^2\,.$$

Fire Ball

example

Dimensions and Duration
↓
Burning Rate
↓
Maximum Surface Emitting Power
↓
Actual Surface Emitting Power
↓
View Factor
↓
Heat Flux

Because there is no time for soot to be collected around the surface of the fire ball,

$$SEP_{act} = SEP_{max} = 342.3 \text{ kW/m}^2.$$

c) View Factor

By direct substitution in Eqs. (C2.44) and (C2.45), one obtains

$$X = \sqrt{H^2 + a^2} = (162.5^2 + 150^2)^{1/2} = 221.1 \text{ m}$$

$$F_{view} = (R / X)^2 = 0.135$$

d) Heat Flux

In order to calculate the heat flux, the atmospheric transmissivity, τ_α (-), from Eq. (C2.47) must first be calculated. In Eq. (C2.47), P_w = 2,320 x 0.7= 1,624 Pa, while X = 224.5 m and

$$\tau_a = 2.02 \left[P_w (X - R)\right]^{-0.09} = 0.67,$$

thus $$q' = SEP_{act} \, F_{view} \, \tau_a = 30.8 \text{ kW/m}^2 .$$

Figure C2.9 shows the heat flux as a function of the distance. It can be observed that at 500 m the heat flux is reduced to 4.9 kW/m^2, which is however still quite a high value (sun radiation is 1 kW/m^2).

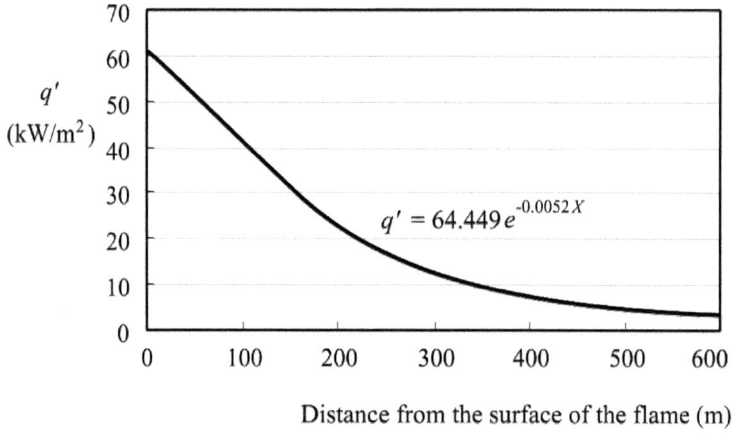

Figure C2.9 Heat flux as a function of the distance.

❐

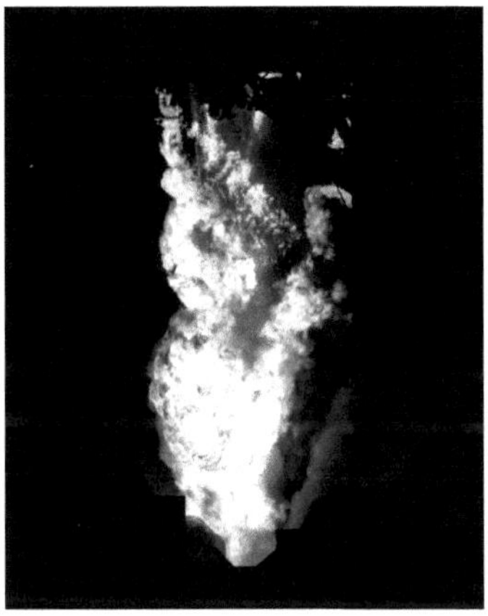

C2.3. Jet Fire

A Jet Fire is defined as a fire of turbulent dispersion resulting from the combustion of flammable substances liberated continuously with considerable momentum in a specific direction. The factor of considerable momentum in a specific direction is what distinguishes a jet fire from a pool fire.

In a jet fire, the following are usually observed:

- The formation of a jet fire does not depend upon meteorological conditions because of the high pressure with which the gas is liberated.
- The duration of a jet fire is not instantaneous, but depends upon the quantity of fuel that is liberated.
- The burning rate is equal to the outflow rate of the flammable gases.
- In contrast to pool fires, the outflow rate is not influenced by the heat of combustion.

Jet fires are common in light hydrocarbons, in natural gas, in gases with flammable condensates, in high-pressure hydrocarbon gases, in fuels, etc. In order to calculate the heat flux, the shape of the jet is considered to be that of a conical frustum.

The algorithm adopted for the calculation of the heat flux in the case of jet fires follows the general methodology, i.e., the dimensions of the fire are first calculated, and then the maximum surface emitting power. Following that, the actual surface emitting power is obtained, and from that, and the view factor, the heat flux is calculated.

The algorithm is based on the model of Chamberlain [Chamberlain 1987], which predicts the flame shape and radiation field of flares from flare stacks and flare booms. Such flares can burn hydrocarbons up to 100 kg/s for a few seconds, creating a 65-m flame and radiating heat of the order of 5,000 MW. This model has been developed in the Thornton Research Center of Shell Oil Company, and was validated with wind tunnel experiments and field tests, both onshore and offshore. The model represents the flame as a frustum of a cone, radiating as a solid body with a uniform surface emissive power.

Jet Fire

calculation procedure

Exit Velocity
↓
Equivalent Diameter
↓
Flame Dimensions
↓
Maximum Surface Emitting Power
↓
Actual Surface Emitting Power
↓
View Factor
↓
Heat Flux

Jet Fire

calculation procedure

Exit Velocity
↓
Equivalent Diameter
↓
Flame Dimensions
↓
Maximum Surface Emitting Power
↓
Actual Surface Emitting Power
↓
View Factor
↓
Heat Flux

C2.3.1. Exit Velocity of the Expanding Jet

The exit velocity, u_j (m/s), of the flammable fuel gases in the expanding jet, is usually expressed in relation to the Mach number, $M_j = u_j/u_{sound}$, where u_{sound} (m/s) represents the velocity of sound. For an ideal gas, the velocity of sound is given by the expression

$$u_{sound} = \sqrt{\frac{\gamma R T}{W_g}} . \tag{C2.49}$$

In the above expression, γ (-) is the ratio of the specific heat capacity at constant pressure C_p (J/kg K), to the specific heat capacity at constant volume C_v (J/kg K), R is the universal gas constant (=$(C_p$-$C_v)W_g$ = 8.3141 J/mol K), T (K) the gas temperature, and W_g (kg/mol) the molecular weight of the gas fuel in the jet. The exit velocity, u_j (m/s), of the flammable fuel gases in the expanding jet is given by the expression

$$u_j = M_j \sqrt{\frac{\gamma R T}{W_g}} . \tag{C2.50}$$

The Mach number, M_j (-), of the jet, can be calculated depending upon the shape of the flow, that is, if it is supersonic or not.

a) Sonic - Supersonic Flow $M_j \geq 1$
In this case, the Mach number, M_j (-) is given by the expression

$$M_j = \sqrt{\frac{(\gamma+1)(P_o / P_a)^{(\gamma-1)/\gamma} - 2}{(\gamma-1)}} . \tag{C2.51}$$

In the above expression, P_a (Pa), represents the ambient pressure and P_o (Pa), the pressure at the point of exit of the jet from the hole, given by the expression

$$P_o = \frac{\rho_o}{W_g} RT = \frac{\dot{m}\, 4}{\pi\, d_o^2} \sqrt{\frac{RT_o}{\gamma W_g}} , \tag{C2.52}$$

where, $\dot{m}$ (kg/s), represent the mass flow rate of the fuel gas in the jet, d_o (m), the diameter of the hole, ρ_o (kg/m^3), the density of the gas at the exit point from the hole, and T_o (K), the temperature at the same point. The temperature, T_o (K), can be calculated from an energy balance between the inside of the vessel (conditions "s") and the hole (conditions "o"), as

$$mC_pT_s + m\frac{u_s^2}{2} = mC_pT_o + m\frac{u_o^2}{2}. \quad \text{(C2.53)}$$

Substituting $u_s = 0$ (point of rest), $C_p = (1/W_g)[\gamma R/(\gamma\text{-}1)]$, and u_o from Eq. (C2.50) for $M_j = 1$, one obtains

$$T_o = \frac{2T_s}{1+\gamma}. \quad \text{(C2.54)}$$

In the case of adiabatic flow (e.g., rapid expansion of gas), the following relations can also be employed for the calculation of the pressure P_o (Pa) and the jet temperature, T_j (K) after the hole

$$P_o = P_s\left(\frac{2}{\gamma+1}\right)^{\frac{\gamma}{\gamma-1}}, \quad \text{(C2.55)}$$

and

$$T_j = T_s\left(\frac{P_a}{P_s}\right)^{\frac{\gamma-1}{\gamma}}. \quad \text{(C2.56)}$$

Eq. (C2.55) is widely employed in cases where the fuel mass flow rate in the jet is not directly known.

b) Subsonic Flow $M_j < 1$

The Mach number, M_j (-) is given as

$$M_j = \sqrt{\frac{\sqrt{1+2(\gamma-1)u_o^2} - 1}{(\gamma-1)}}, \quad \text{(C2.57)}$$

where the velocity u_o (m/s) denotes the velocity of the fuel gas at the exit point from the hole, and can be calculated from the expression

$$u_o = \frac{\dot{m}\,4}{\pi\, d_o^2}\frac{1}{P_a}\sqrt{\frac{RT_s}{\gamma W_g}}. \quad \text{(C2.58)}$$

In Eq. (C2.57), d_o (m) is the diameter of the hole, $\dot{m}$ (kg/s), the mass flow rate of the fuel gas in the jet, P_a (Pa), the ambient pressure and T_s (K) the vessel's temperature.

Jet Fire

calculation procedure

Exit Velocity
↓
Equivalent Diameter
↓
Flame Dimensions
↓
Maximum Surface Emitting Power
↓
Actual Surface Emitting Power
↓
View Factor
↓
Heat Flux

C2.3.2. Source Equivalent Diameter

The use of the equivalent diameter of the source is quite common in simulating combustion processes. The equivalent source diameter, D_s (m), expresses the diameter of a hypothetical nozzle through which flows air of density, ρ_{air} (kg/m^3), with a flow rate $\dot{m}$ (kg/s), and exit velocity u_j (m/s). In this case

$$D_s = \sqrt{\frac{4\,\dot{m}}{\pi\,\rho_{air}\,u_j}}\,. \tag{C2.59}$$

When the source itself is a flow nozzle of diameter d_o (m) and the density ρ_o (kg/m^3) is homogeneous, then the conservation of momentum produces

$$D_s = d_o\sqrt{\frac{\rho_o}{\rho_{air}}}\,. \tag{C2.60}$$

C2.3.3. Flame Dimensions

The length, L_B (m), of the flame can be calculated from the empirical expression [Kalghatgi 1983, 1984]

$$L_B = L_{B0}\left(0.51\,e^{-0.4u_w} + 0.49\right)\left[1 - 0.00607\left(\Theta_{jv} - 90\right)\right], \tag{C2.61}$$

where

$$L_{B0} = Y\,D_s\,, \tag{C2.62}$$

while Y is the root of the equation

$$\left[0.024\left(\frac{gD_s}{u_j^2}\right)^{1/3}\right]Y^{5/3} + 0.2\,Y^{2/3} - \left[\left(\frac{2.85}{W}\right)^{2/3}\right] = 0\,, \tag{C2.63}$$

and

$$W = \frac{W_g}{15.816\,W_g + 0.0395}\,. \tag{C2.64}$$

In the above Eqs. (C2.61)-(C2.64), D_s (m) is the equivalent source diameter, u_j (m/s), the velocity of the fuel in the jet, W_g (kg/mol), the molecular weight of the fuel and Θ_{jv} (°), the angle between the axis of the hole and the horizontal axis in the direction of the wind. Eqs. (C2.61)-(C2.64) produce very good results in hydrocarbon flames.

The angle, α (°), between the axis of the hole and the axis of the flame (see Figure C2.10) can be calculated [Kalghatgi 1983, 1984] as a function of the ratio of the wind velocity to the fuel velocity in the jet,

$$R = \frac{u_{\mathrm{w}}}{u_{\mathrm{j}}}, \tag{C2.65}$$

the angle Θ_{jv} and the Richardson number, *Ri* (-). For small values of the ratio *R*, like $R \leq 0.05$, conditions of the jet prevail, while for large values of the ratio *R*, the angle is more influenced by air. Thus, correspondingly

for $R \leq 0.05$
$$\alpha \, Ri(L_{\mathrm{B0}}) = 8{,}000R + Ri(L_{\mathrm{B0}})\left(\Theta_{\mathrm{jv}} - 90\right)\left(1 - e^{-25.6R}\right) \tag{C2.66}$$

for $R > 0.05$
$$\alpha \, Ri(L_{\mathrm{B0}}) = 1{,}726\sqrt{R - 0.026} + 1{,}334 + Ri(L_{\mathrm{B0}})\left(\Theta_{\mathrm{jv}} - 90\right)\left(1 - e^{-25.6R}\right) \tag{C2.67}$$

where the Richardson number, *Ri*, is given by the relation

$$Ri(L_{\mathrm{B0}}) = L_{\mathrm{B0}} \left(\frac{g}{D_{\mathrm{s}}^2 \, u_{\mathrm{j}}^2} \right)^{1/3}. \tag{C2.68}$$

The Richardson number, *Ri* (-), represents the importance of buoyancy in calculating the size of the flame. It is defined as the cubic root of the ratio of buoyancy to the inflow rate, and it is often used in combustion.

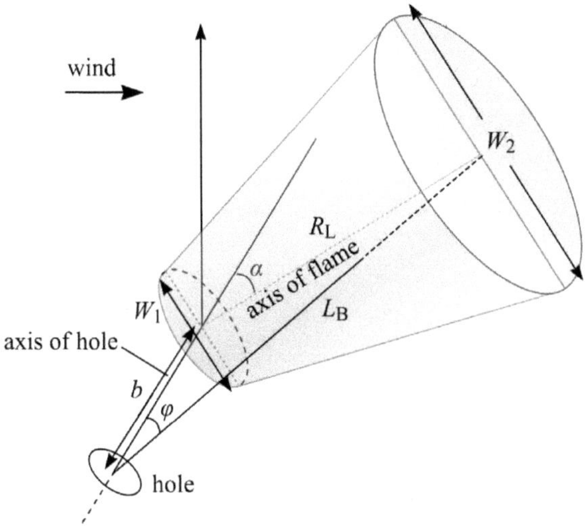

Figure C2.10 Jet fire with flame represented as a conical frustum.

Jet Fire

calculation procedure

Exit Velocity
↓
Equivalent Diameter
↓
Flame Dimensions
↓
Maximum Surface Emitting Power
↓
Actual Surface Emitting Power
↓
View Factor
↓
Heat Flux

Jet Fire

calculation procedure

Exit Velocity
↓
Equivalent Diameter
↓
Flame Dimensions
↓
Maximum Surface Emitting Power
↓
Actual Surface Emitting Power
↓
View Factor
↓
Heat Flux

The lift-off, b (m), of the conical frustum can be calculated from the geometric relation (see Figure C2.10)

$$b = L_{\mathrm{B}} \frac{\sin(\alpha - \varphi)}{\sin \alpha}. \tag{C2.69}$$

The above relation can also be written as

$$b = L_{\mathrm{B}} \frac{\sin(K\alpha)}{\sin \alpha}, \tag{C2.70}$$

where $K = (\alpha - \varphi)/\alpha$. In this case, parameter K (-) is empirically related [Chamberlain 1987], with the velocity ratio R, Eq. (C2.65), as

$$K = 0.187 e^{-20R} + 0.015. \tag{C2.71}$$

When $\alpha = 0°$ or $180°$, the lift-off is reduced to KL_{B}, which in turn equals $0.2L_{\mathrm{B}}$ in no-wind situations ($\alpha = 0°$) or $0.015L_{\mathrm{B}}$ for fires with strong wind present ($\alpha = 180°$).

The length, R_{L} (m), of the frustum can be calculated from the geometric relation (see Figure C2.10)

$$R_{\mathrm{L}} = \sqrt{L_{\mathrm{B}}^2 - b^2 \sin^2 \alpha} - b \cos \alpha. \tag{C2.72}$$

In this equation, the lengths, L_{B} (m) and b (m) are shown in Figure C2.10, while the angle α is the angle between the axis of the hole and the axis of the flame, because of the presence of wind.

The frustum base width, W_1 (m), can be calculated from the empirical expression [Chamberlain 1987]

$$W_1 = D_{\mathrm{s}} \left(13.5 e^{-6R} + 1.5\right) \left(1 - e^{-70 Ri(D_{\mathrm{s}}) CR} \left[1 - \frac{1}{15} \sqrt{\frac{\rho_{\mathrm{air}}}{\rho_{\mathrm{j}}}}\right]\right), \tag{C2.73}$$

where

$$C = 1000 e^{-100R} + 0.8, \tag{C2.74}$$

and

$$Ri(D_{\mathrm{s}}) = D_{\mathrm{s}} \left(\frac{g}{u_{\mathrm{j}}^2 D_{\mathrm{s}}^2}\right)^{1/3}. \tag{C2.75}$$

It can be noted that when the velocity ratio R tends to zero, then the value of the frustum base width, W_1, approaches the value of the equivalent diameter of the source, D_s.

The frustum tip width, W_2 (m), can be calculated from the empirical expression [Chamberlain 1987]

$$W_2 = L_B\left(0.18e^{-1.5R} + 0.31\right)\left(1 - 0.47e^{-25R}\right), \qquad \text{(C2.76)}$$

where the velocity ratio R was defined in Eq. (C2.65). It is noted that when the ratio R tends to zero, then $W_2 = 0.26L_B$, a result which is verified by experimental observations.

The surface area of frustum including end discs, A (m^2), is calculated from the expression

$$A = \frac{\pi}{4}\left(W_1^2 + W_2^2\right) + \frac{\pi}{2}\left(W_1 + W_2\right)\sqrt{R_L^2 + \left(\frac{W_2 - W_1}{2}\right)^2}. \qquad \text{(C2.77)}$$

Fire and explosion on February 18, 2008, in Big Spring Refinery, TX, U.S.A. (Reproduced by kind permission of the Texas Forest Service U.S.A.)

Jet Fire

calculation procedure

Exit Velocity
↓
Equivalent Diameter
↓
Flame Dimensions
↓
Maximum Surface Emitting Power
↓
Actual Surface Emitting Power
↓
View Factor
↓
Heat Flux

Jet Fire

example

Exit Velocity
↓
Equivalent Diameter
↓
Flame Dimensions
↓
Maximum Surface Emitting Power
↓
Actual Surface Emitting Power
↓
View Factor
↓
Heat Flux

EXAMPLE C2.6. *Dimensions of Jet Flame*

Calculate the exit velocity and the dimensions of a jet fire of 30 kg/s mass flow caused inside a high-pressure natural-gas pipeline. The following data are available:

- Initial pressure of pipeline, P_s	:	10	MPa
Wind velocity, u_w	:	5	m/s
Specific heat capacity, C_p	:	1.52	kJ/kg K
Ambient temperature, T_a	:	288	K
Ambient pressure, P_a	:	0.1	MPa
Molecular weight of natural gas, W_g	:	0.0186	kg/mol
Molecular weight of air, W_{air}	:	0.029	kg/mol
Angle between hole axis/wind axis, Θ_{jv}	:	85	o

Assuming supersonic flow, **the Mach number, M_j (-),** is calculated from Eq. (C2.51) as

$$M_j = \sqrt{\frac{(\gamma+1)(P_o / P_a)^{(\gamma-1)/\gamma} - 2}{(\gamma-1)}}$$

where, the pressure P_o, at the exit point of the jet is given by Eq. (C2.55), as

$$P_o = P_s \left(\frac{2}{\gamma+1} \right)^{\frac{\gamma}{\gamma-1}} .$$

where $\gamma = C_p / (C_p - R/W_g) = 1.42$. Using the above two expressions, one obtains $P_o = 5.25$ MPa, and $M_j = 3.71$.

The temperature, T_j (K), and the velocity of the jet, u_j (m/s), are obtained from the Eqs. (C2.56) and (C2.50), respectively, as

$$T_j = T_s \left(\frac{P_a}{P_s} \right)^{\frac{\gamma-1}{\gamma}} = 74.3 \text{ K} \qquad \text{(because } T_s = T_a\text{)},$$

and $$u_j = M_j \sqrt{\frac{\gamma R T_j}{W_g}} = 805.9 \text{ m/s}.$$

The source equivalent diameter, D_s (m), is obtained from Eq. (C2.59) as

$$D_s = \sqrt{\frac{4\,\dot{m}}{\pi\,\rho_{air}\,u_j}} = 0.198 \text{ m}.$$

Having obtained these values, the dimensions of the frustum can be obtained as follows.

The length, L_B (m), of the flame is calculated from the Eqs. (C2.61)-(C2.64), as

$$L_B = L_{B0}\left(0.51\,e^{-0.4u_w} + 0.49\right)\left[1 - 0.00607\left(\Theta_{jv} - 90\right)\right] = 34.6 \text{ m},$$

where $$L_{B0} = Y\,D_s = 60.1 \text{ m},$$

and $Y = 303.6$, is the root of the equation

$$\left[0.024\left(\frac{gD_s}{u_j^2}\right)^{1/3}\right]Y^{5/3} + 0.2\,Y^{2/3} - \left[\left(\frac{2.85}{W}\right)^{2/3}\right] = 0\,,$$

and $$W = \frac{W_g}{15.816\,W_g + 0.0395} = 0.0557.$$

The angle, α (°), between the axis of the hole and the axis of the flame, is calculated from the Eqs. (C2.65)-(C2.68), as

$$R = \frac{u_w}{u_j} = 0.006 \quad \text{and} \quad Ri(L_{B0}) = L_{B0}\left(\frac{g}{D_s^2\,u_j^2}\right)^{1/3} = 4.37,$$

for $R \le 0.05$ $\quad \alpha\,Ri(L_{B0}) = 8{,}000R + Ri(L_{B0})\left(\Theta_{jv} - 90\right)\left(1 - e^{-25.6R}\right) \Rightarrow \alpha = 10.6^\circ$

The lift-off, b (m), of the conical frustum is calculated from the Eqs. (C2.70)-(C2.71), as

$$K = 0.187e^{-20R} + 0.015 = 0.18,$$

Jet Fire

example

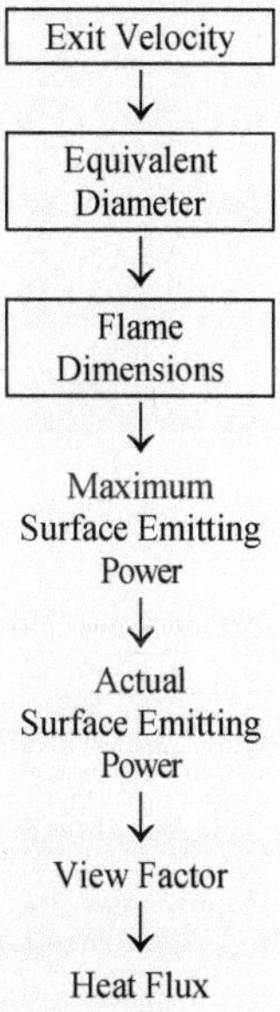

Jet Fire

example

Exit Velocity
↓
Equivalent Diameter
↓
Flame Dimensions
↓
Maximum Surface Emitting Power
↓
Actual Surface Emitting Power
↓
View Factor
↓
Heat Flux

$$b = L_{\mathrm{B}} \frac{\sin(K\alpha)}{\sin\alpha} = 6.27 \text{ m}.$$

The length, R_{L} (m), of the frustum is calculated from Eq. (C2.72) as

$$R_{\mathrm{L}} = \sqrt{L_{\mathrm{B}}^2 - b^2 \sin^2\alpha} - b\cos\alpha = 28.8 \text{ m}.$$

The frustum base width, W_1 (m), is obtained from Eqs. (C2.73)-(C2.75) as

$$W_1 = D_{\mathrm{s}}\left(13.5e^{-6R} + 1.5\right)\left(1 - e^{-70\,Ri(D_{\mathrm{s}})CR}\left[1 - \frac{1}{15}\sqrt{\frac{\rho_{\mathrm{air}}}{\rho_{\mathrm{j}}}}\right]\right) = 2.78 \text{ m},$$

where
$$C = 1000e^{-100\,R} + 0.8 = 538.5,$$

$$Ri(D_{\mathrm{s}}) = D_{\mathrm{s}}\left(\frac{g}{u_{\mathrm{j}}^2 D_{\mathrm{s}}^2}\right)^{1/3} = 0.014,$$

and
$$\frac{\rho_{\mathrm{air}}}{\rho_{\mathrm{j}}} = \frac{T_{\mathrm{j}}}{T_{\mathrm{a}}}\frac{W_{\mathrm{air}}}{W_{\mathrm{g}}} = 0.402.$$

The frustum tip width, W_2 (m), is obtained from Eq. (C2.76)

$$W_2 = L_{\mathrm{B}}\left(0.18e^{-1.5R} + 0.31\right)\left(1 - 0.47e^{-25R}\right) = 10.1 \text{ m}.$$

The surface area of frustum including end discs, A (m²), is calculated from Eq. (C2.77)

$$A = \frac{\pi}{4}\left(W_1^2 + W_2^2\right) + \frac{\pi}{2}\left(W_1 + W_2\right)\sqrt{R_{\mathrm{L}}^2 + \left(\frac{W_2 - W_1}{2}\right)^2} = 665.6 \text{ m}^2.$$

❐

C2.3.4. Maximum Surface Emitting Power

The maximum surface emitting power, SEP_{max} (kW/m^2), expresses the emitting power from the flame's surface if no soot is present. It can be calculated as a function of the heat of combustion, ΔH_c (kJ/kg), the burning rate, m' (kg/s) and the total surface area of frustum including discs, A (m) (see Eq. (C2.9)), as

$$SEP_{max} = F_s \frac{m'}{A} \Delta H_c . \qquad \text{(C2.78)}$$

In the above expression, the radiation fraction F_s (-), represents the fraction of the combustion energy radiated from the flame temperature. For large flames, the radiation fraction is given by an empirical expression [Chamberlain 1987] as a function of the fuel velocity in the jet, as

$$F_s = 0.21 e^{-0.00323\, u_j} + 0.11 \qquad \text{(C2.79)}$$

while u_j (m/s) is the exit velocity of the flammable gases in the expanding jet, defined in Eq. (C2.50). For small fires, the above expression overestimates the radiation fraction F_s.

C2.3.5. Actual Surface Emitting Power

The maximum surface emitting power, SEP_{max} (kW/m^2) (calculated in the previous section), refers to radiation from the flame surface in the case where there is no soot present. As has already been discussed, soot results in a large decrease of this radiation. The actual surface emitting power, SEP_{act} (kW/m^2), takes into consideration the presence of soot and is calculated as

$$SEP_{act} = SEP_{max}(1-s) + SEP_{soot}\, s . \qquad \text{(C2.80)}$$

In the above expression, the term SEP_{soot} (kW/m^2) represents the surface emitting power of soot, while s (-) denotes the fraction of the surface that is covered by soot. For petroleum products, the parameter s in general is taken equal to 80%, and SEP_{soot} = 20 kW/m^2 [Hagglund & Person 1976].

Especially in the case of jet fires, the fraction of the surface that is covered by soot is considered very small and thus s ≈ 0. In this case Eq. (C2.80) becomes $SEP_{act} = SEP_{max}$.

Jet Fire

calculation procedure

Exit Velocity
↓
Equivalent Diameter
↓
Flame Dimensions
↓
Maximum Surface Emitting Power
↓
Actual Surface Emitting Power
↓
View Factor
↓
Heat Flux

Jet Fire

calculation procedure

Exit Velocity → Equivalent Diameter → Flame Dimensions → Maximum Surface Emitting Power → Actual Surface Emitting Power → **View Factor** → Heat Flux

C2.3.6. View Factor

The view factor, F_{view} (-), expresses the fraction of the radiative heat that reaches the receptor (either human or material). In the case of jet fires, the flame is usually considered as a tilted cylinder with a diameter equal to the average of the diameters of the two end discs of the frustum. In this case the equations described in Section C2.1.4 can be employed with the following obvious transformations (that is, one must employ X' in place of X and Θ' in place of Θ):

$$X' = \sqrt{\left(b \sin \Theta_{jv}\right)^2 + \left(X - b \cos \Theta_{jv}\right)^2} \tag{C2.81}$$

$$\Theta' = 90^{o} - \Theta_{jv} + \alpha - \arctan\left(\frac{b \sin \Theta_{jv}}{X - b \cos \Theta_{jv}}\right) \tag{C2.82}$$

$$R = \frac{W_1 + W_2}{4} \tag{C2.83}$$

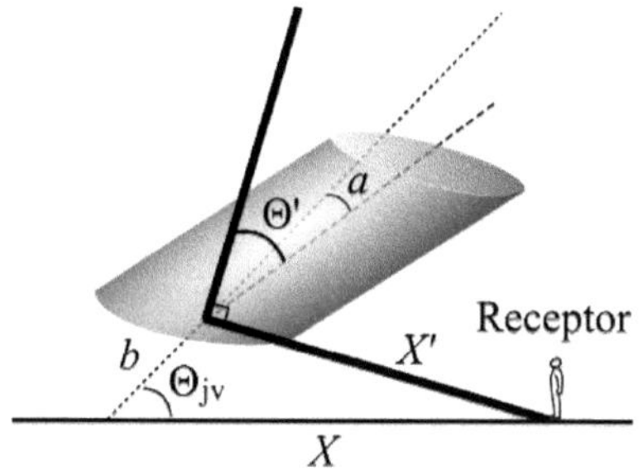

Figure C2.11 Conical frustum as a tilted cylinder.

In relation to Eq. (C2.81), it is reminded that:

- The angle Θ_{jv} (°) is the angle between the axis of the hole and the horizontal axis in the direction of the wind.
- The angle α (°) is the angle between the axis of the hole and the axis of the flame, and
- The angle Θ' (°) is the angle between the axis of the flame and the new y-axis. Therefore it can be obtained (see Figure C2.11) if from 90° one subtracts (a) the angle between X and X' and (b) the angle Θ_{jv}, and adds the angle α; see Eq. (C2.82).

The reader is advised to examine the above figure in relation to Figures C2.10 and C2.4.

C2.3.7. Heat Flux

The heat flux, q' (kW/m^2), in a specific distance from the center of the fire (the distance enters the calculations through the view factor and the atmospheric transmissivity), and is calculated as discussed in the introduction - see Eq. (C2.3) - from the product of the actual surface emitting power, SEP_{act} (kW/m^2), the view factor, F_{view} (-), and the atmospheric transmissivity, τ_α (-), as

$$q' = SEP_{act}\, F_{view}\, \tau_a \qquad \text{(C2.84)}$$

For the calculation of atmospheric transmissivity, τ_α (-), the following empirical expression [Bagster & Pittblado 1989] can be employed

$$\tau_a = c_4 \left[P_w (X - R)\right]^{-0.09} . \qquad \text{(C2.85)}$$

In the above expression, P_w (Pa), denotes the partial water vapor pressure in air while X (m) is the distance of the receptor from the center of the fire of radius R (m). The constant c_4 is equal to 2.02 Pa$^{0.09}$m$^{0.09}$.

The partial water vapor pressure in air can easily be calculated from the saturation vapor pressure, P_w^o (Pa), in air, and the relative humidity, RH (fraction 0-1), as

$$P_w = RH\, P_w^o . \qquad \text{(C2.86)}$$

Jet Fire

calculation procedure

Exit Velocity
↓
Equivalent Diameter
↓
Flame Dimensions
↓
Maximum Surface Emitting Power
↓
Actual Surface Emitting Power
↓
View Factor
↓
Heat Flux

Jet Fire

example

Exit Velocity
↓
Equivalent Diameter
↓
Flame Dimensions
↓
Maximum Surface Emitting Power
↓
Actual Surface Emitting Power
↓
View Factor
↓
Heat Flux

EXAMPLE C2.7. *Heat Flux*

For the jet fire in the high-pressure natural-gas pipeline of Example C2.6, calculate the heat flux in a distance of 50 m from the center of the jet (on the ground). The following data are available:

- Heat of combustion, ΔH_c	:	38	MJ/kg
Saturation water vapor pressure, P_w^o	:	2,320	Pa
Relative humidity, RH	:	0.7	-

The maximum surface emitting power, SEP_{max} (kW/m²), is given from Eq. (C2.78),

$$SEP_{max} = F_s \frac{m'}{A} \Delta H_c = 215 \text{ KW/m}^2,$$

where $$F_s = 0.21 e^{-0.00323\, u_j} + 0.11 = 0.126,$$

assuming $s = 0$, $$SEP_{act} = SEP_{max} = 215 \text{ kW/m}^2 .$$

To calculate the heat flux, one must first calculate the view factor and the atmospheric transmissivity, using

$$X' = \sqrt{\left(b \sin \Theta_{jv}\right)^2 + \left(X - b \cos \Theta_{jv}\right)^2} = 49.84 \text{ m},$$

$$\Theta' = 90^\circ - \Theta_{jv} + \alpha - \arctan\left(\frac{b \sin \Theta_{jv}}{X - b \cos \Theta_{jv}}\right) = 8.41^\circ,$$

$$R = \frac{W_1 + W_2}{4} = 3.22 \text{ m}, \qquad L = R_L = 28.8 \text{ m}.$$

Employing these values (instead of X and Θ) in Eqs. (C2.17)-(C2.23), see Example C2.4, the view factor is obtained as

$$F_{view} = 0.023 \quad (F_v = 0.022, F_h = 0.006).$$

From Eqs. (C2.85)-(C2.86), one obtains $\tau_\alpha = 0.73$, and thus

$$q' = SEP_{act}\, F_{view}\, \tau_a = 3.63 \text{ kW/m}^2.$$

❒

Flash Fire

Reproduced by kind permission of SiamsFX (www.siamsfx.com).

C2.4. Flash Fire

A flash fire can result from the sudden ignition of a cloud of flammable gases, where the flame is not accelerated because of the presence of obstacles or the influence of turbulent dispersion. It is not perfectly understood when a gas outflow will result in a flash fire or a fire ball.

The shock wave that is created from a flash fire is small and as the duration of the fire is also short, its impact on facilities and equipment outside the cloud is of limited nature. Inside the cloud however, buildings and objects will be subsumed by the burning part of the gas cloud, and thus combustible parts will catch fire, and secondary fires may result. Therefore, this kind of fire is examined in risk analysis cases, mostly in relation to consequences in people.

The presence of obstacles in the area where the gases (produced by the combustion) expand can often result in explosions rather than flash fires. The obstacles increase the turbulence in the expansion of the flame resulting in the increase of the burning velocity of the flammable gases, and consequently the probability of explosion, instead of flash fire, rises [see Section C3].

Calculations that usually are carried out in order to estimate the consequences from flash fires use dispersion models to calculate the extent of the fire's expansion. With the use of dispersion models, one can calculate the distance to which the cloud concentration is equal to the low flammability limit (LFL) - see Section D3.2.1 - or more conservatively, the distance to which the cloud concentration is equal to half the LFL.

Some researchers propose the use of the LFL for the calculation of the fire's expansion, although to estimate the consequences from the fire's effects in a distance from the fire, they assume a 50% radiation of the available heat of combustion in the cloud.

A more conservative approach in estimating the consequences from the effects of flash fires can be adopted if the distance to which the cloud concentration is equal to the LFL is increased so much that the expansion of the gases from the combustion is also considered.

Persons inside the flame area will be subjected to fatal injuries (100% fatality), while the consequences for persons outside the flame area can be estimated with the use of probit functions based upon the heat radiation dose [see Section C2.5]. At 22.30 on September 24, 1990, a road tanker carrying LPG was involved in a traffic accident at a busy road junction in the centre of Bangkok. Some 5 tn of LPG was released but did not ignite immediately. When the cloud did ignite, there was evidently a flash fire, but accounts spoke also of an explosion, probably from gas which had entered a nearby building. In the worst-affected area, almost all the shop houses on both sides of the street were destroyed. Reports gave some 68 persons dead and over 100 injured. The vehicle was a flat bed lorry carrying two LPG tanks. The two tanks were interconnected by two lines and it appears that these were severed in the accident. The tanker had no license to carry LPG.

Damages following the explosion on February 8, 2008, at the Imperial Sugar refinery in Port Wentworth, GA, U.S.A. (Reproduced by kind permission of the U.S. Chemical Safety Board.)

C2.5. Effects of Heat Radiation

Injuries caused to people following the onset of a fire are due primarily to the large amounts of heat radiation generated by the combustion of the flammable substances, and secondarily due to the toxic gases produced by the combustion.

Among the usual CO_2 NO_x, SO_X gases produced during combustion, gases of high toxicity, very harmful to people when inhaled, can also be produced. Furthermore, the generation of toxic gases can hinder the efforts of people trying to flee the fire, as well as the efforts of rescue workers and firefighters coming to help, and thus increase injuries and lethality. The dispersion of toxic substances will be examined in Section C5, while in this section the effects of heat radiation to people and materials will be presented.

In Sections C2.1-C2.4 the procedure for calculating the heat flux radiated from a fire was presented. Hence all intermediate algorithms are presented so that the heat flux from a fire can be calculated as a function of the distance from the flame's surface. This relation of the calculated heat flux as a function of the distance will be the basis of this section, while also examining its effects on people and material.

The calculation procedure depends upon the relation of the heat flux to distance, and the use of the radiation dose and the probit functions. From these quantities the probability of injury (1st and 2nd degree burns) or death, will be calculated. Spatial integration of the aforementioned relations will produce the total number of 1st and 2nd degree burns or deaths which result from the particular heat flux of the fire.

It should be emphasized, however, that the results of these calculation procedures are not fully quantitative or binding. Usually their purpose is primarily to estimate the effects and consequences of an event, so that contingency plans of hypothetical scenarios can be drawn, discussed and dealt with.

C2.5.1. Human Skin Burns

Effects on humans are directly related to the intensity of the heat flux and the duration of the person's exposure to it.

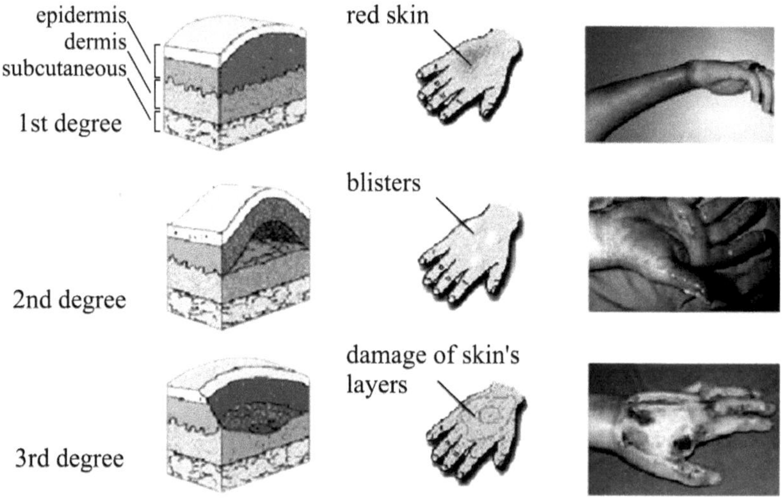

Figure C2.12 Types of burns. (Pictures from http://www.burnsurvivor.com/.)

Heat radiation from a fire has a twofold effect on people. Initially, there is an increase in the heart rate, sweating and rise of the body temperature. The second and more important effect is one that refers to burns caused by the transfer of heat to the skin. Three types of burns can be distinguished (see Figure C2.12):

1st degree : First-degree burns affect only the epidermis or outer layer of skin. The burn site is red, painful, dry, and with no blisters. Mild sunburn is an example. Long-term tissue damage is rare and usually consists of an alteration of the skin color.

2nd degree : Second-degree burns involve the epidermis and part of the dermis layer of skin (0.07 - 0.12 mm depth). The burn site appears red, blistered, and may be swollen and painful.

3rd degree : Third-degree burns destroy the epidermis and dermis. Third-degree burns may also damage the underlying bones, muscles, and tendons. The burn site appears white or charred. There is no sensation in the area since the nerve endings are destroyed.

Effects of Fires

calculation procedure

Human Skin Burns
↓
Thermal Radiation Intensity Limits

Thermal Radiation Dose
↓
Probability of Injury or Death
↓
Overall Effects

Effects on Materials

It should be mentioned that 2nd and 3rd degree burns can lead to permanent disability. Their treatment often requires clinical help in a specialized hospital. A realistic probability of mortality is also present depending on the amount of the skin's surface that the burn covers. It is typically noted that for a burn that spreads over 50% of the skin, a child of 0-9 years old will have an 80% probability of survival, an adult 30-35 years old a 50% probability of survival, while persons over 60 years old, will have no probability of survival. In Table C2.7 average values for some physical properties of skin are given [TNO 1989]:

Table C2.7 Average Values for Properties of Skin of a Man of 70 kg and 1.7 m.

Weight	4	kg
Surface	1.8	m^2
Water content	70-75	% (mass)
Density	110	kg m^{-3}
Thickness	0.05-5	mm
Thermal conductivity	0.65	W $m^{-1}K^{-1}$
Specific heat capacity	34	kJ kg^{-1} K^{-1}

Finally it should be mentioned that skin burns are directly related to the covering of skin with clothes. Provided that clothes do not ignite, they protect the skin. Usually, however, because of the heat radiation, clothes themselves ignite and thus cause worse burns.

C2.5.2. Thermal Radiation Intensity Limits

In order to be able to calculate and consequently avoid the effects of the heat flux, typical thermal radiation intensity limits have often been proposed. In Table C2.8, the thermal radiation intensity limits proposed by the World Bank [World Bank 1988] are shown. These limits are directly related to specific radiation effects to people and materials, and are in full agreement with those proposed by the American Petroleum Institute (API).

The intensity of the sun's radiation is approximately 0.8-1 kW/m^2. When calculating the distances that correspond to specific thermal radiation intensity limits, the intensity of the sun's radiation should also be added. For example, if the calculated from mathematical models heat flux 100 m from the flame's surface is 2 kW/m^2, the actual heat flux to which people are subjected in this distance (taking the sun's radiation into consideration) is about 2.8-3 kW/m^2.

The value for the sun's radiation usually employed in calculating effects on people is equal to 1 kW/m^2.

Table C2.8 Thermal Radiation Intensity Limits [World Bank 1988].

Heat flux ($kW\cdot m^{-2}$)	Effects on materials	Effects on humans
37.5	Equipments damage.	100% lethality in 1 min. 1% lethality in 10 s.
25	Minimum intensity for ignition of wood in prolonged exposure.	100% lethality in 1 min. Serious injuries in 10 s.
12.5	Minimum intensity for ignition, and melting of plastic tubes.	1% lethality in 1 min. 1st degree burns in 10 s.
4		No lethality. 2nd degree burns probable. Pain after exposure of 20 s.
1.6		Acceptable limit for prolonged exposure.

Fire on April 30, 2008, at the Holy Refinery, UT, U.S.A. (Reproduced by kind permission of M. Christensen.)

C2.5.3. Effects on People

In this section the calculation procedure for the effects of the heat flux from fires to people will be described. More precisely, for a specific heat flux in a specific distance or in all the surrounding space, the probability of injury (1st or 2nd degree burns) or death will be calculated. The algorithm is composed of the following three steps:

1. The thermal radiation dose will be calculated as a function of the heat flux obtained at a specific distance (see Sections C2.1.5, or C2.2.6, or C2.3.7), and the escape time, that is the time required by a person to run away from the fire until he is safe.
2. The probability of injury (1st or 2nd degree burns) or death will be calculated, based upon empirical expressions known as probit functions. The probit functions are expressed in relation to the thermal radiation dose.
3. The total number of people suffering 1st or 2nd degree burns, or death, because of the heat flux, will be obtained from the spatial integration of the above probability functions.

a) Thermal Radiation Dose

Burns or deaths are a direct consequence of the intensity of the radiated heat flux from the fire and the exposure time. To determine the number of burns and deaths, the term "thermal radiation dose," D ($W^{4/3} s \cdot m^{-8/3}$), is employed, as it includes both the aforementioned terms being defined from the following equation as:

$$D = t_{\text{eff}} (q')^{4/3}. \qquad \text{(C2.87)}$$

In this equation, q' (W/m^2) is the heat flux calculated (see Sections C2.1.5, or C2.2.6, or C2.3.7) for various types of fires. The time, t_{eff} (s), represents the person's exposure time to this heat flux.

The exposure time of a person to a fire depends upon several factors:

1) the type of fire,
2) the position of the person in relation to the fire, and
3) the time he takes to react and how he reacts when he realizes the event, as well as parameters like his age and physical condition.

The exposure time of a person in the case of fire balls (Section C2.2) and flash fires (Section C2.4) is actually equal to the duration of the fire (<1 min). In case of pool fires (Section C2.1) and jet fires (Section C2.3), the duration of the fire is longer and the person's exposure time to the radiated heat flux depends upon the time he takes to react and run away.

Therefore the person's exposure time is directly related to the escape time from the event. The escape time includes the time of the initial reaction as well as the time required for the person to cross the distance so that he will reach areas protected from the heat flux, or areas where the level of heat flux is low enough for prolonged exposure.

The reaction time depends directly upon the space where the fire broke out as well as the characteristics of the population exposed to it. For example, one would expect different reaction times between people facing a fire in a refinery, and people that happen to be present in a fire following a tanker accident in an inhabited area. In the first case reaction times are expected to be very low, because of the training of the people, their age, and their awareness that such an event may happen. In the second case, reaction times are expected to be much higher, as people exposed to such an event can be much older, of a sensitive group, and certainly not trained or not having similar experiences. The term "sensitive group" can be employed for groups of people who are not expected to react immediately or effectively during an evacuation caused by the appearance of such an emergency, i.e., the elderly, handicapped or children. Reaction times in case of fires, as well as escape velocities proposed by various investigators, are shown in Table C2.9.

Table C2.9 Reaction Times and Escape Velocities.

Investigators	Type of population	Escape velocity (m/s)	Reaction time (s)	Dose for probability of deaths $10^4 \times (W^{4/3} s \cdot m^{-8/3})$ 1%	injuries 50%
Eisenberg *et al.* 1975	-	-	-	960	2,380
Tsao and Perry 1979	-	-	-	420	1,050
H.S.E. 1981	average	2.5	0	1,000	-
	sensitive	1	5	500	-
Hynes 1983		6			
TNO 1989	average	4	5	520	-
	sensitive	-		430	1090

In relation to Table C2.9, some points are worth mentioning. As the world record for running 100 m is about 9.8 s, and this corresponds to a velocity of approximately 10 m/s, then the values for the escape velocities, quoted in the table, as well as the reaction time of 5 s, are logical.

The exposure time, t_{eff} (s), that enters in the definition of the thermal radiation dose, Eq. (C2.87), is usually taken as equal to the sum of an initial reaction time plus the time required to reach a safe point, that is

$$t_{\text{eff}} = t_r + \frac{(x_o - r)}{u}. \tag{C2.88}$$

In the above expression, t_r (s) is the person's reaction time (= 5 s), x_o (m) is the distance between the flame's surface and the position where the intensity of the heat flux is lower than 1 kW/m^2 (low thermal radiation intensity limit), r (m), the distance of the person from the surface of the flame, and u (m/s) the escape velocity - usually taken equal to 4 m/s.

If the exposure time calculated by Eq. (C2.88) is greater than the actual duration of the fire (e.g., in case of fire balls), then the exposure time is taken as equal to the duration of the fire.

b) Probability of Injury or Death

In order to estimate the effects from fires, dose-response curves are usually employed to correlate the magnitude of the consequence to the people with the specified related dose. Dose-response relations have been created from experimental or field data for various parameters like thermal radiation, overpressure, heat, noise, concentration of toxic gases, etc. Many methods for the reproduction of the dose-response curves exist, but the most widely used method today (because of its ease in computer calculations) is the method of probit functions.

For the calculation of the probability, P (-), of injury (1st or 2nd degree burns) or death, as a consequence of a specified dose, D (W$^{4/3}$s·m$^{-8/3}$), the following equations are employed:

$$P = F_{\text{k}} \frac{1}{2}\left[1 + \text{erf}\left(\frac{Pr - 5}{\sqrt{2}}\right)\right], \tag{C2.89}$$

where the probit function, Pr (-), is given by the empirical expression,

$$Pr = c_1 + c_2 \ln D. \tag{C2.90}$$

In the above two equations, the probability, P (-), refers to injury (1st or 2nd degree burns) or death, depending on the values of the coefficients c_1 and c_2 that

Effects of Fires

calculation procedure

Human Skin Burns
↓
Thermal Radiation Intensity Limits

Thermal Radiation Dose
↓
Probability of Injury or Death
↓
Overall Effects on People

Effects on Materials

are chosen from Table C2.10. Values for the error function can easily be found in literature*. The parameter, F_k (-), refers to the influence of clothes in the probability of injury or death, and it constitutes a correction factor, providing there is no ignition of the clothes. Its values range from 0.14 for winter clothes (large coverage of skin area) to 0.95 for summer clothes (small coverage of skin area).

Consequently, for a particular heat flux in a specified distance, the exposure time t_{eff} is calculated from Eq. (C2.88) and the thermal radiation dose from Eq. (C2.87). Having obtained the thermal radiation dose, the corresponding probit function, *Pr,* is obtained from Eq. (C2.90) for injury or death and from this the corresponding probability, *P*, from Eq. (C2.89).

Table C2.10 Coefficients c_1 and c_2 [TNO 1989].

Effect	c_1	c_2
1st degree burn	-39.83	3.0186
2nd degree burn	-43.14	3.0186
Deaths	-36.38	2.56

Based on the values of the coefficients c_1 and c_2 given in Table C2.10, in Table C2.11 the calculated values of the probit functions and probabilities for injury (1st or 2nd degree burns) or death for specific thermal radiation doses are given. The correction factor, F_k (-), that refers to clothing, has been taken equal to unity, i.e., that clothes have no effect.

As an example, we see that for a thermal radiation dose D = 9,000,000 $W^{4/3} s \cdot m^{-8/3}$, the probability of a 1st degree burn is 100%, for a 2nd degree burn is 57.8% and for death is 34.9%. Furthermore, the percentage referring to the probability of death, includes the percentages for 1st and 2nd degree burns (that is, those persons who die from the radiative heat already have 1st and 2nd degree burns). Correspondingly, in the percentage of people with 2nd degree burns, people with 1st degree burns are included. This effect will be clearer in Example C2.8.

* $\mathrm{erf}(-z) = -\mathrm{erf}(z)$. For $z > 0$ and error $< 5 \times 10^{-4}$ [Abramowitz and Stegun 1970].
$\mathrm{erf}(z) = 1 - (1 + 0.278393\, z + 0.230389\, z^2 + 0.000972\, z^3 + 0.078108\, z^4)^{-4}$.

Table C2.11 Probabilities for Specified Doses.

	1st degree		2nd degree		Lethality	
D ($W^{4/3}s{\cdot}m^{-8/3}$)	*Pr* (-)	*P* (%)	*Pr* (-)	*P* (%)	*Pr* (-)	*P* (%)
1×10^6	1.87	0.1				
2×10^6	3.97	15.1	0.66	0.0	0.76	0.0
3×10^6	5.19	57.5	1.88	0.1	1.80	0.1
4×10^6	6.06	85.5	2.75	1.2	2.54	0.7
5×10^6	6.73	95.9	3.42	5.7	3.11	2.9
6×10^6	7.28	98.9	3.97	15.2	3.57	7.7
7×10^6	7.75	99.7	4.44	28.7	3.97	15.2
8×10^6	8.15	99.9	4.84	43.7	4.31	24.5
9×10^6	8.51	100.0	5.20	57.8	4.61	34.9
10×10^6	8.82	100.0	5.51	69.7	4.88	45.3
12×10^6	9.37	100.0	6.06	85.6	5.35	63.7
14×10^6	9.84	100.0	6.53	93.7	5.74	77.1
16×10^6	10.24	100.0	6.93	97.4	6.09	86.1
18×10^6	10.60	100.0	7.29	98.9	6.39	91.7
20×10^6	10.92	100.0	7.61	99.5	6.66	95.1
30×10^6	12.14	100.0	8.83	100.0	7.69	99.6
40×10^6	13.01	100.0	9.70	100.0	8.43	100.0

Note: Values in the table are obtained with $F_k = 1$.

c) Overall Effects

In the previous section it was shown how the probability, *P* (-), of injury (1st or 2nd degree burns) and death can be calculated for a given heat flux at a specific distance. To calculate the overall effects, the previous expressions must be integrated over the whole area around the fire. If one assumes a population density of N_o (persons/m^2) in the vicinity, and the fire has a radius *R* (m), then the total number of persons dying will be equal to:

$$N = \left(N_0 \pi R^2\right) + \int_R^{\infty} P\, N_0\, 2\pi r\, dr\,. \qquad \text{(C2.91)}$$

The first term represents the lethality inside the fire, while the second term (together with the corresponding probit function for lethality), the lethality outside the fire.

For the above expression it has been assumed that the fire is symmetric (spherical) and that the population density is uniform. For calculations of the total number of people with 1st and 2nd degree burns, only the second term should be employed.

Effects of Fires

calculation procedure

Human Skin Burns
↓
Thermal Radiation Intensity Limits

Thermal Radiation Dose
↓
Probability of Injury or Death
↓
Overall Effects on People

Effects on Materials

Effects of Fires

example

Human Skin Burns
↓
Thermal Radiation Intensity Limits

Thermal Radiation Dose
↓
Probability of Injury or Death
↓
Overall Effects on People

Effects on Materials

EXAMPLE C2.8. *Effects on People*

For the pool fire examined in Example C2.4, calculate for the summer season, the following:

a) The probability of injury (1st or 2nd degree burns) and death in a distance of 30 m from the flame's surface.
b) If, in the area where the pool fire was developed, the population density is about 1 person per 20 m^2 (in the whole area), calculate the number of persons with 1st and 2nd degree burns, as well as the number of deaths.

---

a) Probability of Injuries and Deaths in a Distance of 30 m from the Flame's Surface.

From the Example C2.4 (see Figure C2.6), in a distance of 30 m, the heat flux is calculated as $q' = 26.964e^{-0.0238x30} = 13.2$ kW/m^2. From Eq. (C2.88), for $u = 4$ m/s, $x_o = 138.4$ m (at 138.4 m $q'=1$ kW/m^2) and $r = 30$ m, the exposure time is calculated as, t_{eff} (s), equal to 32.1 s.
The thermal radiation dose is subsequently calculated equal to $D = t_{eff}\,(q')^{4/3} = 10x10^6$ W$^{4/3}$s·m$^{-8/3}$.

For the calculation of the probability of deaths, one can use Eq. (C2.89) - (C2.90), that is, for $c_1 = -36.38$ and $c_2 = 2.56$ (Table C2.10)

$$Pr = c_1 + c_2 \ln D \;=\; 4.89,$$

and for $F_k = 0.95$,
$$P = F_k \frac{1}{2}\left[1 + \mathrm{erf}\left(\frac{Pr-5}{\sqrt{2}}\right)\right] = 0.433 \text{ or } 43.3\%.$$

That is, the probability of deaths from the heat flux, at 30 m from the flame's surface is 43.3%.

Employing the corresponding parameters c_1 and c_2 from Table C2.10, one obtains that

- the probability of 1st degree burns is 95%, and
- the probability of 2nd degree burns is 66.4%.

As expected, the above results are in full agreement with those shown in Table C2.11 (note that in the table it was used that $F_k = 1$).

b) Overall Effects

To calculate the total number of deaths, Eq. (C2.91) will be employed,

$$N = \left(N_0 \pi R^2\right) + \int_R^\infty P\, N_0\, 2\pi\, r\, dr\ .$$

- Inside the fire.
  Since $N_o = 0.05$ persons/m^2, and the fire's radius is $R = 21.2$ m (see Example C2.4), the number of deaths inside the fire will be 71 persons.

- Outside the fire.
  The system of Eqs. (C2.87)-(C2.91) must be solved,

$$N_{\text{outside}} = 2\pi\, N_0 \int_R^\infty P\ r\ dr\ ,$$

where
$$P = F_k \frac{1}{2}\left[1 + \text{erf}\left(\frac{Pr - 5}{\sqrt{2}}\right)\right],$$

$$Pr = c_1 + c_2 \ln D\ ,$$

and
$$D = t_{\text{eff}}\left(q'\right)^{4/3} = \left(5 + \frac{(138.4 - r)}{4}\right)\left(26.964\, e^{-0.0238\, r}\right)^{4/3}.$$

The values of the parameters c_1 and c_2 will determine the number of deaths or burns (Table C2.10). An approximate numeric solution of the integral gives:
- 567 persons with 1st degree burns,
- 186 persons with 2nd degree burns,
- 138 deaths.

Therefore, to avoid doublecounting, the probabilities in total show
- 243 persons with 1st degree burns (2nd degree burns and deaths were subtracted),
- 48 persons with 2nd degree burns (deaths were subtracted),
- 209 deaths (71 of them were inside the circle of fire).

❐

C2.5.4. Effects on Materials

Materials' damages from the heat flux can be separated into two categories according to their observed effects:

- 1st degree damages: total destruction or complete material failure,
- 2nd degree damages:superficial, visual decay of materials.

The criterion for these damages is the surface temperature of the examined material. When this surface temperature exceeds a critical temperature, T_{crit} (K), then 1st or 2nd degree damages will be observed. In Table C2.12, values of the critical temperature for various materials are given.

Table C2.12 Critical Temperature and Critical Heat Flux for Various Materials [TNO 1989].

	1st degree damages		2nd degree damages	
	q'_{crit} (kW/m^2)	T_{crit} (K)	q'_{crit} (kW/m^2)	T_{crit} (K)
Stainless Steel		1,400		480
Iron	150-200	770	20-80	470
Wood	15	680	2	370
Aluminum		520	-	-
Glass	4	390	-	-
Synthetic Materials	15	-	2	370

In the same table, typical values of the critical heat flux, q'_{crit} (kW/m^2), are also given. The critical heat flux represents the minimum heat flux required so that the surface of the material attains its critical temperature. It can be observed that except in the case of iron, where (because of its high thermal diffusivity) heat diffuses quickly inside it, the values of the critical heat flux of the other materials are relatively low. Synthetic materials can be reinforced polyester, PVC, Perspex, etc.

To calculate whether the surface temperature, T (K), will reach its critical value, T_{crit} (K), for 1st and 2nd degree damages or not, an energy balance should be performed in the material. That is, the incident heat flux from the fire is equal to (neglecting heat transfer inside the material) the heat emitted from the material by radiation, and the heat removed by convection.

$$aq' - \frac{S_{out}}{S_{in}}\left(\varepsilon\sigma T^4 + h_c(T - T_a)\right) = 0 . \qquad \text{(C2.92)}$$

In Eq. (C.92), α (-), represents the absorption coefficient, q' (W/m^2), the incident heat flux, S_{in} (m), the material's surface that receives the heat flux, S_{out} (m), the material's surface from which heat is emitted, ε (-) the grey-body emissivity and σ the Stephan-Boltzmann constant (= 5.6703×10^{-8} W·m^{-2}·K^{-4}). Furthermore, T (K), denotes the temperature on the material's surface, h_c (W·m^{-2}·K^{-1}), the heat-transfer by convection coefficient and, T_α (K), the ambient temperature. Values for the above coefficients are given in Table C2.13.

Table C2.13 Coefficients of Eq. (C2.92).

	1st degree damages	2nd degree damages
Absorption coefficient, α (-)	1	0.7
Grey-body emissivity, ε (-)	1	
Heat-transfer by convection coefficient, h_c (W·m^{-2}·K^{-1})	2-3 (293 K) 7 (373-473 K)	

Finally, in Figure C2.13 a calculation example (for two different shapes with surfaces h (m^2) and b (m^2), of the surfaces ratio is shown.

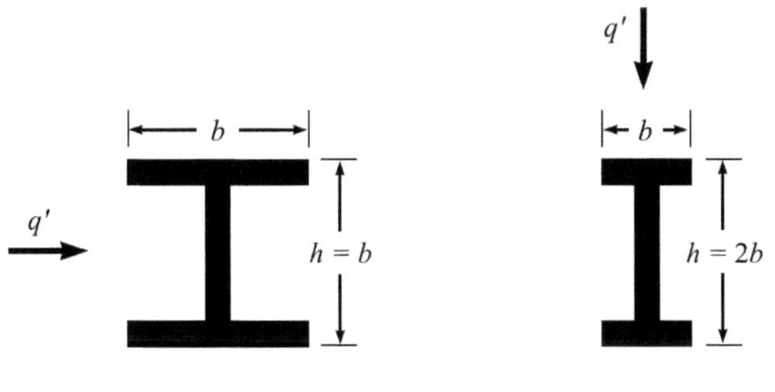

$$\frac{S_{out}}{S_{in}} = \frac{2(h+b)}{h} = 4 \qquad \frac{S_{out}}{S_{in}} = \frac{2(b+2b)}{b} = 6$$

Figure C2.13 Calculation example of surfaces ratio S_{out}/S_{in}.

Hence from Eq. (C2.92) the material's surface temperature for the specific heat flux is calculated. By comparing this value with the corresponding critical temperatures shown in Table C2.13, the expected damage can be estimated.

EXAMPLE C2.9. *Effects on Materials*

For the fire ball examined in Example C2.5, calculate the material damages in a warehouse 50 m from the flame's surface. The warehouse is wooden with large glass openings and an iron supporting structure.

From the diagram in Figure C2.9 (or the equation of the curve), it can be calculated that at 50 m, the heat flux is equal to 50,000 W/m^2.

In order to calculate the surface temperature, T (K), Eq. (C2.92) must be solved,

$$aq' - \frac{S_{\text{out}}}{S_{\text{in}}}\left(\varepsilon\,\sigma T^4 + h_c(T - T_a)\right) = 0\,.$$

For 1st degree damages, from Table C2.13:
$\alpha = 1$, $\varepsilon = 1$ and $h_c = 7$ W·m^{-2}·K^{-1} and $\sigma = 5.6703\times10^{-8}$ W·m^{-2}·K^{-4}.

- For wood and glass, $S_{out}/S_{in} = 2$. Therefore $T = 786$ K
- For iron, $S_{out}/S_{in} = 6$. Therefore $T = 579$ K

For 2nd degree damages, from Table C2.13:
$\alpha = 0.7$, $\varepsilon = 1$ and $h_c = 7$ W·m^{-2}·K^{-1} and $\sigma = 5.6703\times10^{-8}$ W·m^{-2}·K^{-4}.

- For iron, $S_{out}/S_{in} = 6$. Therefore $T = 523$ K

Thus, according to the critical temperatures shown in Table C2.13,

- The iron support structure will not suffer serious damages, but superficial decays (1st degree damages $T_{crit} = 770$ K> $T = 579$ K while for the 2nd degree damages $T = 523$ K > $T_{crit} = 470$ K).
- The wooden walls will be completely destroyed ($T = 786$ K > $T_{crit} = 680$ K).
- The glass openings will also be completely destroyed ($T = 786$ K > $T_{crit} = 390$ K).

❐

C2.6. Examples

In this section initially a case study will be discussed aiming to demonstrate the significance of the magnitude of the variables involved in calculating a fire and its effects. Following this, a list of major industrial accidents caused by fire is presented.

C2.6.1. Case Study: Fire in a Gasoline Tanker in a City

A passenger car collides with a gasoline tanker whilst traveling on a busy city central avenue (see Figure C2.14). The tanker has a total capacity of 20 m^3, and at the time was one-quarter full. As a result of the accident, all the content flow out creating a pool of 3 cm depth, which immediately ignites.

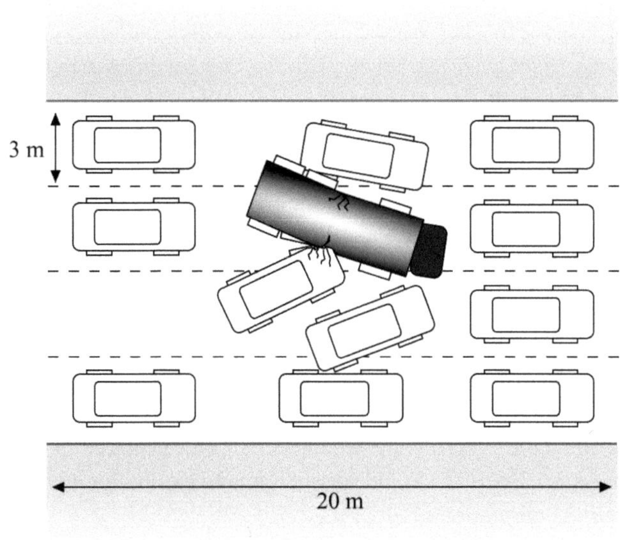

Figure C2.14. Collision on a busy city central avenue.

Calculate the statistically expected number of injuries (1st and 2nd degree burns), and deaths.

Case Study

fire in gasoline tanker

Burning Rate
↓
Maximum Surface Emitting Power
↓
Actual Surface Emitting Power
↓
View Factor
↓
Heat Flux
↓
Overall Effects on People

Case Study

fire in gasoline tanker

Burning Rate
↓
Maximum Surface Emitting Power
↓
Actual Surface Emitting Power
↓
View Factor
↓
Heat Flux
↓
Overall Effects on People

The following data are available:

- Density of gasoline, ρ_L	:	875	kg/m^3
Heat of combustion, ΔH_c	:	45	MJ/kg
- Soot surface emitting power, SEP_{soot}	:	20	kW/m^2
Soot surface fraction, s	:	0.8	-
- Wind velocity, u_w	:	0	m/s
Density of air, ρ_{air}	:	1.21	kg/m^3
- Saturation water vapor pressure, P_w^o	:	2,320	Pa
Relative humidity, RH	:	70	%
- Population density in the square, N_o	:	0.5	persons/m^2

Since the gasoline tanker was one-quarter full, 5 m^3 of gasoline leaked to the street. The pool formed had a depth, d (m), of about 0.03 m and thus the radius, R (m), of the lake was

$$R = \sqrt{V/(\pi\, d)} = 7.28 \text{ m},$$

and surface A = 166.7 m^2. In Figure C2.15 the pool area is shown.

Calculation of heat flux, *q'*

The statistically expected number of injuries and deaths is a function of the heat flux, q' (kW/m^2), that in turn is calculated from the product of the actual surface emitting, SEP_{act} (kW/m^2), the view factor, F_{view} (-), and the atmospheric transmissivity, τ_α (-). Hence, first these three terms will be calculated.

The burning rate, m' (kg/m^{2}s), of gasoline is obtained from Table C2.2, as

$$m' = 0.055 \text{ kg/m}^2\text{s} .$$

The mean flame's length (zero wind velocity) will be calculated from the Pritchard -Binding expression, Eq. (C2.13), because Thomas's expression underestimates the flame's length (see subsection C2.1.2), as

$$L/D = 10.615\left[\frac{m'}{\rho_{air}\sqrt{g\,D}}\right]^{0.305}(u^*)^{-0.03} \quad => \quad L = 28.26 \text{ m} \qquad (u^* = 1)$$

If the Thomas expression, Eq. (C2.12) was employed, the length would have been found equal to 19.16 m, which is significantly shorter.

Case Study

fire in gasoline tanker

Burning Rate
↓
Maximum Surface Emitting Power
↓
Actual Surface Emitting Power
↓
View Factor
↓
Heat Flux
↓
Overall Effects on People

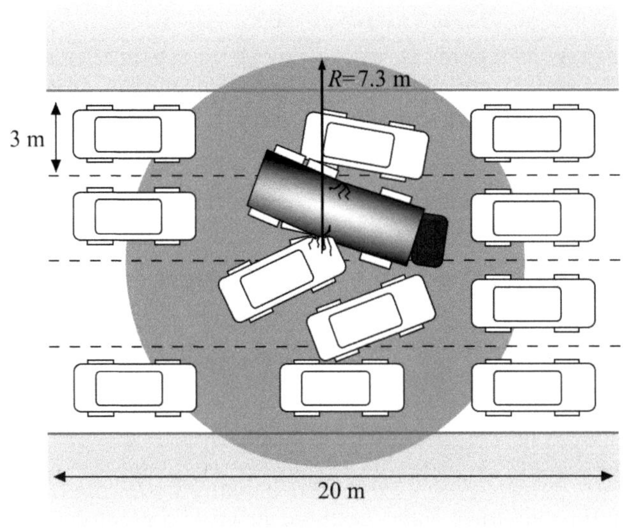

Figure C2.15. Pool formed following the collision.

Having calculated the length of the flame, the maximum surface emitting power, SEP_{max} (kW/m^2), can be calculated from Eq. (C2.9), as

$$SEP_{max} = F_s \frac{1}{1+4(L/D)} m' \, \Delta H_c = 56.5 \text{ kW/m}^2.$$

The actual surface emitting power, SEP_{act} (kW/m^2), is calculated from Eq. (C2.14),

$$SEP_{act} = SEP_{max}(1-s) + SEP_{soot}\, s = 27.4 \text{ kW/m}^2.$$

If the flame's length obtained from the expression of Thomas, Eq. (C2.12), was employed, then the actual surface emitting power would have been, SEP_{act} = 31.8 kW/m^2 (which value is quite near to the above one, as the flame's length does not strongly influence the surface emitting power).

Case Study

fire in gasoline tanker

Burning Rate
↓
Maximum Surface Emitting Power
↓
Actual Surface Emitting Power
↓
View Factor
↓
Heat Flux
↓
Overall Effects on People

Table C2.14. Heat Flux as a Function of Distance.

X (m)	α (-)	β (-)	F_v (-)	F_h (-)	F_{view} (-)	τ_a (-)	q' (kW/m^2)
7.28	3.88	1.00	0.500	0.496	0.704	1.00	19.23
9.0	3.88	1.24	0.405	0.298	0.503	0.99	13.58
10.0	3.88	1.37	0.364	0.256	0.445	0.95	11.53
12.5	3.88	1.72	0.290	0.189	0.346	0.90	8.46
15.0	3.88	2.06	0.240	0.147	0.282	0.86	6.64
20.0	3.88	2.75	0.175	0.096	0.200	0.83	4.50
30.0	3.88	4.12	0.105	0.045	0.115	0.78	2.45
40.0	3.88	5.49	0.069	0.024	0.073	0.76	1.51
50.0	3.88	6.86	0.048	0.014	0.049	0.74	1.00
60.0	3.88	8.24	0.034	0.008	0.035	0.73	0.70
100.0	3.88	13.73	0.013	0.002	0.013	0.69	0.25

The view factor, F_{view} (-), and the atmospheric transmissivity, τ_a (-), are calculated as a function of the distance X (m) from the center of the fire, from Eqs. (C2.26)-(C2.29) and Eqs. (C2.30)-(C2.32) correspondingly. Results are shown in Table C2.14. In the same table, parameters α (-) and β (-) employed in the view factor calculations and defined by Eq. (C2.20) are also presented. Also the two components, F_v (-) and F_h (-), of the view factor F_{view} (-), the atmospheric transmissivity, τ_a (-) and the heat flux, q' (kW/m^2), are given in the same table.

In Figure C2.16, the heat flux as a function of the distance X (m) is shown.

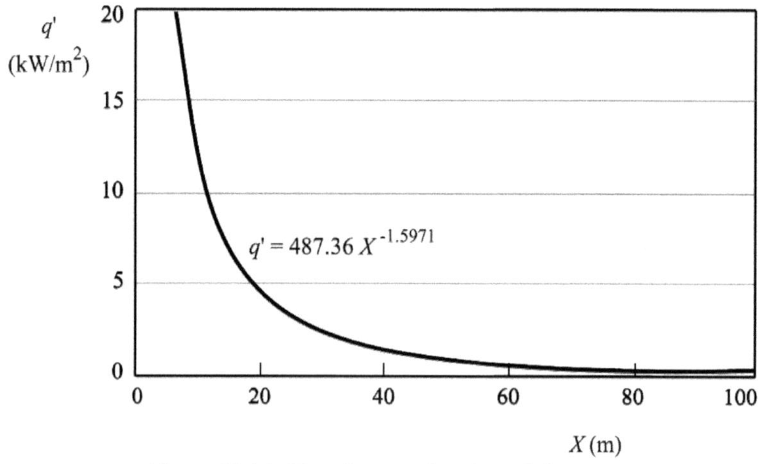

Figure C2.16. Heat flux as a function of distance.

Total number of deaths

To calculate the total number of deaths, one must first calculate the deaths inside the fire itself, and then those outside the fire.

It will be assumed that there are no further explosions, resulting in a "domino" effect.

- Inside the fire

  According to Figure C2.15, inside the pool fire, in addition to the gasoline tanker, there were 4 whole cars and 4 halves. Assuming on average 1.5 persons per car, the number of people that died immediately (because they were inside the fire) is 9. Adding the tanker's driver, the total number of deaths inside the fire is 10 persons.

- Outside the fire

  The following system of Eqs. (C2.87)-(C2.91) must be solved:

$$N_{\varepsilon\kappa\tau\acute{o}\varsigma} = 2\pi \, N_0 \int_R^\infty P \; r \; dr \; ,$$

where
$$P = F_k \, \frac{1}{2}\left[1 + \mathrm{erf}\left(\frac{Pr - 5}{\sqrt{2}}\right)\right] \qquad (F_k = 1),$$

$$Pr = c_1 + c_2 \ln D \; ,$$

and
$$D = t_{eff}\,(q')^{4/3} = \left(5 + \frac{(50 - r)}{4}\right)\left(487.36\, r^{-1.5971}\right)^{4/3} .$$

In the above expression, the distance 50 m represents the position from the center of the fire where the heat flux has dropped to 1 kW/m^2.

The values of the parameters c_1 and c_2 determine the number of injuries or deaths (Table C2.10). An approximate numeric solution of the integral produces:

- 276 persons with 1st degree burns
- 97 persons with 2nd degree burns
- 56 deaths

Case Study

fire in gasoline tanker

Burning Rate
↓
Maximum Surface Emitting Power
↓
Actual Surface Emitting Power
↓
Heat Flux
↓
Overall Effects on People

Case Study

fire in gasoline tanker

Burning Rate
↓
Maximum Surface Emitting Power
↓
Actual Surface Emitting Power
↓
Heat Flux
↓
Overall Effects on People

Therefore, to avoid doublecounting, the probabilities in total show
- 123 persons with 1st degree burns (2nd degree burns and deaths were subtracted)
- 41 persons with 2nd degree burns (deaths were subtracted)
- 66 deaths (10 were inside the fire)

In reality, obstacles (e.g., walls, buildings, etc.) will normally be present at some distance, which act to slow the expansion of the heat flux. Thus results must be altered accordingly in order to take into account those obstacles.

❐

C2.6.2. Major Industrial Accidents Caused by Fires

In Table C2.15 an indicative list of major industrial accidents caused by fires is given. The list is part of the database MinA - Major Industrial Accidents due to Fires, Explosions and Toxic-Gas Releases [MinA 2006].

Table C2.15. Major Industrial Accidents Caused by Fires.

Date	Country	City/State	D	I	E[a]	Chemical	Unit	M$[b]
2001								
21.9	USA	Lake Charles, LA		3		Hydrocarbons	Refinery	62
14.8	USA	Lemont, IL				Hydrocarbons	Refinery	43
16.5	UK	Birkenhead				Hydrocarbons	Petrochemicals	129
28.4	USA	Wood River, IL				Hydrocarbons	Refinery	81
23.4	USA	Carson City, CA				Hydrocarbons	Refinery	147
10.4	Aruba	Aruba				Oil	Refinery	159
11.2	USA	St James, LA				Styrene	Petrochemicals	21
1999								
11.3	India	Bombay High				Natural Gas	Oil Platform	42
19.2	Greece	Thessaloniki				Hydrocarbons	Refinery	51
1998								
6.10	France	Berre l'Etang				Hydrocarbons	Refinery	27
10.5	Egypt	Ras Gharib				Oil	Distribution	38
1997								
14.10	India	Wishakaptnam	34	31	150,000		Refinery	
21.1	USA	Martinez, CA	1	46		Hydrocarbons	Refinery	
7.10	Taiwan	Kaoshiung	10	47		MEKPO	Warehouse	
23.6	India	Chennai	3	4		LPG		
1996								
6.8	USA	Heilliecourt, PA					Warehouse	
1995								
16.10	USA	Rouseville, PA				Hydrocarbons	Refinery	54
9.10	UK	Wilton				Propylene		25
24.7	France	Blotzheim		1		Plastics		
27.4	Russia	Ukhta	12			Natural Gas		
24.1	Indonesia	Cilapcap				Natural Gas	Refinery	
1994								
4.11	Nigeria	Onitsha	60			Oil	Transport	
2.11	Egypt	Donca	410			Oil	Warehouse	

[a] D I E - Number of Deaths, number of Injuries, number of people Evacuated.

[b] M$: Material Damage in million $ (in 2005 prices).

Note: LPG, liquefied propane gas; MEKPO, methyl ethyl ketone peroxide.

- Continued

Table C2.15 (cont.). Major Industrial Accidents Caused by Fires.

Date	Country	City/State	D	I	E[a]	Chemical	Unit	M$[b]
1994								
8.8	USA	Baton Rouge, LA				Ethylene	Petrochemicals	34
25.7	UK	Pembroke		27		Hydrocarbons	Refinery	108
1993								
25.2	Japan	Kawasaki				Hydrocarbons	Refinery	49
24.8	France	Mirande				Plastics		
1.7	Canada	St Terese		36		Dioxins		
23.5	Germany	Munich	2					
2.4	USA	Baton Rouge, LA				Hydrocarbons	Refinery	92
7.1	South Korea	Chongju	27	50		LPG		
1992								
27.11	Belgium	Diest				Dioxins	Petrochemicals	
10.9	USA	Camden, NJ		12		HCl	Petrochemicals	
1991								
3.11	USA	Beaumont, TX				Hydrocarbons	Refinery	21
7.9	Israel	Haifa					Chemicals	20
1.9	China	Shaxi	30	650		Pesticides		
1.5	USA	Sterlington, LA				Hydrocarbons	Petrochemicals	153
12.1	USA	Port Arthur, TX				Hydrocarbons	Refinery	37
1990								
30.11	Saudi Arabia	Ras Tanura				Hydrocarbons	Refinery	47
25.11	USA	Denver, CO				Oil	Distribution	47
15.11	Portugal	Porto Leixoes	14	76		Propane	Refinery	26
26.7	Lebanon	Chtaura		45		Oil		
10.7	Brazil	Rio de Janeiro	3			Hydrocarbons		
1.4	Australia	Sydney			10,000		Warehouse	
11.5	India	Thane	35	10		Hydrocarbons		
1989								
3.10	USA	Sabine Pass, TX				Natural Gas	Pipeline	
5.9	USA	Martinez, CA				Hydrogen	Refinery	73
25.5	USA	San Bernadino, CA	2	3		Natural Gas	Pipeline	
10.4	USA	Richmond, CA				Hydrogen	Refinery	133
1988								
9.11	India	Bombay	35	16		Oil	Refinery	
22.9	UK	North Sea				Natural Gas	Oil Platform	116
8.6	USA	Port Arthur, TX				Propane	Warehouse	23
8.6	France	Tours		3	200,000			
4.5	USA	Henderson, TX	2	350		Ammonia	Petrochemicals	
24.4	Brazil	Enchova				Natural Gas	Oil Platform	546

[a] D I E - Number of Deaths, number of Injuries, number of people Evacuated.

[b] M$: Material Damage in million $ (in 2005 prices).

Note: HCl, hydrochloric acid; LPG, liquefied propane gas.

- Continued

Table C2.15 (cont.). Major Industrial Accidents Caused by Fires.

Date	Country	City/State	D	I	E[a]	Chemical	Unit	M$[b]
1987								
20.12	USA	Cook Inlet, AK				Natural Gas	Oil Platform	202
12.12	India	Maharashtra	25	23		Naphtha	Refinery	
24.11	USA	Torrance, CA				Hydrocarbons		22
11.10	Canada	Ft. McMurray, AL				Hydrocarbons		52
23.8	China	Lanzhou	5			Natural Gas	Train Wagon	
23.7	Canada	Mississauga				Hydrocarbons		29
2.6	France	Port Herriot	2	8		Oil	Warehouse	
24.3	USA	Natchitoke, MS				Sulfuric acid		
1986								
1.11	Switzerland	Basel				Pesticides	Warehouse	
8.7	USA	Miamisburg, OH		400	40,000	Acid	Train Wagon	
15.6	USA	Pascagoula, MS	3	76		Aniline		
24.2	Greece	Thessaloniki						368
21.2	USA	Lancaster, KY	3			Natural Gas	Pipeline	
1985								
21.12	Italy	Naples				Oil	Distribution	71
1.11	India	Padaval	43	82		Natural Gas		
22.6	USA	Anaheim, CA		12	10,000	Pesticides	Warehouse	
19.5	Italy	Priolo	23	11		Ethylene	Petrochemicals	110
27.4	USA	Beaumont, KY	5	3		Natural Gas	Pipeline	
20.3	USA	Plainfield, NJ				Hydrogen		
26.2	USA	Coachella, CA		236	2,000	Pesticides		
1984								
15.8	Canada	Ft. McMurray, AL				Hydrocarbons	Refinery	129
13.12	Venezuela	Las Piedras				Hydrocarbons	Refinery	105
25.2	Brazil	Sao Paolo	50	221	80	Natural Gas		
24.2	Brazil	Cubatao		150		Natural Gas	Pipeline	
1983								
31.8	Brazil	Pojuca	42	100	>1,000	Natural Gas		
30.8	UK	Milford Haven		6		Oil	Refinery	22
30.7	USA	Baton Rouge, LA				VCM	Train Wagon	
27.7	UK	Dursley		5		Ammonia		20
1.7	USA	Port Arthur, TX					Chemicals	23
26.5	USA	Prudhoe Bay, AK				LNG		53
7.4	USA	Avon, CA				Hydrocarbons		86
1982								
19.12	Venezuela	Tacos, Caracas	150	500	40,000	Oil		90
4.10	USA	Freeport, TX				Oil		23
3.5	USA	Oakland, CA	7			Natural Gas	Transport	

[a] D I E - Number of Deaths, number of Injuries, number of people Evacuated.

[b] M$: Material Damage in million $ (in 2005 prices).

Note: LNG, liquefied natural gas; VCM, vinyl chloride monomer.

- Continued

Table C2.15 (cont.). Major Industrial Accidents Caused by Fires.

Date	Country	City/State	D	I	E[a]	Chemical	Unit	M$[b]
1982								
18.4	Canada	Edmonton, AB				Ethylene	Petrochemicals	39
31.3	Japan	Kashima				Hydrocarbons		21
13.2	UK	Morley					Warehouse	
20.1	Canada	Ft. McMurray, AL				Hydrogen	Compressors	32
1981								
19.12	Venezuela	Caracas	153	500		Oil	Warehouse	
20.8	Kuwait	Shuaiba				Hydrocarbons	Refinery	86
1980								
31.12	USA	Corpus Christi, TX				Hydrocarbons		31
25.11	USA	Kenner, LA	7	6		Natural Gas	Transport	
26.7	USA	Muldraugh, KY		4		VCM	Train Wagon	
15.6	USA	New Orleans, LA				Natural Gas	Pipeline	
5.6	Malaysia	Port Kelang	3	200	>3,000		Transport	
17.5	USA	Deer Park, TX				Acid	Petrochemicals	42
11.2	UK	Longport				LPG	Warehouse	
1.1	UK	Barking					Warehouse	
1979								
11.12	Puerto Rico	Ponce				Hydrocarbons	Petrochemicals	36
11.12	Australia	Geelong				Oil		23
30.8	USA	Good Hope, LA	12	25		Butane	Transport	
8.1	Ireland	Banty Bay				Petrol	Distribution	50
1978								
14.7	Taiwan	Kaoshiung	33	49		MEKPO	Warehouse	
30.5	USA	Texas City, TX	7	11		Butane	Refinery	142
29.5	USA	Lewisville, AR		2		VCM	Train Wagon	
29.5	Mexico	Santa Cruz	52	88		Propylene	Transport	
1977								
28.12	USA	Goldonna, LA	2	9		LPG	Train Wagon	
24.9	USA	Romeoville, IL				Natural Gas	Transport	19
24.9	USA	Beattyville, KY	7	6		Natural Gas	Transport	
8.7	USA	Fairbanks, AK				Oil	Pipeline	94
1976								
31.8	USA	Gadsden, AL	3	24		Natural Gas		
1975								
4.12	USA	Seattle, WA					Transport	
13.5	USA	Devers, TX	4			LPG	Pipeline	
30.4	USA	Eagle Pass, TX	16	7		Propane	Transport	

[a] D I E - Number of Deaths, number of Injuries, number of people Evacuated.
[b] M$: Material Damage in million $ (in 2005 prices).
Note: LPG, liquefied propane gas; MEKPO, methyl ethyl ketone peroxide; VCM, vinyl chloride monomer.

- *Continued*

Table C2.15 (cont.). Major Industrial Accidents Caused by Fires.

Date	Country	City/State	D	I	E[a]	Chemical	Unit	M$[b]
1975								
16.3	USA	Avon, CA				Oil	Refinery	28
31.1	USA	Marcus Hook, PA				Oil	Transport	21
17.1	USA	Lima, OH				Oil	Distribution	
1974								
13.9	USA	Griffith, IN				Propane	Warehouse	
9.6	USA	Bealeton, VA				Natural Gas	Pipeline	
17.1	UK	Aberdeen				Butane	Transport	
1973								
6.9	Germany	Gladbeck					Chemicals	
24.8	Virgin Islands	St. Croix				Hydrocarbons		37
8.7	Japan	Tokuyama	1	16		Ethylene	Petrochemicals	57
5.7	USA	Kingman, AZ	13	89		Propane	Transport	
10.2	USA	Staten Island, NY	40			LNG		
1972								
21.9	USA	Turnpike, NJ	2	22		Propane	Transport	
4.8	Italy	Trieste				Oil	Distribution	39
9.5	USA	Lynchburg, VA	2	3		Propane	Transport	
6.4	USA	Doraville, GA	2	161		Natural Gas		
9.2	USA	Tewksbury, MA	2	21			Warehouse	
1970								
12.11	USA	Hudson, OH				LPG	Transport	
21.6	USA	Crescent City, IL				Propane	Transport	
1969								
14.5	UK	Wilton Teeside	2	23		Cyclohexane		
6.3	Venezuela	Puerto La Cruz				Hydrocarbons		20
1968								
5.12	USA	Yutan, NE	5			LPG	Pipeline	
1967								
19.12	USA	El Segundo, CA	13	5		Oil		
1966								
4.1	France	Feyzin	18	83		Propane	Refinery	114
1964								
16.6	Japan	Niigata				Hydrocarbons	Refinery	114
1962								
4.8	Saudi Arabia	Ras Tanura	1	111		Propane		

[a] D I E - Number of Deaths, number of Injuries, number of people Evacuated.

[b] M$: Material Damage in million $ (in 2005 prices).

Note: LNG, liquefied natural gas; LPG, liquefied propane gas.

- Continued

Table C2.15 (cont.). Major Industrial Accidents Caused by Fires.

Date	Country	City/State	D	I	E[a]	Chemical	Unit	M$[b]
1959								
2.6	USA	Deer Lake, PA	11	10		LPG	Train Wagon	
1958								
22.5	USA	Signal Hills, CA	2	34		Oil		66
1956								
29.7	USA	Amarillo, TX	20	>32		Oil	Warehouse	
1949								
10.10	USA	Winthrop, MO	1			LPG	Train Wagon	
1948								
13.10	USA	Sacramento, CA	2			Butane	Transport	
1947								
16.4	USA	City Port, TX	552	3,000		Ammonia	Transport	
1944								
21.11	USA	Denison, TX	10	45		Butane		
20.10	USA	Cleveland, OH	128	300		LNG	Warehouse	
14.4	India	Bombay	350	1,800	50,000	Explosives	Transport	
1943								
18.1	USA	Los Angeles, CA	5	>25		Butane	Transport	
1915								
27.9	USA	Ardmore, OK	40			Hydrocarbons	Train Wagon	

[a] D I E - Number of Deaths, number of Injuries, number of people Evacuated.

[b] M$: Material Damage in million $ (in 2005 prices).

Note: LNG, liquefied natural gas; LPG, liquefied propane gas.

C3
Vapor Cloud Explosions (VCEs)

If a flammable cloud is formed during the leakage of flammable gases, its direct ignition can sometimes lead to a flash fire - see Section C2.4. If, however, its ignition is for some reason delayed (5-10 min), then a vapour cloud explosion (VCE) is the probable outcome. For ignition to take place, the composition of the flammable gases in some part of the vapor cloud must be between the flammability limits, while at the same time a source able to supply the required energy (usually of the order of 10 J), must be available. The variables that influence the evolution and the intensity of an explosion are:

On September 21, 2001, an explosion in the fertilizer production unit of AZF (TotalFinaElf) in Toulouse, France, had the force of a 3.2 earthquake on the Richter scale. The explosion totally destroyed the unit and the surrounding area, killing 30 persons and injuring 3,000. (Reproduced by kind permission of E. Grimault.)

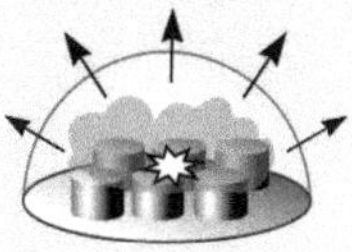

VCE

introduction

- The type and the quantity of the flammable substance.
- The time span from the onset of the leakage until the ignition.
- The configuration of the space where the leakage took place.
- The position and the number of ignition sources in relation to the place of leak.

The damaging effects of a vapor cloud explosion are mostly due to the overpressure (shock wave) that is created from the fast expansion of the combustion products. The shock wave is the most important cause of damage to people, equipment and facilities.

To simulate or predict the effects of vapor cloud explosions, the following types of models are usually employed [Lea & Ledin 2002]:

a) **Empirical analytic models** that are usually based on a single curve (or family of curves) of overpressure as a function of the distance, with parameters like
   - the part of the energy that is released as shock wave (TNT method),
   - the strength of the explosion (Multi-Energy method), or
   - the flame expansion speed (Baker-Strehlow method).
b) **Numerical 3D CFD computer models of finite volumes** which are usually based upon
   - turbulent analysis (CFX, EXSIM, NEWT, REACFLOW, etc.), or
   - empirical relations (FLACS, AutoReaGas, COBRA, etc.).

In this work only empirical analytic models will be presented and with these models the overpressure will be calculated as a function of the distance. From the magnitude of the overpressure, the effects on persons and materials will be calculated (see Section C3.5). The case study (see Section C3.6.1) will demonstrate the applicability and the advantages of these models.

The preference for presenting empirical rather than numerical models follows the aim of this handbook (as has already been mentioned) which is to be a useful reference tool for the experienced or the new engineer, by presenting a series of quick and simple algorithms, in contrast to the complex commercial packages, whose very high cost makes them inaccessible to most users.

It should also be mentioned that most of the algorithms presented in this handbook have been derived from large series of experiments and observations. This ensures their validity to a certain degree, especially when used to analyze cases/incidents with similar conditions.

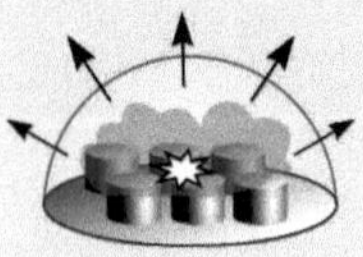

C3.1. Cloud Expansion Mechanism

VCE

expansion mechanism

The expansion mechanism of a vapor cloud explosion can be analyzed in the following stages:

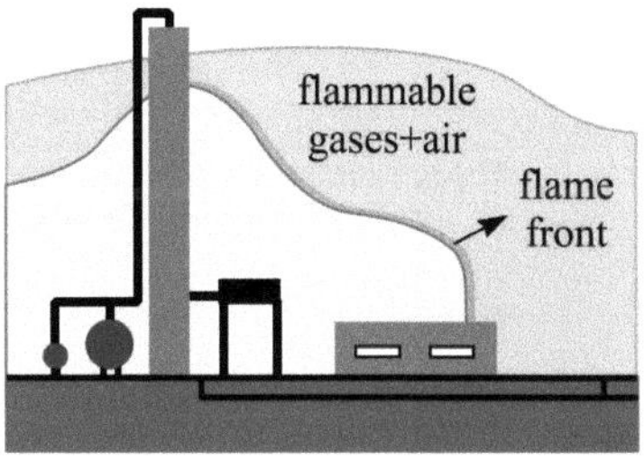

Stage 1.

When the cloud ignites, the flame starts to propagate away from the point of ignition, with a speed proportional to the developed overpressure. The flammable mixture of gas-air is pushed in front of the flame. In this stage the flow of combustion products can be considered as laminar.

Flame speed about 5-30 m/s.
Overpressure, very low.

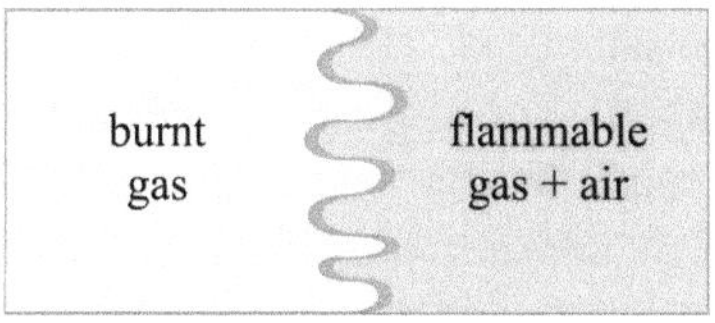

Stage 2.

Because of the unstable nature of the flame and large turbulent eddies, a wrinkled-frame front appears, resulting in an increase of the flame surface, thus an increase in its burning rate and consequently its speed.

Deflagration.
Flame speed 30-500 m/s.
Overpressure up to 2-3 mbar.

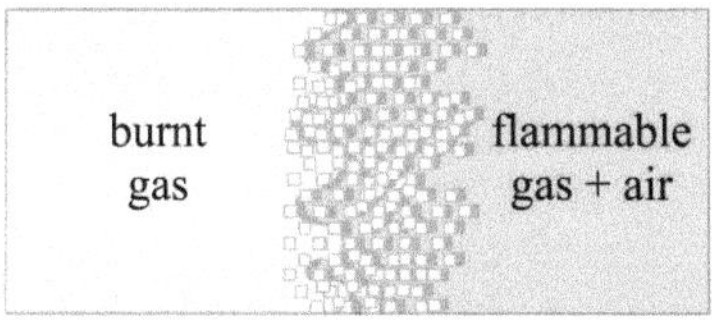

Stage 3.

The presence of obstacles in the flow results in a further increase of the flame speed. The flow becomes turbulent, and the burning front is changed into a zone where flammables and combustion products coexist. This surface increase produces a further increase of the burning rate.

Deflagration.
Flame speed 500-1,000 m/s.
Overpressure up to 1 bar.

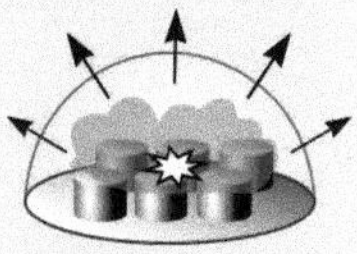

VCE

expansion mechanism

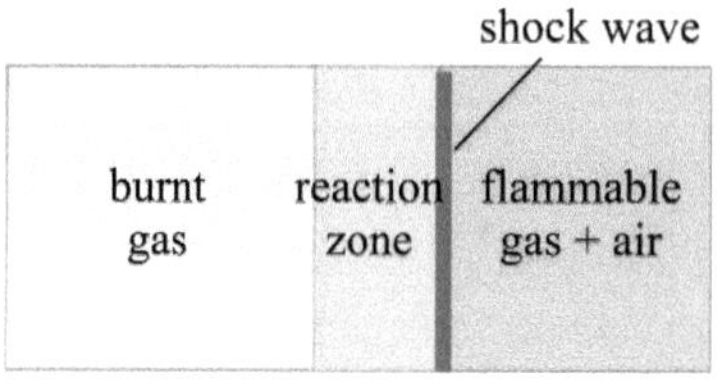

Stage 4.

The flame speed continues to increase, and the reactive mixture in front of the zone of turbulent combustion is subjected to compression and heat because of mixing with combustion products. Hence, temperatures higher than the self-ignition point are reached, resulting in the creation of a shock wave.

Detonation.
Flame speed up to 2,200 m/s.
Overpressure up to 20 bar.

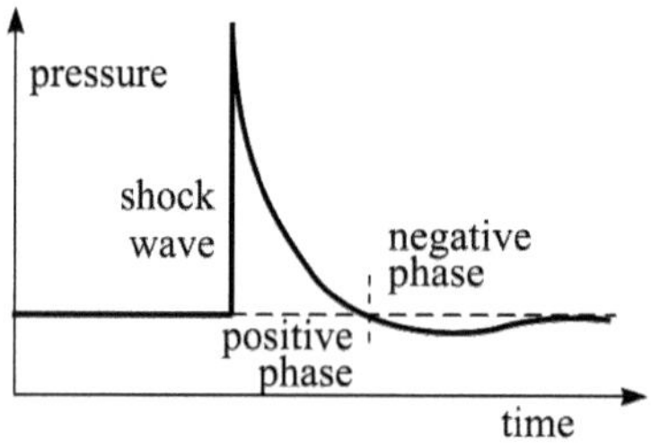

Stage 5.

Chemical energy is transformed into mechanical energy via shock wave (40% transformation). The shock wave induces a drastic change in the properties of the surrounding space (pressure, density, molecular velocity). These properties are instantaneously increased (positive phase), then decreased with slower rates to values lower than the ambient ones (negative phase), to return slowly afterwards to their ambient values.

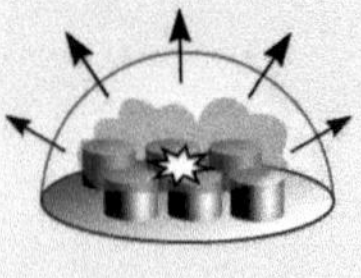

C3.2. Equivalent TNT Mass Method

Equivalent TNT Mass Method

calculation procedure

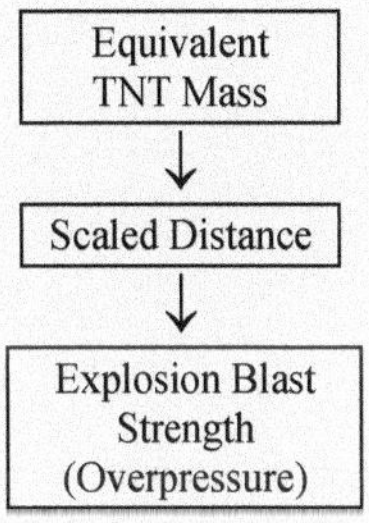

According to this method, the power of the vapor cloud explosion equates to an equivalent mass of TNT (tri-nitrotoluene) that would produce the same explosive power. First, the mass of the flammable gas in the cloud with concentrations between the lower and the upper flammability limits (LFL and UFL) is estimated. This mass is consequently multiplied by the heat of combustion to obtain the total available energy of combustion. This energy is multiplied by a parameter (0 to 1) that accounts for the non-ideality of the explosion, and then divided with the heat of combustion of TNT, in order to calculate the equivalent TNT mass. The equivalent TNT mass is employed for the calculation of the shock wave in a specific distance from the source.

This method is particularly easy to use and there is an abundance of data for the characterization of TNT explosions. However, VCEs behave significantly differently than explosions of TNT or similar explosives, and thus calculations with this method usually overestimate the effects of such an explosion. In general, the TNT method is employed today only as a first estimate in the determination of the effects of an explosion.

C3.2.1. Equivalent TNT Mass and Overpressure

The method is based on the empirical diagram of Brasie & Simpson [Brasie & Simpson 1968] (Figure C3.1), and produces the overpressure, P_S (kPa), as a function of a scaled distance, Z (m/kg$^{1/3}$), defined by the equation

$$Z = \frac{x}{M_{TNT}^{1/3}}, \qquad \text{(C3.1)}$$

where x (m) is the distance from the center of the explosion and M_{TNT} (kg) denotes the equivalent TNT mass, obtained from the expression

$$M_{TNT} = \frac{f_E \, \Delta H_c \, M_G}{\Delta H_{TNT}}. \qquad \text{(C3.2)}$$

In the above expression, M_G (kg) denotes the mass of the flammable gas that takes part in the explosion, while ΔH_c (kJ/kg) and ΔH_{TNT} (kJ/kg) are the heat of combustion of the flammable gas and the heat of combustion of TNT (= 4,760 kJ/kg), respectively. The coefficient, f_E (-), denotes the fraction of the energy released as shock wave (usually value between 0.01 and 0.1). Eq. (C3.1) is based on the well-known cubic-root law of scaling which most powerful explosives

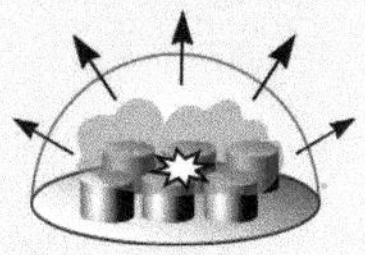

Equivalent TNT Mass
↓
Scaled Distance
↓
Explosion Blast Strength (Overpressure)

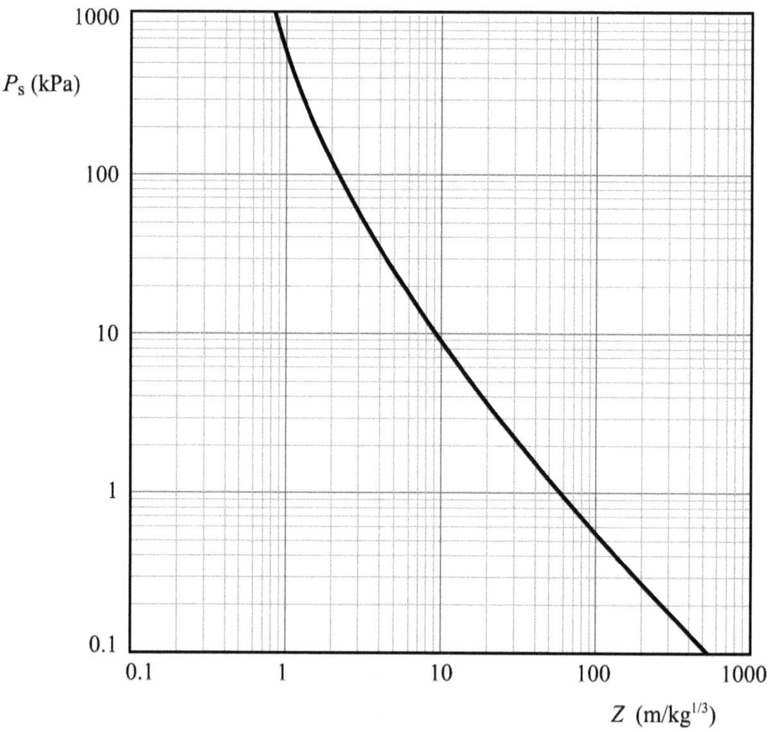

Figure C3.1. Overpressure as a function of the scaling distance [Brasie & Simpson 1968].

(dynamite, TNT, nitroglycerin and others) follow. According to the cubic-root law, the distance from the center of the explosion to a specific distance is proportional to the cubic-root of the mass of the explosive employed.

In place of the diagram of Figure C3.1, one can also use the following more recent analytical expression for the overpressure, P_s (kPa), of the shock wave

$$P_s = \frac{80{,}800\left(1+\left[\frac{Z}{4.5}\right]^2\right)}{\sqrt{1+\left[\frac{Z}{0.048}\right]^2}\sqrt{1+\left[\frac{Z}{0.32}\right]^2}\sqrt{1+\left[\frac{Z}{1.35}\right]^2}}. \qquad \text{(C3.3)}$$

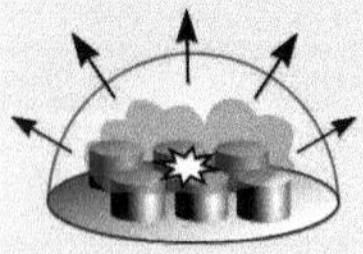

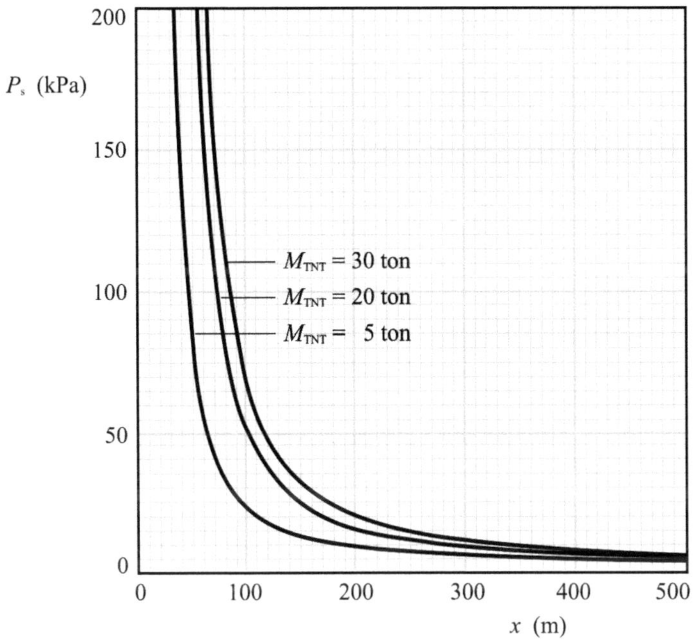

Figure C3.2. Overpressure as a function of distance - Eq. (C3.3).

Values obtained by Eq. (C3.3) are in very close agreement with those obtained by Figure C3.1. Differences are observed only in very small distances where the increase in pressure is quite steep. Characteristic values of the overpressure of the shock wave as a function of the distance for three equivalent TNT masses obtained from the solution of Eqs. (C3.1)-(C3.3) are shown in Figure C3.2.

The main advantage of the TNT method is its great simplicity of use. For this reason it is widely employed in the calculation of the overpressure of an explosion, but also in its characterization. The most important disadvantages of the method are the following:

a) The TNT method calculates the overpressure of an explosion without taking into consideration the space configuration where the explosion takes place. As will be shown in the next section, an explosion in the middle of an area full of equipment, or in a closed space, will exhibit different power from an equivalent one in an open space.
b) Parameter f_E in most cases is unknown, and greatly influences the prediction.
c) The method does not calculate the time evolution of the explosion.

Equivalent TNT Mass Method

calculation procedure

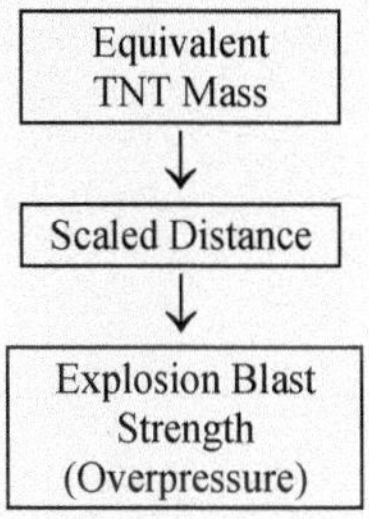

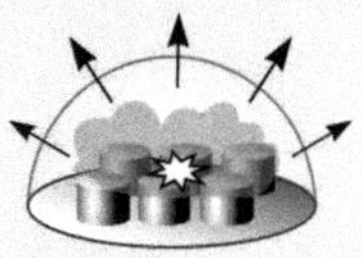

Equivalent TNT Mass Method

example

Equivalent TNT Mass
↓
Scaled Distance
↓
Explosion Blast Strength (Overpressure)

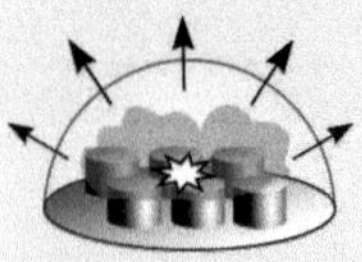

EXAMPLE C3.1. *Calculation of the Overpressure*

The PEMEX LPG (Liquefied Propane Gas) Terminal, of 16,000 m^3 capacity, in San Juan Ixhuatepec in the outskirts of Mexico City, was regularly supplied by 3 refineries. At 5:35 am of November 19, 1984, the control room noticed a pressure drop in the pumping station, without however being able to find its cause. An 8-in diameter pipeline between a spherical storage tank and a group of cylindrical vessels was leaking. The leak lasted between 5-10 min, while a 0.4 m/s wind was in the area. A large vapor cloud was formed, followed by a VCE. The resulting casualties included 550 deaths and more than 6,400 wounded. Material damages were estimated at $34,000,000 (in 2005 prices). The facilities dated from 1961-1962 and thus had already been in operation for 20 years. During this time, the area surrounding the terminal was inhabited, and there were houses within a distance of 130 m. The terminal's capacity was 16,000 m^3 (29,760 kg), and it is estimated that about 4,750 kg of propane leaked and evaporated into the atmosphere.

In Figures C3.3 and C3.4, the area layout as well as photographs before and after the explosion are shown. Local observations showed that in a distance of 200 m, the overpressure was over 0.3 bar. Calculate the overpressure in distances of 25, 75, 125 and 200 m, and draw the overpressure curve as a function of the distance from the center of the explosion. The following data are available:

- Heat of combustion of propane, ΔH_c : 46,010 kJ/kg
  Heat of combustion of TNT, ΔH_{TNT} : 4,760 kJ/kg

First, one must calculate the equivalent TNT mass, M_{TNT} (kg), from Eq. (C3.2). Since the fraction of energy released as a shock wave is not known, it is arbitrarily assumed that $f_E = 0.05$ (it takes values between 0.01 and 0.1)

$$M_{TNT} = \frac{f_E\ \Delta H_c\ M_G}{\Delta H_{TNT}} = \frac{0.05 \times 46{,}010 \times 4{,}750}{4{,}760} = 2{,}296 \text{ kg}.$$

Hence for a distance $x = 25$ m, $Z = x / M_{TNT}^{1/3} = 1.90\ \text{m kg}^{-1/3}$

and from Eq. (C3.3)

$$P_S = \frac{80{,}800\left(1 + \left[\frac{Z}{4.5}\right]^2\right)}{\sqrt{1 + \left[\frac{Z}{0.048}\right]^2}\ \sqrt{1 + \left[\frac{Z}{0.32}\right]^2}\ \sqrt{1 + \left[\frac{Z}{1.35}\right]^2}} = 233 \text{ kPa or 2.3 bar.}$$

Equivalent TNT Mass Method

example

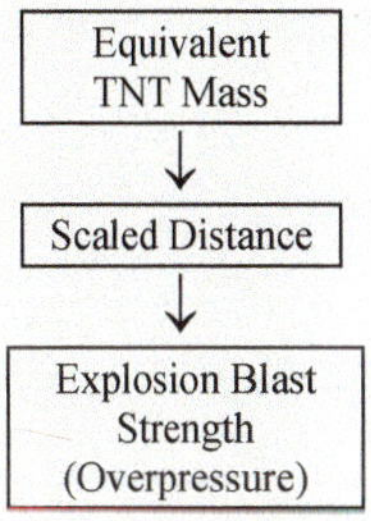

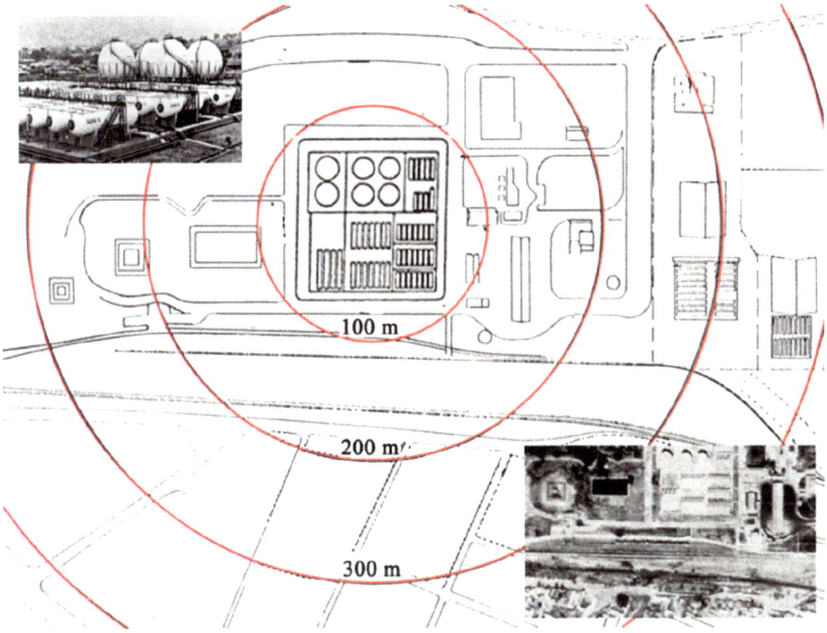

Figure C3.3. Area layout and photographs before the explosion [Lerdo de 1985, Pietersen 1988].

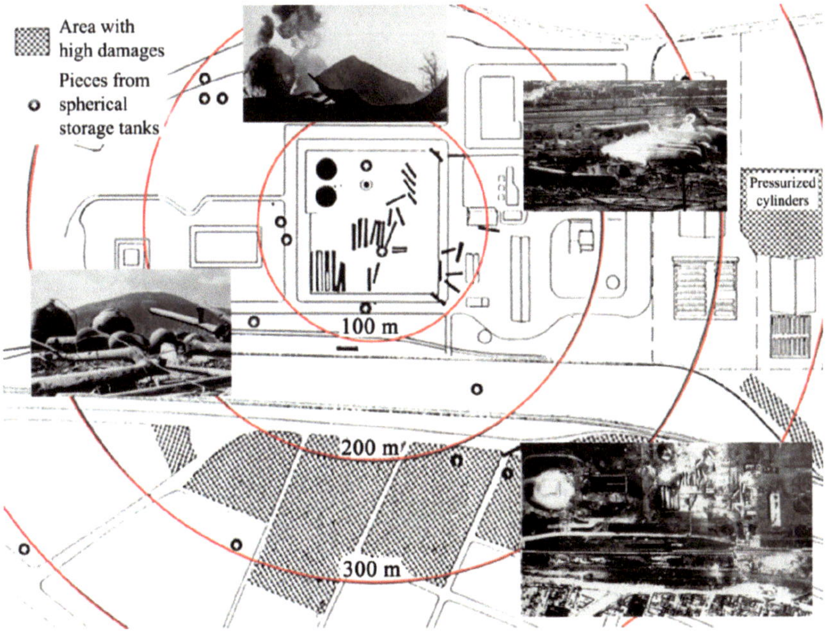

Figure C3.4. Area layout and photographs after the explosion [Arturson 1987, Lerdo de 1985, Pietersen 1988].

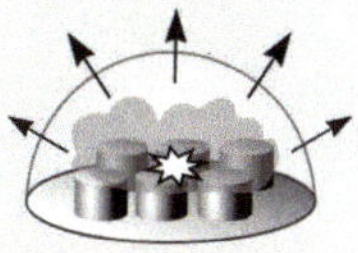

Equivalent TNT Mass Method

example

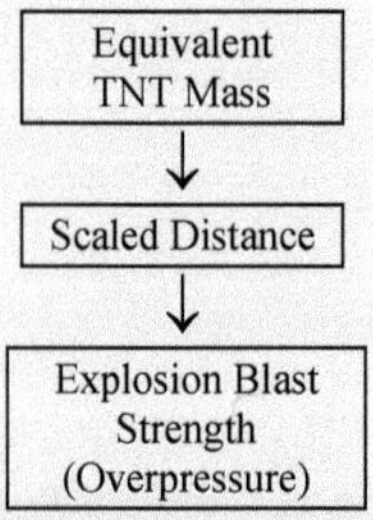

Therefore in a distance of 25 m from the center of the explosion, the overpressure is 2.33 bar. In Table C3.1, the values calculated from Eq. (C3.3), as well as the values calculated using the diagram of Figure C3.1, are shown. There is a good agreement between the values (differences are present only in small distances where the slope of the overpressure curve is very steep).

Table C3.1. Overpressure as a Function of Distance.

x (m)	Z (m/kg$^{1/3}$)	P_s - Eq. (C3.3) (bar)	P_s - Figure C3.1. (bar)
25	1.90	2.33	1.20
75	5.69	0.23	0.21
125	9.49	0.11	0.09
200	15.16	0.06	0.05

In Figure C3.5, the whole curve calculated by Eq. (C3.3) is shown. In relation to the observation (based upon the level of damages in that distance), and also on the destruction at a radius of 50 m (see photographs), it seems that the method estimates a quite lower overpressure. In the same diagram, the influence of parameter f_E, is also shown - quite small in this particular case.

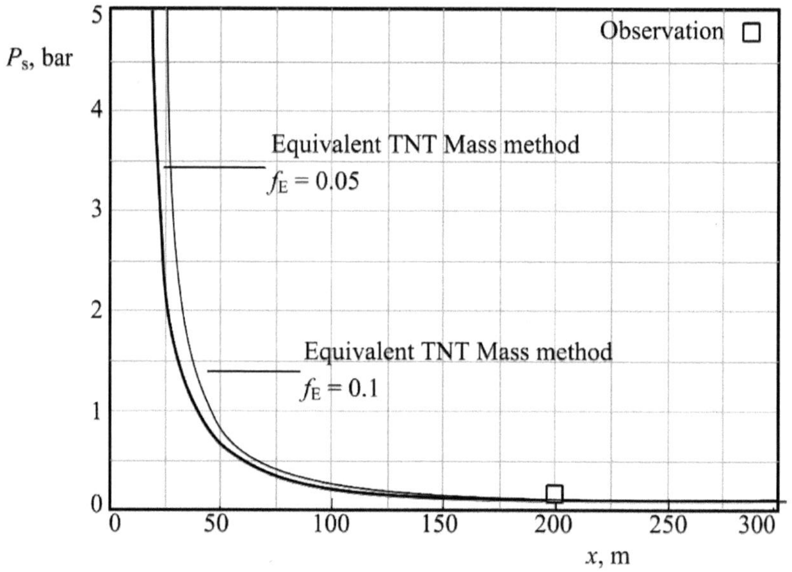

Figure C3.5. Overpressure as a function of the distance (TNT method).

❐

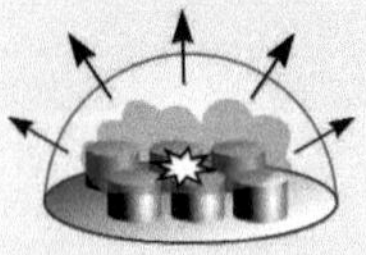

C3.3. Multi-Energy Method

Multi-Energy Method

calculation procedure

Cloud Dimensions
↓
Obstructed Regions
↓
Explosion Blast Strength (Overpressure)
↓
Positive Phase Duration

In contrast to the usual simulation methods, where the vapor cloud explosion is regarded as an entity, the Multi-Energy method assumes that the vapor cloud explosion is composed of a number of sub-explosions taking place inside specific areas of the cloud, see Figure C3.6, corresponding to the various sources of blast that exist in the cloud [Berg 1985].

The most important assumption of the method is that the strength of the explosion blast, and thus the overpressure developed, depends upon the layout of the space where the cloud is spreading. More precisely, only the obstructed or partially obstructed regions (regions with high equipment density) will contribute to a high strength explosion blast. The remaining parts of the cloud will slowly burn, without a serious contribution to the strength of the blast [Berg 1985, Berg & Lannoy 1993, Mercx *et al.* 2000].

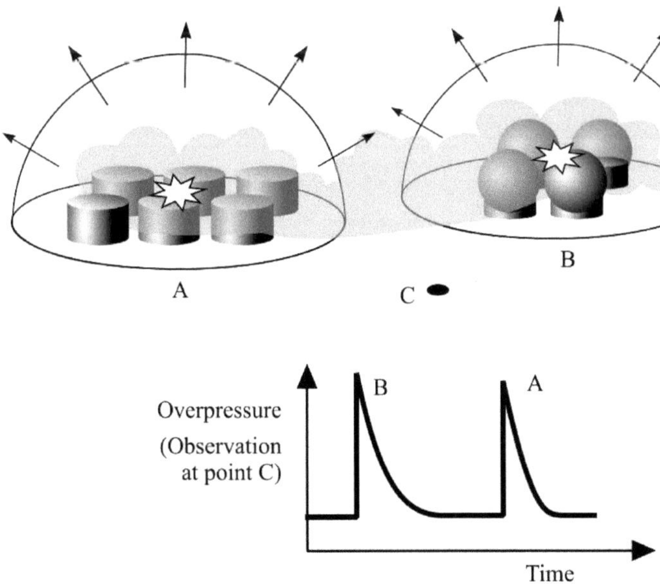

Figure C3.6. Sub-explosions taking place inside the cloud.

Initially the dimensions of the cloud based upon the amount of leaked flammable gas must be estimated, and the probable explosion sources must be identified. Following this, a series of empirical criteria are employed in order to identify the obstructed regions, to calculate the volume they occupy and thus to obtain the space left free for the vapor cloud to spread.

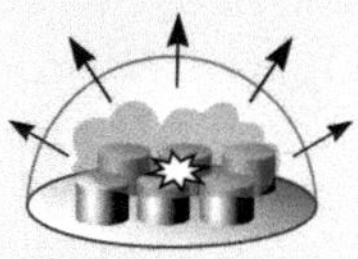

Multi-Energy Method

calculation procedure

Consequently the energy of the explosion is calculated, and thus from empirical expressions the resulting overpressure in specific distances from the explosion center, as well as the duration of the positive phase, are obtained. The calculation of the overpressure is directly dependent upon the type of region where the explosion took place, i.e., if it is an obstructed region or not.

From the overpressure, the effects to humans or material damages can easily be calculated (see Section C3.5).

Cloud Dimensions
↓
Obstructed Regions
↓
Explosion Blast Strength (Overpressure)
↓
Positive Phase Duration

C3.3.1. Cloud Dimensions

The volume, V (m^3) of the resulting vapor cloud (composed of flammable gas and air) is calculated from the reaction's stoichiometry, from which the volume of the oxygen required is obtained and therefore the volume of the required air.

In the case that the leaked fluid is in the liquid state (pool of flammable liquid), then the liquid's evaporation rate must be multiplied with the time until the explosion, so as to obtain the total amount of vapor that participated in the development of the vapor cloud.

The radius of the resulting cloud, R (m), is derived from the volume, V (m^3) of the cloud, being considered as a hemisphere, as

$$R = \left(\frac{3V}{2\pi} \right)^{1/3} . \qquad \text{(C3.4)}$$

C3.3.2. Obstructed Regions

A non-obstructed region is a region that does not include any kind of obstacles, and therefore the cloud can be evenly distributed, i.e., the strength of the explosion blast is very low. On the contrary, an obstructed region is a region of high density of obstacles (equipment, walls, buildings, etc.) resulting in the increase of the spreading velocity of the cloud, as flow changes from laminar to turbulent and thus the strength of the explosion blast becomes very high. Hence, the area surrounding the explosion's center must be separated into obstructed and non-obstructed regions [Berg 1985, Berg & Lannoy 1993, Mercx *et al.* 2000].

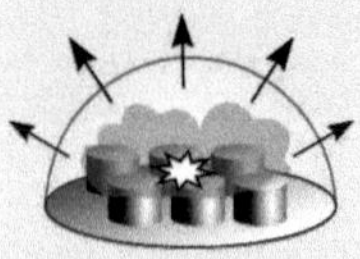

The procedure of "building-up" an obstructed region is based on the effect obstacles have on the generation of turbulence in the expansion flow ahead of the flame. The space around the explosion's center is separated into obstructed and non-obstructed regions. The cloud can only spread in the free space of every obstructed region (i.e., in between obstacles).

Two empirical rules have been proposed [Berg 1985] in order to include or not a new object in an obstructed region. More specifically, the distance, X (m), between each new object and its previous one must satisfy the following two conditions

$$\text{a)}\quad X < 25\,\text{m} \qquad \text{(C3.5)}$$

$$\text{b)}\quad X < 10\,D_1 \quad \text{or} \quad X < 1.5\,D_2\,, \qquad \text{(C3.6)}$$

where, D_1 (m), is the smallest dimension of the object on the plane perpendicular to the direction of the flame's propagation, and D_2 (m), is the dimension of the object parallel to the direction of the flame's propagation. To fully understand the application of the above rules, see Example C3.2.

EXAMPLE C3.2. *Obstructed Regions*

If the explosion takes place in the first storage tank, analyze whether all three storage tanks shown in Figure C3.7 should be included in the same obstructed region.

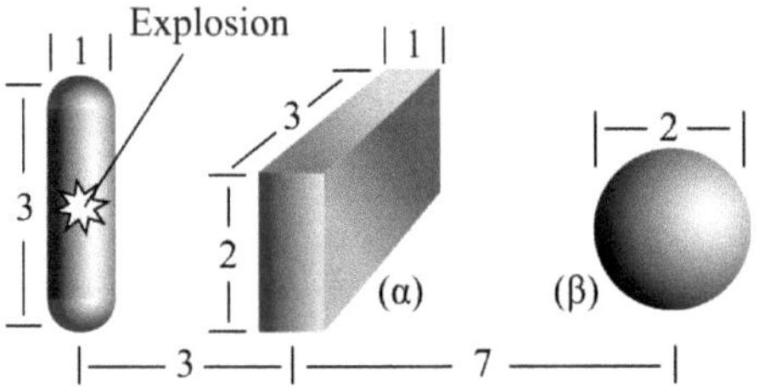

Figure C3.7. Explosion in storage tank.

In order to include storage tank (a) in the obstructed region, the validity of the above two criteria, Eqs. (C3.5) and (C3.6) will be examined:

a) $3 < 25$ m (valid)

b) $3 < 10\,D_1 = 10 \times 2 = 20$ (valid) or $3 < 1.5\,D_2 = 1.5 \times 1 = 1.5$ (not valid).

Therefore both criteria are satisfied, and thus tank (a) is included in the same obstructed region with the tank where the explosion occurred.

The same analysis is carried out for storage tank (b),

a) $7 < 25$ m (valid)

b) $7 < 10\,D_1 = 10 \times 2 = 20$ (valid) or $7 < 1.5\,D_2 = 1.5 \times 2 = 3$ (not valid).

Hence, in this case both criteria are also satisfied, and thus all three storage tanks are included in the same obstructed region.

❒

Cloud Dimensions
↓
Obstructed Regions
↓
Explosion Blast Strength (Overpressure)
↓
Positive Phase Duration

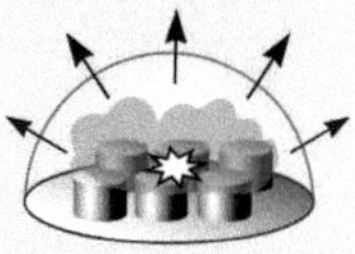

Multi-Energy Method

calculation procedure

Cloud Dimensions
↓
Obstructed Regions
↓
Explosion Blast Strength (Overpressure)
↓
Positive Phase Duration

In this way, starting from the point of the explosion, space is separated into obstructed and non-obstructed regions. If the volume of the cloud is larger than the free space in the obstructed region, then calculations are continued in two parts:

- calculations with a cloud volume equal to the free space of the obstructed region, and a high strength of explosion blast (corresponds to high overpressure), and
- calculations based on the remaining volume of the cloud, with low strength of explosion blast (corresponds to low overpressure).

If the volume of the cloud is smaller than the available free space in the obstructed region, then further calculations are based upon this volume and not upon the free volume of the obstructed region. In the next section the significance of the strength of the explosion blast will be discussed.

C3.3.3. Strength of Explosion Blast and Overpressure

The coefficient of the strength of the explosion blast characterizes, as already mentioned, the strength of the explosion blast. In the diagram of Figure C3.8, the scaled overpressure, P_S' (-), is given, as a function of the scaled distance, r' (-). Both these quantities are defined [Berg 1985] as

$$P_S' = \frac{P_S}{P_a}, \tag{C3.7}$$

and

$$r' = x\left(\frac{E}{P_a}\right)^{-1/3}. \tag{C3.8}$$

The parameter of these curves is the coefficient of the strength of the explosion blast, as mentioned above. A coefficient of 10 refers to a high strength explosion with very high overpressure, etc.

In Eqs. (C3.7) and (C3.8), P_S (MPa) denotes the overpressure caused by the explosion, P_a (MPa), the ambient pressure (= 0.1 MPa), x (m), the distance from the center of the explosion and E (MJ), the total energy released by the explosion.

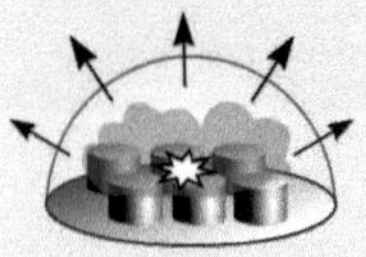

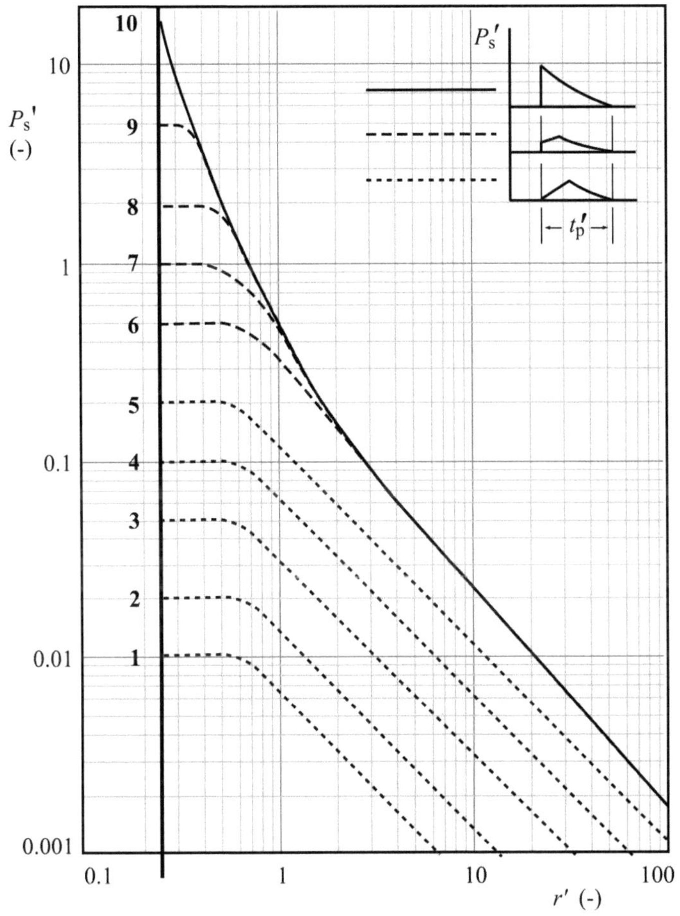

Figure C3.8. Scaled overpressure as a function of scaled distance, with parameter (values 1-10) the coefficient of strength of the explosion blast [Berg 1985].

The calculation of the overpressure is carried out according to the following algorithm:

1) For every region (obstructed or not) the strength of the blast is chosen (high obstacle density will result in a high strength blast).
2) The total energy, E, released during the explosion in this region is calculated.
3) Following that, for a specific distance x, the scaled distance r' is calculated.
4) From the diagram in Figure C3.8, the scaled overpressure that corresponds to this scaled distance is obtained, and from that the overpressure, P_s (MPa), of the explosion at that distance.

Multi-Energy Method

calculation procedure

Cloud Dimensions
↓
Obstructed Regions
↓
Explosion Blast Strength (Overpressure)
↓
Positive Phase Duration

Multi-Energy Method

calculation procedure

Cloud Dimensions
↓
Obstructed Regions
↓
Explosion Blast Strength (Overpressure)
↓
Positive Phase Duration

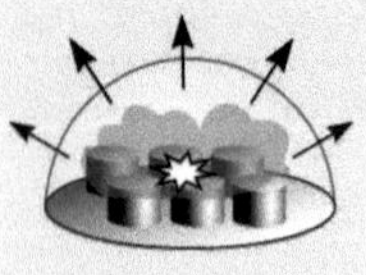

While in the Equivalent TNT Mass method the unknown parameter was the fraction of energy released as shock wave (coefficient f_E), in the Multi-Energy method the unknown parameter is the coefficient of strength of the explosion blast. This must be estimated according to the equipment density in the surrounding area. If the equipment density is high in the area, then the value of the coefficient of strength will have a large value (it takes values from 1 to 10; see diagram in Figure C3.8).

The type of line in Figure C3.8 characterizes the shape of the shock wave that will follow, according to the shapes in the upper right corner of the diagram. It is obvious that a high strength explosion will result in a very sudden shock wave.

For computer applications, for the two cases of blast strength 10 and 3, the following equation can be used

$$P_S = 10^{-b \log_{10} r' - c} \qquad \text{(C3.9)}$$

where P_s (bar) is the overpressure and coefficients b and c are given in Table C3.2.

Table C3.2. Coefficients b and c of Eq. (C3.9).

Coefficient of Strength of Explosion Blast	Range of r'	b	c
10	$0.15 < r' < 1.0$	2.3721	0.3372
	$1.0 \le r' \le 2.5$	1.5236	0.3372
	$r' > 2.5$	1.1188	0.5120
3	$r' \le 0.6$	0	1.3010
	$r' > 0.6$	0.9621	1.5145

C3.3.4. Positive Phase Duration

One of the advantages of the Multi-Energy method is that it also predicts, empirically, the duration, t_p (s), of the positive phase of the explosion. In Figure C3.9, the scaled positive phase duration, t'_p, as a function of the scaled distance r', Eq. (C3.8), is shown. The duration, t_p (s), of the positive phase of the explosion can be obtained from the expression

$$t_p = \frac{t'_p}{C_s}\left(\frac{E}{P_a}\right)^{1/3} . \tag{C3.10}$$

In the above expression, P_a (MPa), denotes the ambient pressure (= 0.1 MPa), E (MJ), the total energy released, and C_s (m/s), the velocity of sound (= 340 m/s).

Hence, in every distance from the center of the explosion, a value of the overpressure, Figure C3.8, and a value of the duration of the positive phase, Figure C3.9, can be obtained; see Example C3.3.

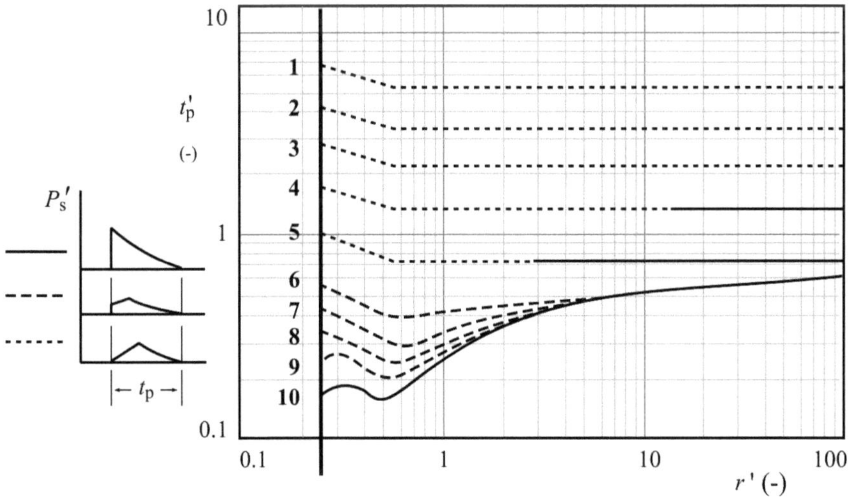

Figure C3.9. Duration of the positive phase, with parameter (values 1-10) the coefficient of strength of the explosion blast [Berg 1985].

The basic advantage of the Multi-Energy method is that it takes into consideration the surrounding area of the explosion, and it calculates the overpressure according to the obstacle density. It also predicts the duration of the positive phase of the explosion.

Multi-Energy Method

calculation procedure

Cloud Dimensions
↓
Obstructed Regions
↓
Explosion Blast Strength (Overpressure)
↓
Positive Phase Duration

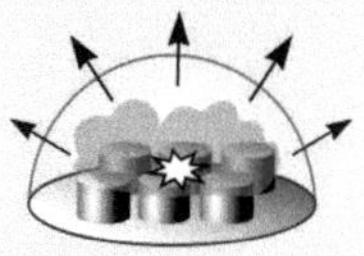

Multi-Energy Method

example

Cloud Dimensions
↓
Obstructed Regions
↓
Explosion Blast Strength (Overpressure)
↓
Positive Phase Duration

EXAMPLE C3.3. *Overpressure Calculation*

In Example C3.1 the events around the explosion of November 19, 1984, in the PEMEX terminal outside Mexico City, were presented. The facility had a 16,000 m^3 capacity and 4,750 kg of propane gas leaked. The final casualty count was 550 deaths, more than 6,400 wounded and material damages of $34,000,000 in 2005 prices [MInA 2006]. In that example, the overpressure as a function of the distance was calculated with the Equivalent TNT Mass method.

Using the Multi-Energy method, calculate the resulting overpressure at distances of 25, 75, 125 and 200 m, and plot the overpressure as a function of distance curve. Compare these results with those obtained by the Equivalent TNT Mass method. In Figure C3.10 a detail of the area of the explosion is shown. The following data are available:

- Heat of combustion for propane, ΔH_c : 46,010 kJ/kg
  Density of propane, ρ (15 °C) : 1.86 kg/m^3

a) Cloud dimensions

The density of propane is 1.86 kg/m^3 (15°C, 0.1 MPa). Hence, the volume of propane in the cloud is $4{,}750/1.86 = 2{,}554\ m^3$.

The combustion reaction is: $C_3H_8 + 5\ O_2 \rightarrow 3CO_2 + 4H_2O$

Hence, the ratio of propane:oxygen : 1:5
and the ratio of propane:air : 1:25 (as air contains 21% O_2)

- Therefore, the volume of the whole cloud $= 2{,}554 \times 25 = 63{,}884\ m^3$.
  Since the cloud's shape is a hemisphere, its radius can be obtained from the expression $(2\pi R^3/3) = 63{,}884$, and hence $R = 31.2$ m.

This radius determines the areas where the cloud will disperse. The equipment shown in the figure has an approximate dimension of 100 x 100 m^2; hence, it certainly takes part in the determination of the obstructed regions (a larger area of the facility is always considered, as a large part of the empty space is occupied by the equipment).

b) Obstructed regions

The above radius determines the part of the facility where the cloud of propane will disperse. In Figure C3.10, the storage tanks, their dimensions and the center of the explosion are shown. For a new object to be part of an obstructed region, its distance from the previous one must satisfy Eqs. (C3.5) and (C3.6):

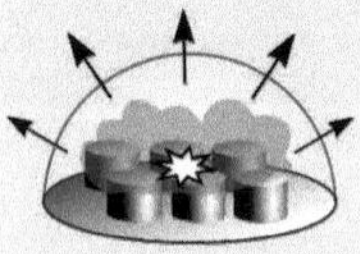

Multi-Energy Method

example

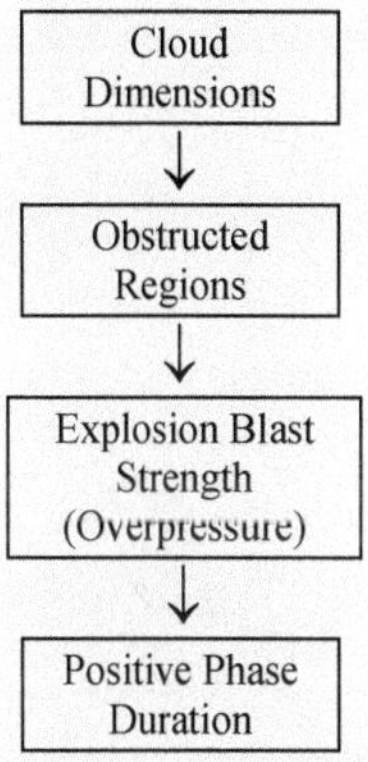

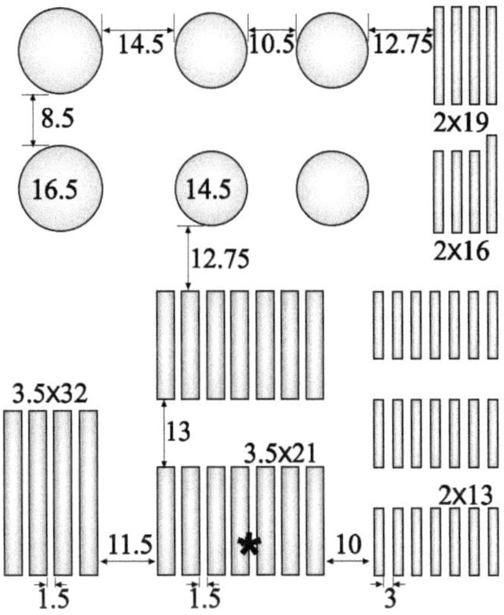

Figure C3.10. Detail of PEMEX facility (dimensions in m).

- 2 spherical storage tanks diameter 16.5 m.
  4 spherical storage tanks diameter 14.5 m.
- 5 cylindrical storage tanks diameter 2 m, length 19 m.
  3 cylindrical storage tanks diameter 2 m, length 16 m.
- 21 cylindrical storage tanks diameter 2 m, length 13 m.
- 14 cylindrical storage tanks diameter 3.5 m, length 21 m.
- 4 cylindrical storage tanks diameter 3.5 m, length 32 m.

Minimum height beneath storage tanks = 2 m
Height of tubing over storage tanks = 0.5 m

a) $X < 25$ m

b) $X < 10\,D_1$ or $X < 1.5\,D_2$

where, D_1 (m) is the smallest dimension of the object perpendicular to the direction of the flame's propagation, and D_2 (m) is the object's dimension that is parallel to the direction of the flame's propagation.

1) Starting from the point of the explosion (*), the first cylindrical tank on the right is considered. It is at a distance of 1.5 m from the previous one (<25 m) and smaller than $10\,D_1$ = 10 x 3.5 = 35. The same follows for the next two storage tanks. Correspondingly, the distance of the first one from the (2x13) group, from the last of the group (3.5x21), is 10 m < 25 m but also < $10\,D_1$ = 10 x 2 = 20. Correspondingly, it can easily be shown that all tanks in the figure are part of the same obstructed region.

2) Although all storage tanks belong to the same obstructed region (according to the two aforementioned criteria), the empty space in the center left (about 30 X 33 m^2) does not. There are also different amounts of empty space over each group of storage tanks. Hence it is preferred to separate the obstructed region into four distinct regions, as shown in Figure C3.11. In these regions the blast explosion's coefficient will be high, while in the remaining area will be low.

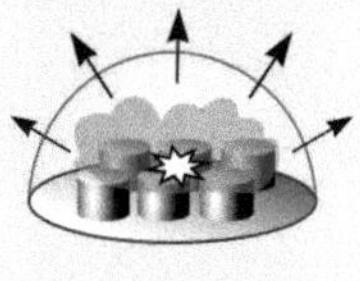

Multi-Energy Method

example

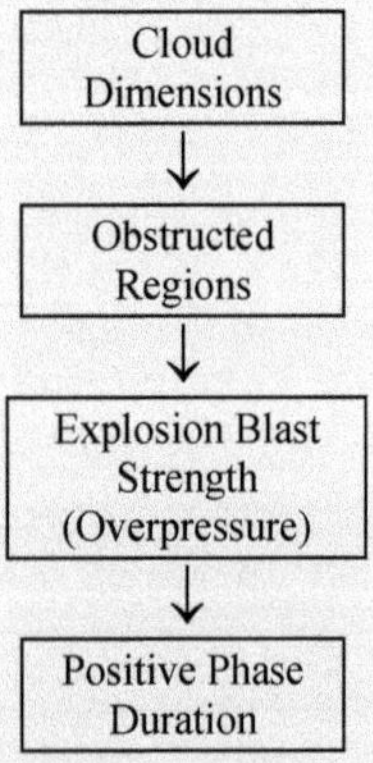

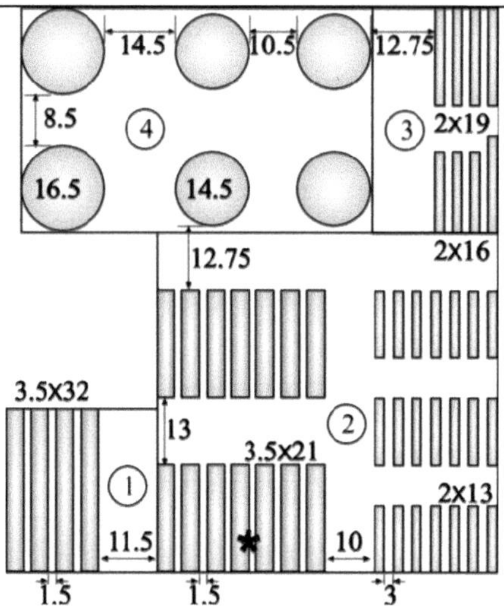

Figure C3.11. Obstructed regions of PEMEX facility (dimensions in m).

- 2 spherical storage tanks diameter 16.5 m.
  4 spherical storage tanks diameter 14.5 m.
- 5 cylindrical storage tanks diameter 2 m, length 19 m.
  3 cylindrical storage tanks diameter 2 m, length 16 m.
- 21 cylindrical storage tanks diameter 2 m, length 13 m.
- 14 cylindrical storage tanks diameter 3.5 m, length 21 m.
- 4 cylindrical storage tanks diameter 3.5 m, length 32 m.

Minimum height beneath storage tanks = 2 m
Height of tubing over storage tanks = 0.5 m

The volume, V (m^3), of these regions is:

(1) $V_1 = 32 \times (3.5\times4+1.5\times3+11.5) \times (3.5+2+0.5) = 5{,}760\ \text{m}^3$

(2) $V_2 = (42+13+12.75)\times(7\times3.5+6\times1.5+10+7\times2+6\times3) \times (3.5+2+0.5) = 30{,}691\ \text{m}^3$

(3) $V_3 = (33+8.5)\times(4\times2+3\times3+12.75)\times(2+2+0.5) = 5{,}555\ \text{m}^3$

(4) $V_4 = (33+8.5)\times(16.5+29+14.5+10.5)\times(16.5+2+0.5) = 55{,}590\ \text{m}^3$

Total volume of obstructed region $= 97{,}596\ \text{m}^3$

Hence, subtracting the storage tanks capacity, the empty space available for the dispersion of the cloud is $= 97{,}596 - 16{,}000 = 81{,}596\ \text{m}^3$.

3) The space available for the dispersion of the cloud inside the obstructed region was found to be equal to 81,596 m^3. This volume is larger than the volume of the cloud, which is 63,884 m^3. Hence the entire cloud will be dispersed inside the obstructed region. A coefficient of an explosion blast of 10 will be assumed (high strength - worst case).
If the cloud was larger than the available empty space, then the calculations would have been carried out as follows:
- for the volume of cloud equal to available empty space, calculation would have been carried out with a large coefficient of explosion blast (e.g., 10),
- for the remaining volume of cloud that is outside the obstructed region, a low coefficient of explosion blast would have been used (e.g., 3).

Multi-Energy Method

example

c) Energy of the explosion

The energy, E (MJ), released by the explosion, can be calculated from the heat of combustion, ΔH_c (MJ/kg), the volume of the cloud, V (m^3), the density ρ (kg/m^3) of propane, and the stoichiometry of the reaction (ratio of propane:air = 1:25), as

$$E = V[\Delta H_c \times \rho \times (1/25)] = 63{,}884\,[46.01 \times 1.86 \times (1/25)] = 218 \text{ GJ}.$$

d) Overpressure

The overpressure can be obtained from Figure C3.8. More analytically, at a distance of 75 m (the 25 m distance is still inside the radius of the explosion), from Eq. (C3.8),

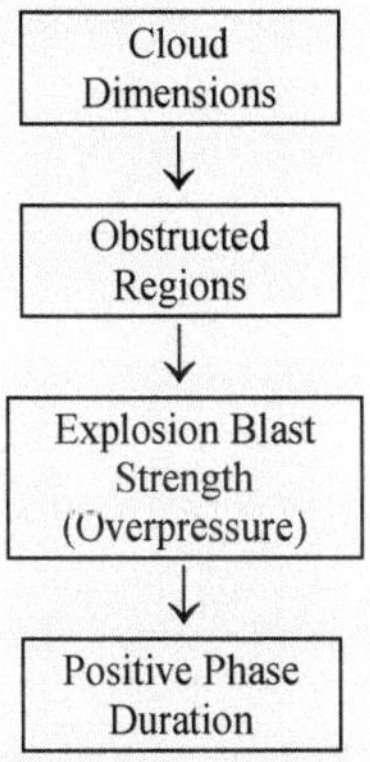

$$r' = x\left(\frac{E}{P_a}\right)^{-1/3} = 75\left(\frac{218{,}000}{0.1}\right)^{-1/3} = 0.58\,.$$

For this value, and for a coefficient of explosion blast equal to 10, from Figure C3.8 or from Eq. (C3.9), the overpressure is calculated as 1.69 bar. The other results can be seen in Table C3.3 and in Figure C3.12.

Table C3.3. Overpressure as a Function of Distance.

	Equivalent TNT Mass*		Multi-Energy		
x (m)	Z (m/kg$^{1/3}$)	P_s (bar)	r'	P_s (bar)	t_p (ms)
25	1.90	2.33	-	-	-
75	5.69	0.23	0.58	1.69	58
125	9.49	0.11	0.96	0.50	83
200	15.16	0.06	1.54	0.24	112

* Eq. (C3.3).

e) Time duration of the explosion's positive phase

As calculated in the previous step, the distance of $x = 75$ m corresponds to a scaled distance $r' = 0.58$. From the diagram in Figure C3.9 and for a coefficient of explosion blast equal to 10, the scaled positive phase is 0.15, which corresponds from Eq. (C3.10) to time $t_p = 58$ ms. That is, the duration of the overpressure of 0.23 bar, registered at a distance of 75 m was 58 ms.

The remaining results are shown in Table C3.3 and plotted in the diagram of Figure C3.12.

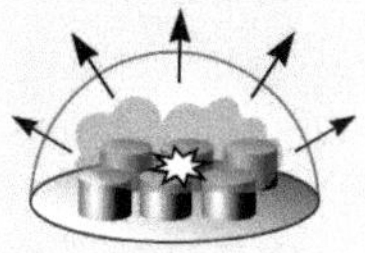

Multi-Energy Method

example

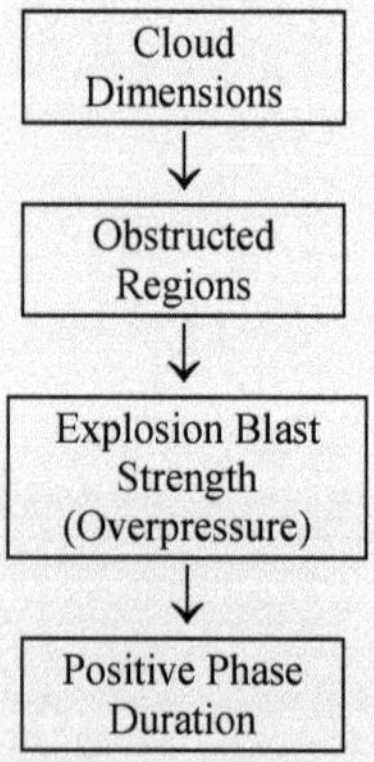

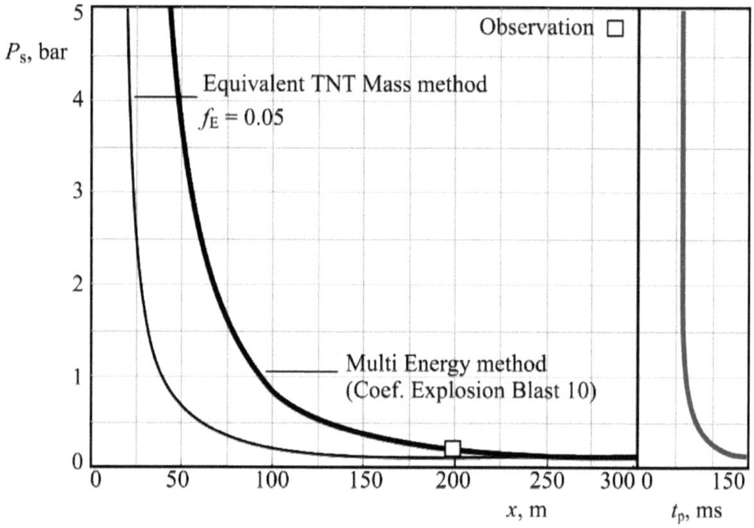

Figure C3.12. Comparative results of the overpressure as a function of distance.

f) Critical evaluation

In Figure C3.12 the overpressure vs. distance curve obtained by the Multi-Energy method as well as the one obtained by the Equivalent TNT method (Example C3.1) are plotted. The following are noted:

- Values obtained by the Multi-Energy method are higher than those obtained by the Equivalent TNT method.
- There is no value of the coefficient f_E of the Equivalent TNT method that will result in a superimposition of the curves. Only if the leaked mass was increased from 4,750 kg to 30,000 kg, which is not logical, would the TNT curve increase so that the two curves could coincide. Furthermore, even if the coefficient of the explosion blast was decreased from 10 to 7, the two curves would still be far apart.
- Values produced by the Multi-Energy method are nearer to the actual values observed based upon the damages that occurred from the explosion.
- The Multi-Energy method can also predict the time duration of the positive phase of the explosion.

In general, one can observe that although the Equivalent TNT method can quickly produce an answer in calculating the overpressure as a function of the distance, the values obtained by the Multi-Energy method are closer to actual conditions.

❒

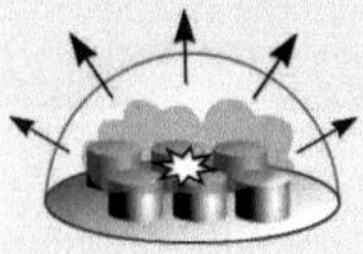

C3.4. Baker-Strehlow Method

Baker-Strehlow Method

calculation procedure

The Baker-Strehlow method [Baker *et al.* 1996, Baker *et al.* 1998] was first published in 1996 and is based upon the same idea of obstructed regions that were initially put forward by the Multi-Energy method. In both these methods, the presence of obstacles in the expansion of the flame causes vapor cloud explosions of higher intensity.

- In the Multi-Energy method, obstructed regions are determined. In these regions the explosive blast is of a higher intensity, and is characterized by the explosion's blast coefficient.

- In the Baker-Strehlow method, the important parameter in the selection of the intensity of the explosion blast is the flame propagation speed. This is determined by
  a) the way the flame front propagates,
  b) the reactivity of the fuel, and
  c) the density of the obstacles.

Cloud Dimensions
↓
Flame Speed
↓
Explosion Blast Strength (Overpressure)

First the dimensions of the cloud will be determined and then the energy of the explosion exactly as it was carried out in the case of the Multi-Energy method. Following that the overpressure will be calculated as a function of the scaled distance, with the flame speed as a parameter. The effects of the shock wave to people and materials will be calculated in Section C3.5, as a function of the overpressure at a given distance.

C3.4.1. Cloud Dimensions

The volume, V (m^3), of the resulting vapor cloud (composed of flammable gas and air) is calculated from the reaction's stoichiometry, from which the volume of the required oxygen is obtained and therefore the volume of required air.

In the case that the leaked fluid is in a liquid state (pool of flammable liquid), then the liquid's evaporation rate must be multiplied with the time until the explosion, so as to obtain the total amount of vapor which participated in the development of the vapor cloud.

The radius of the resulting cloud, R (m), is derived from the volume, V (m^3), of the cloud, considered as a hemisphere, as

$$R = \left(\frac{3V}{2\pi}\right)^{1/3}. \qquad \text{(C3.11)}$$

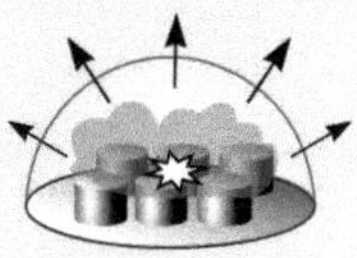

Baker-Strehlow Method

calculation procedure

Cloud Dimensions
↓
Flame Speed
↓
Explosion Blast Strength (Overpressure)

C3.4.2. Flame Speed

As mentioned, the flame speed is the important parameter in the determination of the intensity of the explosion blast. The presence of obstacles to the expansion of the flame produces large turbulent eddies and thus a wrinkled-frame front appears resulting in an increase of the overall flame surface. The increased surface requires higher fuel consumption per unit time, and hence the rate of emitted heat is increased and consequently the flame speed.

The parameters which influence the flame speed according to the Baker-Strehlow method [Baker *et al.* 1996, Baker *et al.* 1998], are the way the flame expands, the reactivity of the fuel and the obstacle density. More analytically, see Table C3.4:

- **The way the flame expands** can be classified in three categories.

  3-D: When a flame is free to expand in three dimensions (spherically or hemispherically), the total surface area of the flame is increased analogous to the square of the distance from the center of the explosion. The flame flow field can weaken freely in the three dimensions. For this reason field velocities are low, and thus the disturbances to the field because of obstacles are small. Furthermore, the effect of these disturbances to the flame speed is small, as they affect only a small part of the initial surface of the flame.

  2-D: In a two-dimension expansion (e.g., expansion of cylindrical flame under a concrete floor or deck) the total surface of the flame is analogous to the distance from the center of the explosion. Hence disturbances of the flame surface will have a greater effect than in a three-dimensional expansion, and disturbances in the flow field will also be stronger.

Table C3.4. Flame Speed in Mach Numbers (M_j).

Flame Expansion	Fuel Reactivity	Obstacle Density: High	Obstacle Density: Medium	Obstacle Density: Low
1-D	High	5.2	5.2	5.2
	Medium	2.27	1.77	1.03
	Low	2.27	1.03	0.294
2-D	High	1.77	1.03	0.588
	Medium	1.24	0.662	0.118
	Low	0.662	0.471	0.079
3-D	High	0.588	0.153	0.071
	Medium	0.206	0.100	0.037
	Low	0.147	0.100	0.037

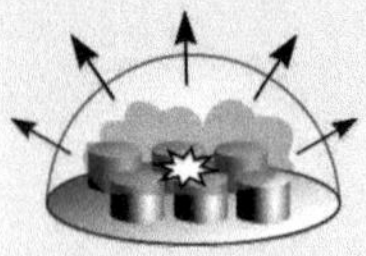

1-D: During a one-dimensional expansion (e.g., expansion of a flame inside a cylinder), the flame's surface is stable, there is almost no weakening of the field, and there is no deviation from the existing flow field. Thus, a very powerful feedback mechanism is created for the increase of the flame's acceleration.

For the above reasons, one can note in Table C3.4 that the highest speeds appear in one-dimensional expansions.

- **Fuel reactivity** is also separated into three categories. Low reactivity is shown by gases like methane and carbon monoxide. High reactivity is exhibited only by hydrogen, acetylene, ethylene oxide, propylene oxide and ethylene (the flame speed in laminar flow of these fuels exceeds the value of 0.8 m/s). All remaining fuels are considered as fuels of middle reactivity.

- **The obstacles density** is the most difficult parameter to be quantified (there have been many attempts to quantify this parameter, which are well outsidc the scope of this book). Hence the correct use of this parameter relies on the logic and judgment of the reader.

According to the aforementioned discussion, the flame speed can be approximately selected from the Table C3.4.

C3.4.3. Strength of Explosion Blast and Overpressure

In order to calculate the strength of the explosion blast, it was proposed in 1996 [Baker *et al.* 1996, Baker *et al.* 1998] to employ the diagram of scaled overpressure vs. scaled distance, shown in Figure C3.13.

In 1999, this curve was substituted by a family of curves [Tang & Baker 1999, Tang & Baker 2000] and the method was consequently renamed as the Baker-Strehlow-Tang method. In this handbook, the initial diagram of the method, shown in Figure C3.13, will be employed, due to its simplicity.

Cloud Dimensions
↓
Flame Speed
↓
Explosion Blast Strength (Overpressure)

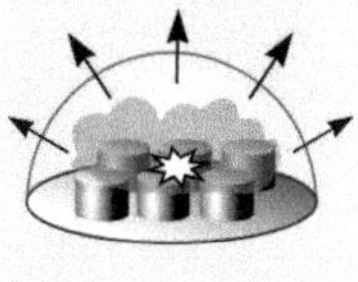

Baker-Strehlow Method

calculation procedure

Cloud Dimensions
↓
Flame Speed
↓
Explosion Blast Strength (Overpressure)

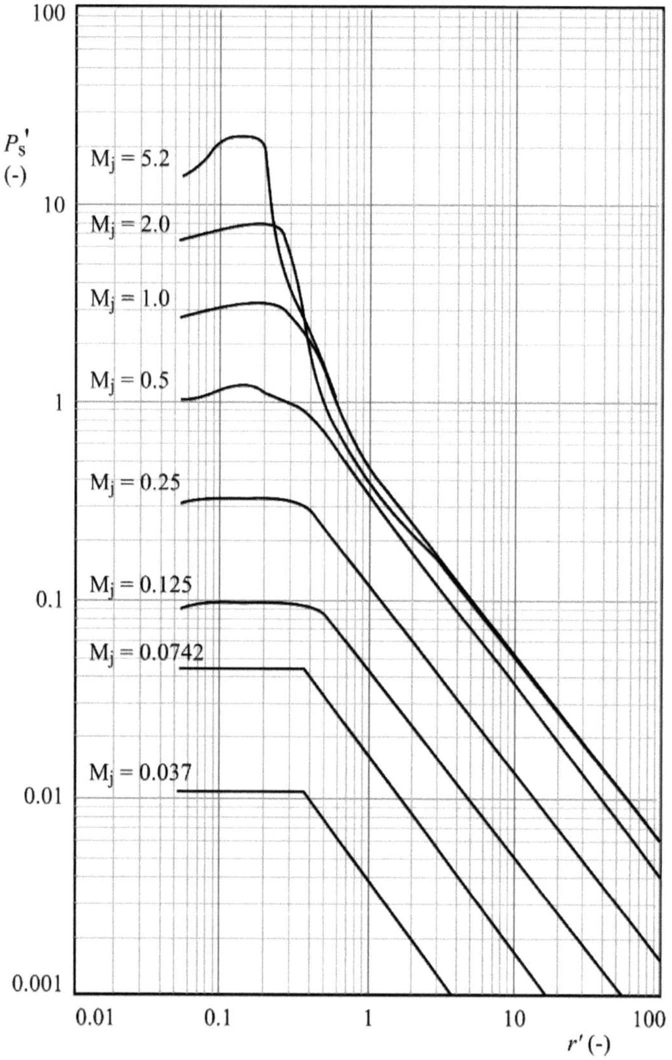

Figure C3.13. Scaled overpressure as a function of the scaled distance (Baker-Strehlow method).

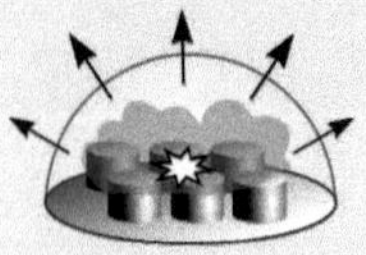

Baker-Strehlow Method

calculation procedure

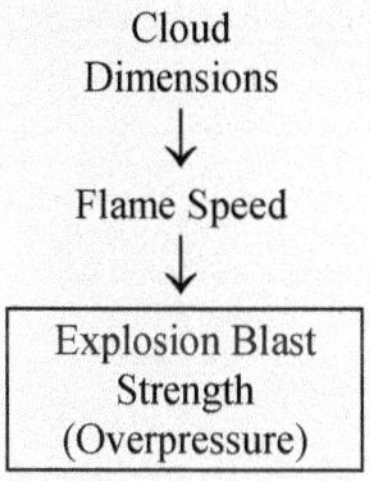

Similar to the Multi-Energy method, in the Baker-Strehlow method a scaled pressure and a scaled distance is defined as

$$P_s' = \frac{P_s}{P_a}, \tag{C3.12}$$

and

$$r' = x\left(\frac{E}{P_a}\right)^{-1/3}, \tag{C3.13}$$

where P_s (MPa) denotes the overpressure caused by the explosion, P_a (MPa) is the ambient pressure (= 0.1 MPa), x (m) the distance from the center of the explosion, and E (MJ) the total energy that is released.

The calculation procedure is carried out in the following steps:

1) For every region the flame speed is selected (Table C3.4).
2) The energy, E, released by the explosion in this region is then calculated.
3) Following that, for a specific distance, x, the scaled distance r', is calculated.
4) From the diagram in Figure C3.13, the scaled overpressure that corresponds to this scaled distance is obtained, and from this the overpressure, P_s (MPa), of the explosion.

Today, for the calculation of the overpressure of an explosion, the tendency is to employ CFD packages that simulate the whole industrial facility in three dimensions. However, this procedure is very complex, time consuming and very expensive. For these reasons the three methods presented here are widely employed. The Equivalent TNT Mass method produces a good first estimate, while the other two methods, the Multi-Energy and the Baker-Strehlow method, yield a good description of the explosion and the overpressure.

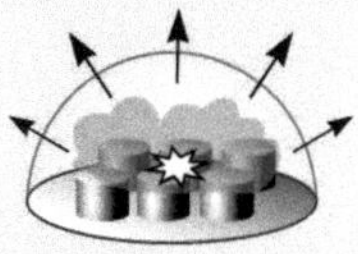

Baker-Strehlow Method

example

Cloud Dimensions

↓

Flame Speed

↓

Explosion Blast Strength (Overpressure)

EXAMPLE C3.4. *Calculation of the Overpressure*

In Example C3.1 the events around the explosion of November 19, 1984, in the PEMEX terminal outside Mexico City, were presented. The facility had a 16,000 m^3 capacity and 4,750 kg of propane gas leaked. The final casualty count was 550 deaths, more than 6,400 wounded and material damages of $34,000,000 in 2005 prices [MInA 2006]. In Example C3.1, the overpressure as a function of the distance was calculated with the Equivalent TNT Mass method, while in Example C3.3, the overpressure as a function of the distance was calculated with the Multi-Energy method.

Calculate using the Baker-Strehlow method, the resulting overpressure at distances of 25, 75, 125 and 200 m and plot the overpressure as a function of distance from the center of the explosion. Compare these results with those obtained by the Equivalent TNT Mass method and the Multi-Energy method. In Figure C3.10 a detail of the area of the explosion was given. The following data are available:

- Heat of combustion for propane, ΔH_c : 46,010 kJ/kg
  Density of propane, ρ (15°C) : 1.86 kg/m^3

---

a) Cloud Dimensions

The cloud dimensions were calculated in Example C3.3, where it was found that

- The total volume of the cloud will be = 2,554 x 25 = 63,884 m^3.
  Since the cloud's shape is hemispherical, its radius can be obtained from the expression $(2\pi R^3/3) = 63{,}884$, and therefore $R = 31.2$ m.

This radius determines the areas where the cloud will disperse. The equipment shown in the figure has the approximate dimensions of 100 x 100 m^2; hence, it certainly takes part in the determination of the obstructed regions (a larger area of the facility is always considered, as a large part of the empty space is occupied by the equipment).

All the equipment constitute a single region from which the empty space is deducted, according to the argument put forward in Example C3.3. Hence it was calculated that

Total volume of region including all equipment = 97,596 m^3.

Therefore, subtracting the capacity of the storage tanks, the free space available for the expansion of the cloud is = 97,596 - 16,000 = 81,596 m^3.

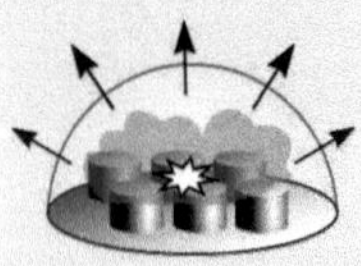

Hence the free space available for the expansion of the cloud inside the obstructed region is equal to 81,596 m^3. This volume is larger than the volume of the cloud that is 63,884 m^3. Therefore, the whole cloud will fill in the obstructed region.

Baker-Strehlow Method

example

Cloud Dimensions
↓
Flame Speed
↓
Explosion Blast Strength (Overpressure)

b) Flame Speed

According to the discussion of Section C3.4.1, propane will be considered as a fuel of high reactivity. The expansion of the flame in this case can be considered as a 2-D propagation, as the layout of the storage tanks will facilitate such a type of expansion. Hence the flame speed, as seen from Table C3.4. will be equal to 1.77.

c) Energy of Explosion

The energy, E (MJ), released by the explosion, can be found from the heat of combustion, ΔH_c (MJ/kg), the volume of the cloud, V (m^3), the density ρ (kg/m^3) of the propane, and the stoichiometry of the reaction (ratio of propane:air = 1:25), as

$$E = V[\Delta H_c \times \rho \times (1/25)] = 63{,}884\,[46.01 \times 1.86 \times (1/25)] = 218 \text{ GJ}.$$

d) Overpressure

The overpressure can be found from the diagram of Figure C3.13. More analytically, at a distance of 75 m (the 25 m distance is still inside the radius of the explosion), from Eq. (C3.13),

$$r' = x\left(\frac{E}{P_a}\right)^{-1/3} = 75\left(\frac{218{,}000}{0.1}\right)^{-1/3} = 0.58$$

For this value, and for a flame speed of 1.77, from the diagram of Figure C3.13 the overpressure is obtained as 1.1 bar. The other results are shown in Table C3.5 and in Figure C3.14.

Table C3.5. Overpressure as a Function of Distance.

	TNT Method Eq. (C3.3)			Multi-Energy Method		Baker-Strehlow Method
x (m)	Z ($m/kg^{1/3}$)	P_s (bar)	r'	P_s (bar)	t_p (ms)	P_s (bar)
25	1.90	**2.33**	-	-	-	-
75	5.69	**0.23**	0.58	**1.69**	58	**1.10**
125	9.49	**0.11**	0.96	**0.50**	83	**0.40**
200	15.16	**0.06**	1.54	**0.24**	112	**0.25**

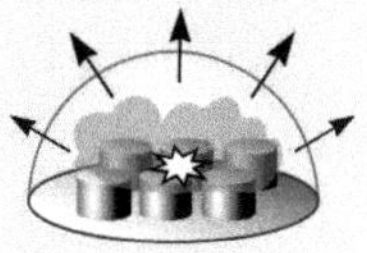

Baker-Strehlow Method

example

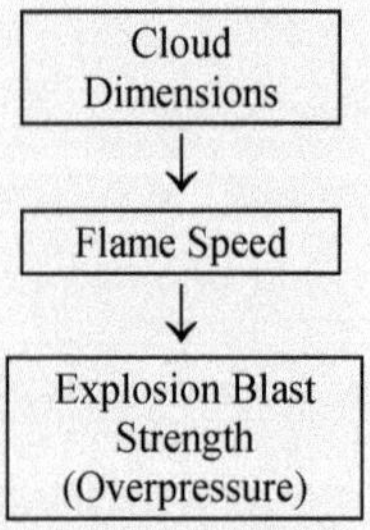

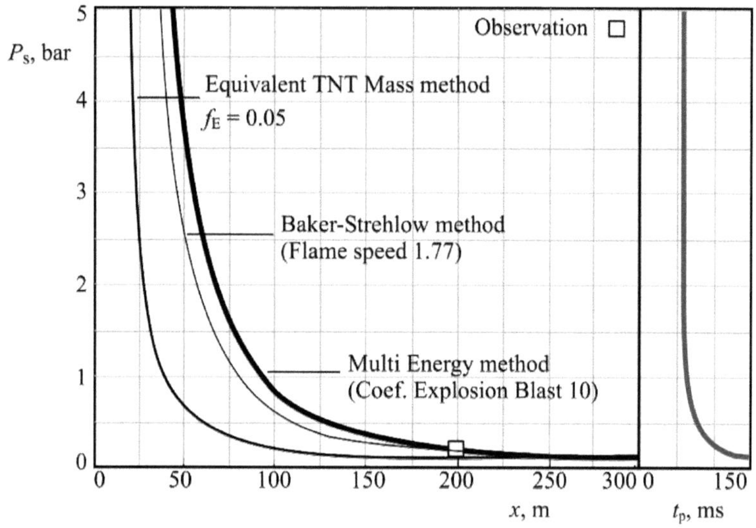

Figure C3.14. Overpressure as a Function of Distance, for the Three Methods Examined.

Figure C3.14 shows the overpressure as a function of the distance for the three methods. The reader must first study the photographs in Figures C3.3 and C3.4, where high damages are observed in houses at distances of 200 to 300 m. This fact indicates that at a distance of 200 m, the overpressure was at least 0.3 bar.

Studying the diagram in Figure C3.14 (and Table C3.5) one observes that the methods of Baker-Strehlow and Multi-Energy predict overpressure curves which are quite close to the actual damages. The Equivalent TNT Mass method underestimates the overpressure curve.

It should, however, be pointed out that these results are characteristic of this example and must not be generalized to all cases.

❒

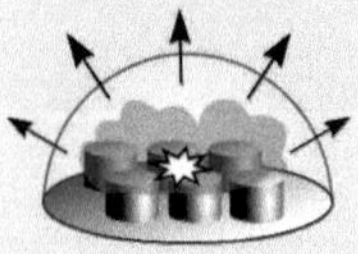

C3.5. Effects of Explosions

In the previous sections, three methods describing vapor cloud explosions (VCEs) were presented. It was thus shown how to calculate

- the overpressure, resulting from a shock wave, as a function of the distance from the center of the explosion, and
- the time duration of the positive phase of the explosion, at a given distance.

It is certain that the effects of an explosion on people and structures will be a function of the overpressure generated by the explosion, as well as its positive-phase duration. In the next two sections these effects will be studied.

In Table C3.6, a series of empirical criteria, which connect the overpressure with structural and other damages, are shown. These values are empirical average values that do not take into consideration the duration of the positive phase of the explosion (two common alternative units of pressure are employed in the table).

Table C3.6. Extent of Damage as a Function of the Overpressure of the Shock Wave.

	P_s	
	(kPa)	(bar)
Glass Panes		
Fracture 5%	0.7 - 1	0.007 - 0.01
Fracture 50%	1.4 - 3	0.014 - 0.03
Fracture 90%	3 - 6	0.03 - 0.06
Buildings		
Movement of tiles	3 - 5	0.03 - 0.05
Destruction of doors and window frames	6 - 9	0.06 - 0.09
Destruction of wall construction 50-70%	35 - 80	0.35 - 0.8
Near total demolition	80 - 260	0.8 - 2.6
Pillars		
Destruction	80 - 260	0.8 - 2.6
Large trees		
Destruction	50 - 100	0.5 - 1.0
Railway cars		
Derailment limit	80 - 190	0.8 - 1.9

The most important characteristics of a shock wave are the peak overpressure P_s (MPa), the positive phase duration, t_p (s) - see Section C3.3.4 - and the so-called "impulse." The impulse, i_s (MPa·s), relates the peak overpressure and the positive phase duration as

$$i_s = \int_{t_p} \left(P_s(t) - P_a\right) dt\,, \qquad \text{(C3.14)}$$

where P_a (MPa) denotes the ambient pressure. The above expression will be extensively employed in the following sections in calculating the effects of an explosion.

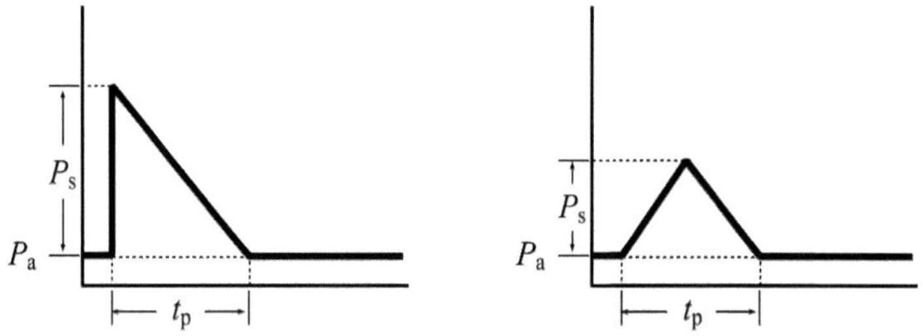

Figure C3.15. Simplified overpressure-time curves
(only the left one is characteristic of a shock wave).

Examining the simplified overpressure vs. time curves shown in Figure C3.15 (but also the corresponding discussions of Figures C3.8 and C3.9), the impulse, approximately, is given by the expression

$$i_s = \frac{1}{2} P_s\, t_p\,. \qquad \text{(C3.15)}$$

The human body can easily adapt to quite large changes in pressure, if these are applied gradually. Sudden large changes in pressure, however, can lead to dramatic and possibly fatal damage to vital human organs which contain air, most notably the lungs. Additionally, the ear, although less vital, is most sensitive to such pressure changes.

The analysis that will follow is based upon the calculation of the probability, P (-), of injury or death because of the explosion, and the following expressions are employed (similar to Section C2.5, for the calculation of the effects of fires):

$$P = \frac{1}{2}\left[1 + \operatorname{erf}\left(\frac{Pr - 5}{\sqrt{2}}\right)\right], \qquad \text{(C3.16)}$$

where the probit function, *Pr* (-), is given by the expression,

$$Pr = c_1 + c_2 \ln S \,. \qquad \text{(C3.17)}$$

In the above expression, variable S is defined according to the type of the effect, with corresponding values for the coefficients c_1 and c_2. Values for the error function are easily found in literature*. It should be noted, although obvious, that the probability of surviving an injury is $(1 - P)$.

C3.5.1. Effects on People

The effects from explosions on humans are usually divided into the following categories:

1) Direct or primary effects
   The overpressure from the explosion's blast, can cause injury to sensitive human organs, or death.
2) Indirect effects
   The indirect effects are sub-divided into two categories
   - secondary effects refer to injuries or death caused by fragments or debris thrown by explosion's blast, or
   - tertiary effects that refer to injuries or death caused by whole-body displacement and collision with stationary objects or structures, as a consequence of the explosion's blast and the associated wind.

One of the reasons for the distinction between direct and indirect effects is that, for direct effects, it is **assured** that a person will be subjected to the pressure increase. Indirect effects, on the other hand, present only a **possibility** that a person might be hit by fragments or debris, or that a person who undergoes a whole-body displacement may collide with an obstacle.

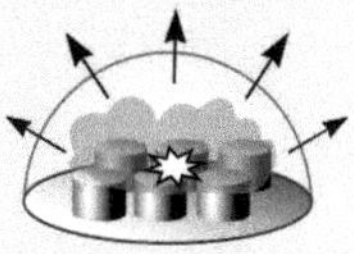

* $\operatorname{erf}(-z) = -\operatorname{erf}(z)$. For $z > 0$ and error $< 5\times10^{-4}$ [Abramowitz and Stegun 1970].
$\operatorname{erf}(z) = 1 - (1 + 0.278393\,z + 0.230389\,z^2 + 0.000972\,z^3 + 0.078108\,z^4)^{-4}$.

Effects of Explosions
calculation procedure

Effects on People
Lung Damage
Ear-Drum Rupture
Head Impact
Whole-Body Displacement Impact
Fragments and Debris
Effects on Structures

a) Lung Damage

The explosion can cause a sudden extreme pressure differential between the inside and outside of the lungs, as the pressure to which the human body is subjected, suddenly increases. As a consequence, the thorax is pressed inwards, causing lung damage and possible death. Since the inward pressure process is associated with a finite time, in addition to the value of the overpressure, its duration is also important.

In order to calculate lung damages, Bowen [Bowen *et al.* 1968] proposed a probit function that took into consideration the position of the person in space. The probit function was defined as a function of the scaled pressure, $\overline{P}$ (-), and the scaled impulse, $\bar{i}$ ($Pa^{1/2}\cdot s\cdot kg^{-1/3}$), both defined as

$$\overline{P} = \frac{P'}{P_a} \quad \text{and} \quad \bar{i} = \frac{i}{m^{1/3}\sqrt{P_a}} = \frac{1}{m^{1/3}\sqrt{P_a}}\left(\frac{1}{2}P'\,t_p\right). \tag{C3.18}$$

In Eq. (C3.18), P' (Pa) denotes the exerted total overpressure due to the position of the person, and P_a (Pa), the ambient pressure, respectively. The total overpressure, P' (Pa), can be higher than the peak overpressure, P_s (Pa), of the shock wave, as it can include reflected pressure depending on the position of the human body. The mass, m (kg), refers to the mass of the human body. Usually it is taken as:

- 5 kg for babies,
- 25 kg for small children,
- 55 kg for adult women,
- 75 kg for adult men.

In Table C3.7, empirical expressions for the calculation of the total overpressure, P' (Pa), for various positions of the human body in relation to the propagation of the shock wave, are given in [Bowen *et al.* 1968].

The calculation procedure for the probability of death, P (-), because of lung damage caused by the exertion of the blast pressure difference, can be as follows:

- In a specific distance from the center of the explosion, the overpressure P_s (Pa) and the positive-phase duration, t_p (s), are calculated.
- From Table C3.7, and for a specific position of the human body, the total overpressure, P' (Pa), is calculated.
- The mass of the human body is selected, and thus the scaled impulse, $\bar{i}$ ($Pa^{1/2}\cdot s\cdot kg^{-1/3}$), and the scaled pressure, $\overline{P}$ (-), are calculated from Eq. (C3.18).
- The value of the probit function is calculated,

$$Pr = 5.0 - 5.74 \ln S\,, \tag{C3.19}$$

Table C3.7. Position of the Human Body in Relation to the Shock Wave.

Position of the human body	Total overpressure P' (Pa)
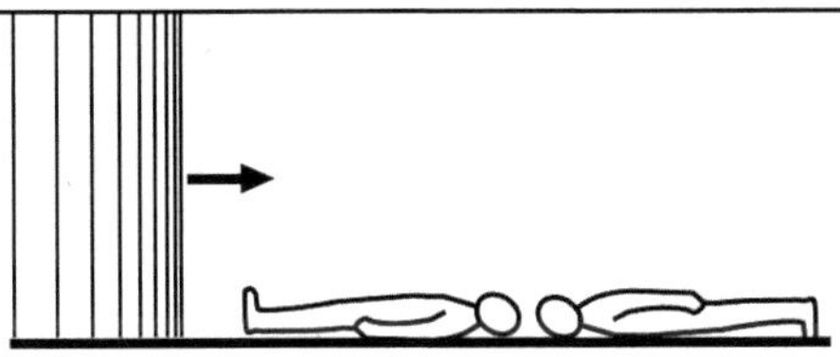 a) No obstruction of shock wave due to the human body.	$P' = P_s$
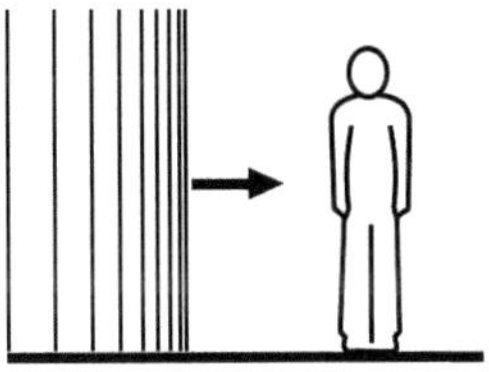 b) Shock wave flows around the human body.	$P' = P_s + \dfrac{5P_s^2}{2P_s + 14 \times 10^5}$
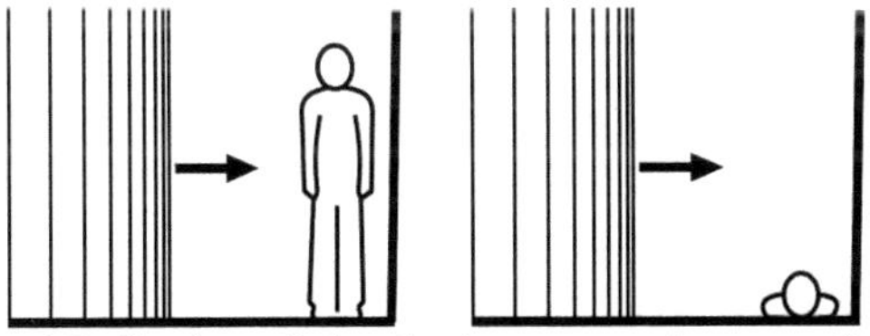 c) Reflection of the shock wave against a surface in the immediate surroundings of the person.	$P' = \dfrac{8P_s^2 + P_s 14 \times 10^5}{P_s + 7 \times 10^5}$

where

$$S = \frac{4.2}{\overline{P}} + \frac{1.3}{\overline{i}}. \tag{C3.20}$$

- The probability of death, P (-), because of lung damage caused by the exertion of the blast pressure difference, is calculated from Eq. (C3.16), from the value of the probit function.

To calculate the probability of death of all the persons affected by the overpressure, Eq. (C3.16) must be integrated spatially according to the algorithm described in subsection (c) of Section C2.5.3.

b) Eardrum Rupture

Hirsch in 1968 [Hirsch 1968] proposed that the probability, P (-), of eardrum rupture, can be calculated from Eq. (C3.16), and a probit function with the peak overpressure, P_s (Pa), as parameter, as

$$Pr = -12.6 + 1.524 \ln P_s \,. \tag{C3.21}$$

c) Head Impact

Although lung damage or ear-drum rupture are direct effects of the shock wave, the impact of the head is a tertiary indirect effect. The shock wave can push the head of a person backwards, resulting in skull rupture or fracture, or even the collision of the head with another stationary or non-stationary object. The possibility of death, P (-), after head impact, can be calculated from Eq. (C3.16), and a corresponding probit function [Baker *et al.* 1983], as

$$Pr = 5.0 - 8.49 \ln S \,, \tag{C3.22}$$

where

$$S = \frac{2.43 \times 10^3}{P_s} + \frac{4 \times 10^8}{P_s \, i_s} \,. \tag{C3.23}$$

d) Whole-Body Displacement Impact

The shock wave can also throw back the whole body backwards, causing injuries because of its impact with other objects. Such injuries are also tertiary indirect effects.

The probability of death, P (-), following such a whole-body displacement, can be calculated from Eq. (C3.16), and a corresponding probit function [Baker *et al.* 1983], as

$$Pr = 5.0 - 2.44 \ln S \,, \tag{C3.24}$$

where

$$S = \frac{7.38 \times 10^3}{P_s} + \frac{1.3 \times 10^9}{P_s \, i_s} \,. \tag{C3.25}$$

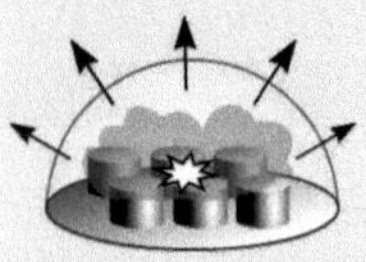

It is noted that in the above equation, P_s (Pa) denotes the overpressure and i_s (Pa·s) the impulse.

e) Fragments and Debris

Injuries or lethal injuries caused by fragments or debris constitute a secondary indirect effect. In this case the probability of death, *P* (-), following impact with a fragment, can be calculated from Eq. (C3.16), as a function of a corresponding probit function [Baker *et al.* 1983], the velocity, *u* (m/s), of the fragment and its mass, *m* (kg), as

for $0.001 \le m \le 0.1$ kg $$Pr = -29.15 + 2.10 \ln\left(m u^{5.115}\right) \qquad (C3.26)$$

for $0.1 < m \le 4.5$ kg $$Pr = -17.65 + 5.30 \ln\left(\frac{1}{2} m u^2\right) \qquad (C3.27)$$

for $m > 4.5$ kg $$Pr = -13.19 + 10.54 \ln(u) \qquad (C3.28)$$

The probit functions require the knowledge of the velocity of the fragment. It is noted that for an impact with a fragment of mass 4.5 kg that moves with a velocity of 7 m/s, the probability of survival is negligible. For further information, the reader is referred to the literature [Baker *et al.* 1983].

In March 21, 2008, at the BP Texas City Refinery, TX, U.S.A., a massive explosion killed 15 and inured 180 persons. (Reproduced by kind permission of the U.S. Chemical Safety Board.)

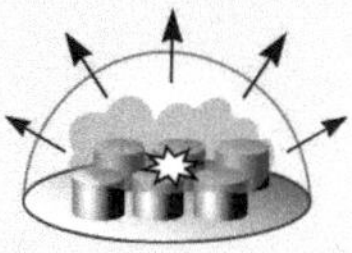

Effects of Explosions

example

Effects on People

Lung Damage

Ear-Drum Rupture

Head Impact

Whole-Body Displacement Impact

Fragments and Debris

Effects on Structures

EXAMPLE C3.5. *Effects of Explosions on People*

In Example C3.3, the Multi-Energy method was employed to calculate the overpressure and the positive-phase duration, as a function of the distance, for the explosion that took place in the PEMEX Terminal in the outskirts of Mexico City at November 19, 1984. Calculate the probability of death of a person weighing 70 kg, who is at a distance of 60 m from the center of the explosion.

From Figure C3.12, at 60 m, $P_s = 0.3$ MPa, and $t_p = 0.06$ s.

a) Lung Damage

In relation to the three positions of the human body shown in Table C3.7, one can calculate

1) $P' = 0.3$ MPa.
   Eq. (C3.18): $\overline{P} = 3.0$, $\overline{i} = 6.91\ \mathrm{Pa^{1/2} \cdot s \cdot kg^{-1/3}}$
   Eq. (C3.20): $S = 1.59$. Eq. (C3.19): $Pr = 2.34$ (erf = -0.992)
   Hence, the probability of death due to lung damage is 0.4%.

2) $P' = 0.525$ MPa.
   Eq. (C3.18): $\overline{P} = 5.25$, $\overline{i} = 12.1\ \mathrm{Pa^{1/2} \cdot s \cdot kg^{-1/3}}$
   Eq. (C3.20): $S = 0.91$. Eq. (C3.18): $Pr = 5.56$ (erf = 0.423)
   Hence, the probability of death due to lung damage is 71.1%.

3) $P' = 1.14$ MPa.
   Eq. (C3.18): $\overline{P} = 11.4$, $\overline{i} = 26.4\ \mathrm{Pa^{1/2} \cdot s \cdot kg^{-1/3}}$
   Eq. (C3.20): $S = 0.42$, Eq. (C3.19): $Pr = 10.0$ (erf = 1.00)
   Hence, the probability of death due to lung damage is 100%.

b) Eardrum Rupture

$P_s = 3.0$ MPa. Eq. (C3.19): $Pr = 6.62$ (erf = 0.895)
Hence, the probability of eardrum rupture is 94.8%.

d) Head Impact

$P_s = 0.3$ MPa. Eq. (C3.23): $S = 0.16$. Eq. (C3.22): $Pr = 20.76$ (erf = 1.000)
Hence, the probability of death due to head impact is 100%.

e) Whole-Body Displacement Impact

$P_s = 0.3$ MPa. Eq. (C3.25): $S = 0.51$. Eq. (C3.24): $Pr = 6.66$ (erf = 0.904)
Hence, the probability of death due to whole-body displacement is 95.2%.

❐

C3.5.2. Effects on Structures

The calculation of the effects of the shock wave on buildings will be carried out with the use of probit functions. The probability of collapse, P (-), of the building is given, as previously, from Eq. (C3.16). The probability is obtained as a function of the peak overpressure, P_s (Pa) of the explosion blast, and the impulse i_s (Pa·s), given by Eq. (C3.15).

The following cases that refer to buildings of up to four floors [TNO 1989], are presented.

a) Building Collapse

$$Pr = 5.0 - 0.22 \ln S , \qquad \text{(C3.29)}$$

where

$$S = \left(\frac{40{,}000}{P_s}\right)^{7.4} + \left(\frac{460}{i_s}\right)^{11.3} . \qquad \text{(C3.30)}$$

b) Major Structural Damage

$$Pr = 5.0 - 0.26 \ln VS , \qquad \text{(C3.31)}$$

where

$$V = \left(\frac{17{,}500}{P_s}\right)^{8.4} + \left(\frac{290}{i_s}\right)^{9.3} . \qquad \text{(C3.32)}$$

c) Minor Damages

$$Pr = 5.0 - 0.26 \ln S , \qquad \text{(C3.33)}$$

where

$$V = \left(\frac{4{,}600}{P_s}\right)^{3.9} + \left(\frac{110}{i_s}\right)^{5.0} . \qquad \text{(C3.34)}$$

d) Breakage of Window Panes

$$Pr = -16.58 + 2.53 \ln P_S . \qquad \text{(C3.35)}$$

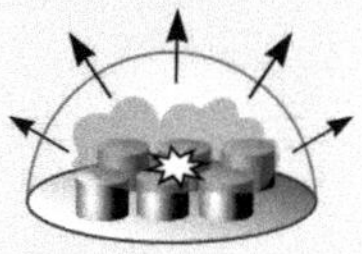

Effects of Explosions

example

Effects on People

Lung Damage

Ear-Drum Rupture

Head Impact

Whole-Body Displacement Impact

Fragments and Debris

Effects on Structures

EXAMPLE C3.6. *Effects of Explosions on Structures*

In Example C3.3, the Multi-Energy method was employed to calculate the overpressure and the positive-phase duration, as a function of the distance, for the explosion that took place in the PEMEX Terminal in the outskirts of Mexico City at November 19, 1984. Calculate the probability of damages in a three-floor building that is situated 200 m from the center of the explosion.

From Table C3.3, at 200 m, $P_s = 0.24$ bar, and $t_p = 0.112$ s.
Hence the impulse is $i_s = 1300$ Pa·s

a) Building Collapse

From Eq. (C3.30): $S = 43.8$.
From Eq. (C3.29): $Pr = 4.17$ (erf = -0.594).
Hence, the probability of building collapse is 20.3%.

b) Major Structural Damage

From Eq. (C3.32): $S = 0.07$.
From Eq. (C3.31): $Pr = 5.69$ (erf = 0.510).
Hence, the probability of major structural damage is 75.5%.

c) Minor Damages

From Eq. (C3.34): $S = 0.0016$.
From Eq. (C3.33): $Pr = 6.67$ (erf = 0.906).
Hence, the probability of minor damages is 95.3%.

d) Breakage of Window Panes

From Eq. (C3.35): $Pr = 8.94$ (erf = 1.000).
Hence, the probability of breakage of window panes is 100%.

It can be noticed that, in this case, the above results agree logically with the corresponding simplified observations of Table C3.6, where for the overpressure of 0.24 bar, the following effects would have been predicted:

- breakage of all window panes,
- destruction of doors and window frames, movement of tiles.
- but no collapse of the building (it would require an overpressure of 2 bar).

We note that the equations are based on average observations and may differ in practice according to the quality of construction.

❐

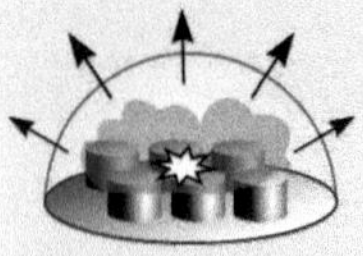

C3.6. Examples

In this section the 1974 disaster at Flixborough will be presented as a case study. Following that, other accidents caused by vapor cloud explosions will be listed.

C3.6.1. Case Study: the Flixborough Accident

On June 1, 1974, a large explosion destroyed the industrial complex of Nypro (UK) Ltd. in Flixborough, North Lincolnshire, England (Figures C3.16-C3.18). The explosion resulted in the deaths of 28 persons and the injuries of 76 others.

The plant produced caprolactam, a precursor chemical used in the manufacture of Nylon 6. The first stage in the production of caprolactam involves the oxidation of cyclohexane to cyclohexanone. One of the six reactor vessels (reactor 5) showed cracks and had to be placed out of operation, bypassing it with a temporary 20 cm diameter pipe. Unfortunately the engineer in charge of design had just left the company and thus the bypass was designed by engineers who were not experienced in high-pressure pipework. When operation started, the bypass pipe broke down immediately, resulting in the release in the atmosphere of 40 tons of cyclohexane at 150°C and 10 atm (approximately the volume of two reactor vessels). This accident was one of the most important industrial accidents in recent history and led to a widespread public outcry over industrial plant safety, and significant tightening of the U.K. government's regulations (adoption of the Control of Industrial Major Accident Hazard Act - CIMAH).

Figure C3.16. Flixborough 1974 after the explosion. (HSE 1975, reproduced under the terms of Crown Copyright Policy Guidance issued by HMSO.)

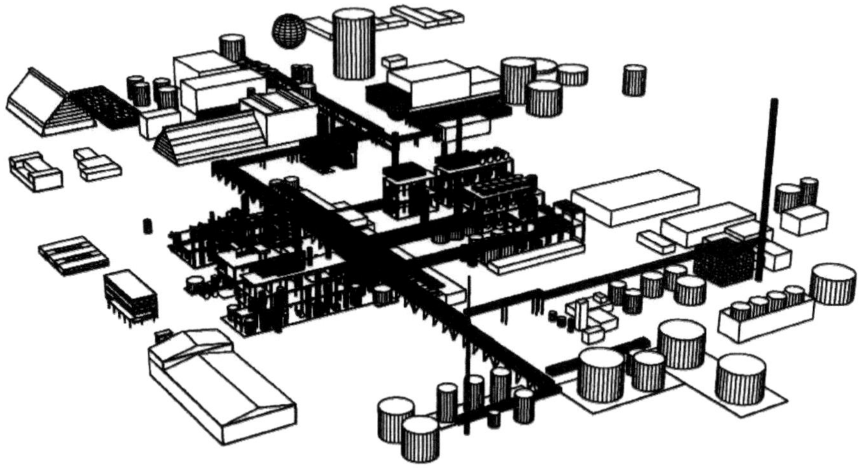

Figure C3.17. Flixborough 1974 [Hoiset *et al.* 2000].

It can be assumed that

a) 70% of the 40 tons of cyclohexane that was released contributed to the formation of the flammable cloud (based on local observations),

b) The temperature of the cloud was slightly above the boiling point of cyclohexane (80.3°C), and

c) There was a stoichiometric ratio of cyclohexane:air in the cloud.

Data for cyclohexane:

Heat of combustion for cyclohexane, ΔH_c	:	43,850	kJ/kg
Density of cyclohexane, ρ (80.3 °C, 0.1 MPa)	:	3	kg/m^3

a) Equivalent TNT Mass Method

To start, the equivalent TNT mass, M_{TNT} (kg), must be calculated from Eq. (C3.2). Since from the 40 tons of cyclohexane, 70% contributed to the flammable cloud formation, the mass of the cloud should be 40,000 x 0.7 = 28,000 kg.

And if $f_E = 0.09$

then $$M_{TNT} = \frac{f_E \, \Delta H_c \, M_G}{\Delta H_{TNT}} = \frac{0.09 \times 43{,}850 \times 28{,}000}{4{,}760} = 23{,}215 \text{ kg} .$$

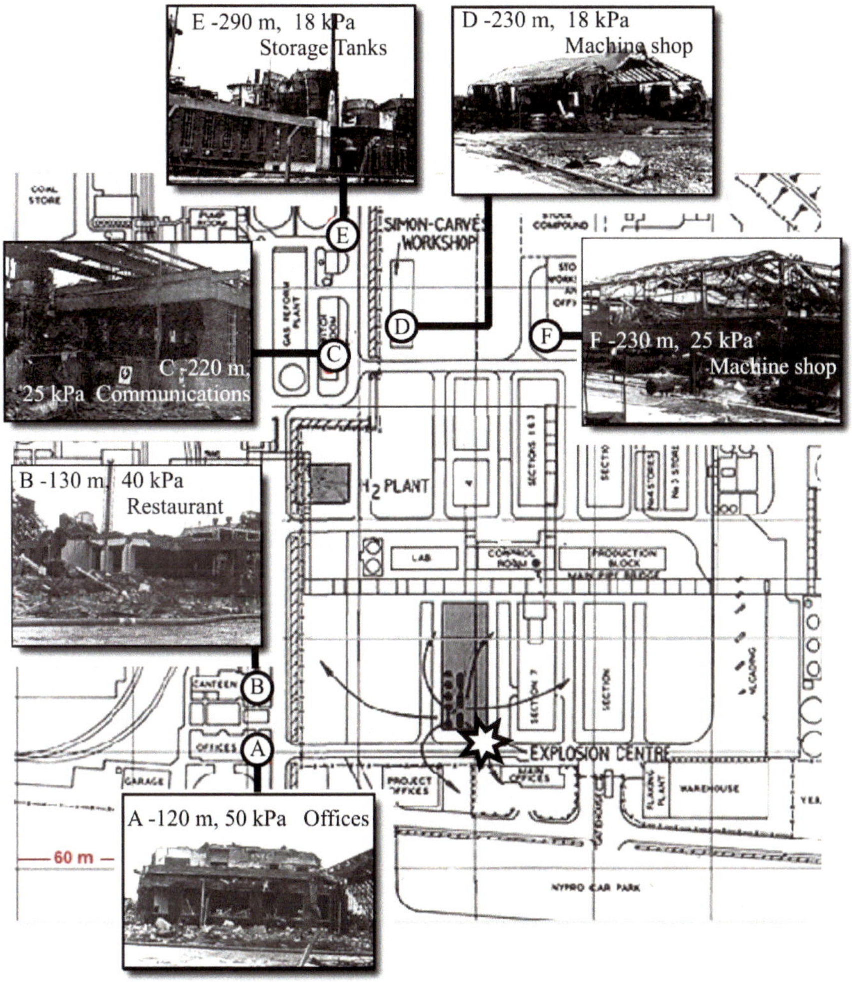

Figure C3.18. Flixborough 1974: Ground plan and disaster photographs [Sadee *et al.* 1976].

Hence, for a distance $x = 50$ m, $Z = x / M_{\mathrm{TNT}}^{1/3} = 2\ \mathrm{m\,kg^{-1/3}}$, and from Eq. (C3.3)

$$P_\mathrm{S} = \frac{80{,}800\left(1+\left[\dfrac{Z}{4.5}\right]^2\right)}{\sqrt{1+\left[\dfrac{Z}{0.048}\right]^2}\sqrt{1+\left[\dfrac{Z}{0.32}\right]^2}\sqrt{1+\left[\dfrac{Z}{1.35}\right]^2}} = 279\ \mathrm{kPa}\ \text{ or }\ 2.8\ \mathrm{bar}.$$

At 40, 100 and 150 m distances, the overpressure is 4.68, 0.58 and 0.26 bar, respectively. The overpressure curve for the TNT method is shown in Figure C3.20.

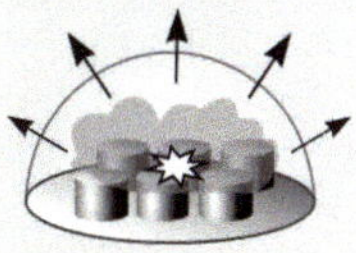

Case Study

Flixborough 1974

Equivalent TNT Mass Method

Multi-Energy Method

Baker-Strehlow Method

Critical Evaluation

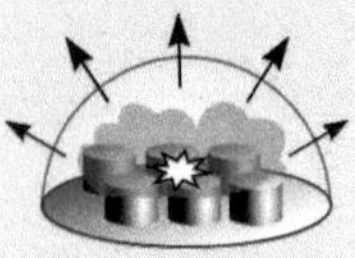

b) Multi-Energy Method

1) Cloud dimensions

Since the ratio of cyclohexane:air is 1:43, the volume, V (m^3), of the cloud will be equal to (28,000 kg/3 kg/m^3)x(43/1) = 401,334 m^3.

Observers testified that the height of the cloud was less than 10 m. Simulating the cloud with a cylinder of 10 m height, its radius, R (m), can be calculated from 401,334 = 10 (πR^2) and thus R = 113 m.

2) Obstructed regions

The radius of 113 m determines the part of the facilities in which the cyclohexane cloud was spread (a larger part of the facilities than that determined by this radius is always considered, as the equipment takes up a large part of the empty space). These facilities, except the restaurant and the nearby offices (see Section C3.19), are separated by a distance of less than 25 m, and thus the criterion <10 D_1 or <1.5 D_2 is valid. That is, all these sections constitute one obstructed region (the explosion took place in Section 25A, Section 7 is less than 25 m apart, while

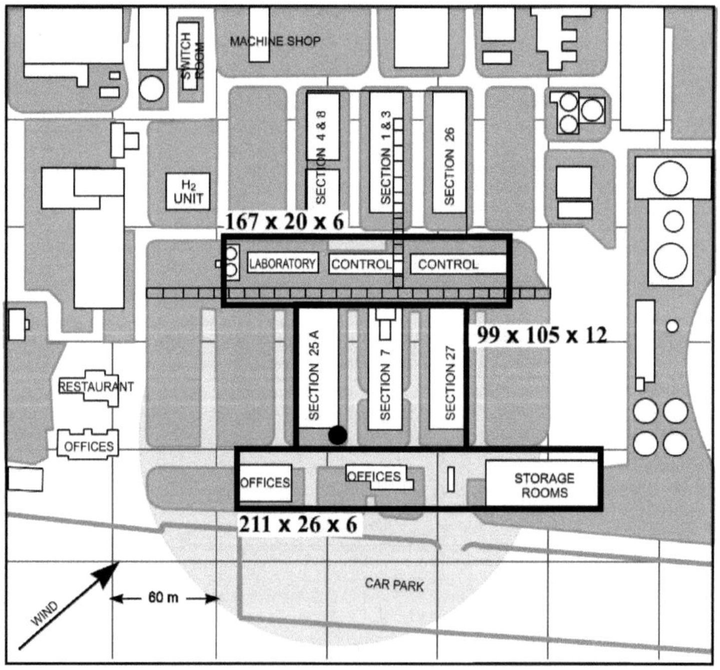

Figure C3.19. Detail of the explosion region.

Section 27 is less than 25 m from Section 7, etc.), and thus it will be assumed that there is a single source of explosion. Since however, this virtual box that would surround these regions will also include a lot of empty space, it is preferable to define three sub-regions, whose dimensions are shown in Figure C3.19 (length x width x height). The volume of the obstructed region is simply the sum of the volumes of the three sub-regions

$$(167 \times 20 \times 6) + (99 \times 105 \times 12) + (211 \times 26 \times 6) = 177{,}700 \text{ m}^3$$

It can further be assumed that 50% of this space is occupied by equipment (vessels, pipes, etc.) that is 0.5 void fraction. Therefore only 50% of this volume (88,850 m^3) is available for the cloud to expand. Explosion inside this space will have a much higher blast (coefficient of explosion blast equal to 10), than the explosion in the remaining unobstructed regions (coefficient of explosion blast equal to 3). Hence, the remaining volume of 312,484 m^3 is ignored (coefficient of explosion blast equal to 3).

3) Energy of explosion

The reaction that takes place is $C_6H_{12} + 9O_2 \rightarrow 6CO_2 + 6H_2O$

Therefore the ratio of cyclohexane:air is 1:43.

The energy, E (GJ), of the explosion is

$$E = 88{,}850 \times (3/43) \times 43.85 = 272 \text{ GJ.}$$

4) Overpressure

We can assume a coefficient of explosion's blast equal to 10, due to the very large damages that took place. For a distance x = 100 m from Eq. (C3.8) r'= 0.72 and from the diagram in Figure C3.8 one obtains the overpressure P_s = 1.01 bar. In 150 m distance, the overpressure will be 0.41 bar, respectively. The overpressure curve is shown in Figure C3.20.

c) Baker-Strehlow Method

The dimensions of the cloud, as well as its energy, will be considered identical to those obtained for the Multi-Energy method. The important parameter in the Baker-Strehlow method is the flame speed. It will be assumed that the expansion takes place in two dimensions, that the equipment density is high and that the cyclohexane is very reactive. These assumptions, from Table C3.4, produce a flame speed equal to 1.77.

Based on this flame speed, for a distance of 100 m and r' = 0.72, from the diagram of Figure C3.13, one obtains P_s = 0.55 bar.

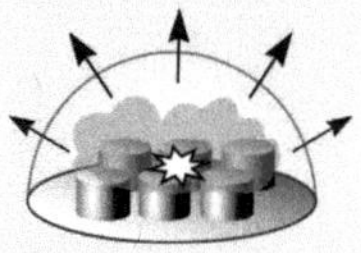

Case Study

Flixborough 1974

Equivalent TNT Mass Method

Multi-Energy Method

Baker-Strehlow Method

Critical Evaluation

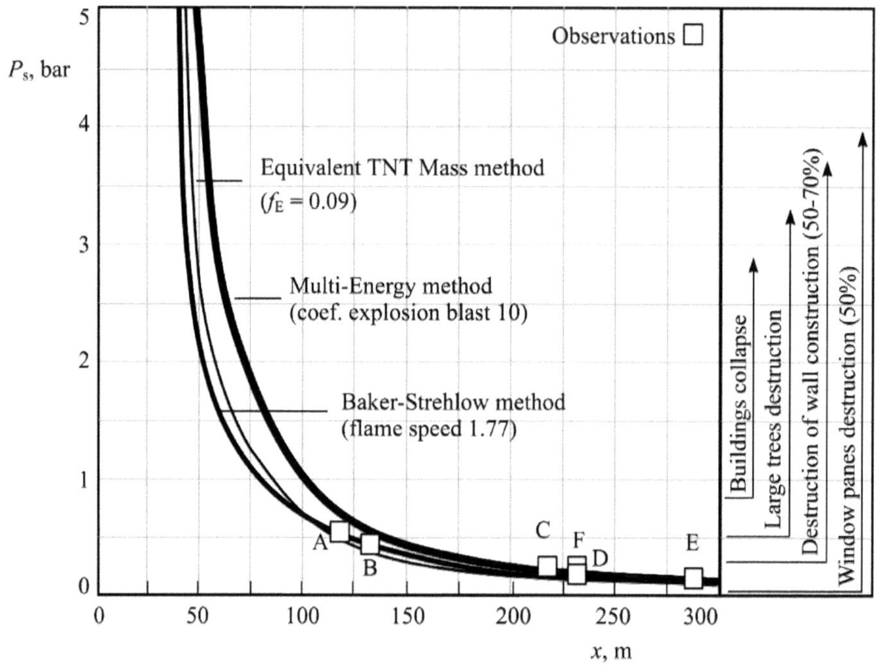

Figure C3.20. Overpressure vs. distance curves.

d) Results Comparison

Examining the diagram in Figure C3.20, one can deduce that in this specific example all three methods produced comparative results. Even if the results of the Multi-Energy method may be considered slightly higher, a small change of the coefficient of explosion's blast will bring them in full agreement with the other methods.

The local observations, shown in the diagram, are based on the level of damages in specific distances as seen in the photographs of Figure C3.18. It can also be observed according to a diagram, that up to the distance of 75 m from the explosion center, there will be total destruction of houses, etc.

Finally it should be emphasized that the whole analysis so far referred to the simulation of explosions that did take place, that is, the effort concentrated to changing the parameters (fraction of energy released, coefficient of explosion blast, flame speed) so that the predicted effects will match the observed ones. The interesting (and of course the most difficult) part is to predict the overpressure-distance curves for different scenarios, when no precise data are available.

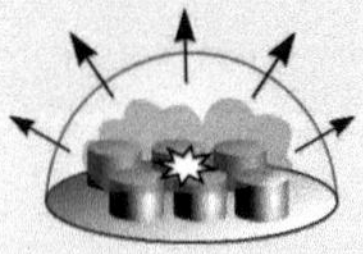

C3.6.2. Major Industrial Accidents Caused by Explosions

In Table C3.8 an indicative list of major industrial accidents caused by explosions (any kind of explosion as usually the type of explosion is not specified) is given. The list is part of the database MinA - Major Industrial Accidents due to Fires, Explosions and Toxic-Gas Releases [MinA 2006].

Table C3.8. Major Industrial Accidents Caused by Explosions.

Date	Country	City/State	D	I	E[a]	Chemical	Unit	M$[b]
2006								
29.5	Serbia	Baric	4	4	0			
2005								
11.12	UK	Hertfordshire		43	>2,000	Oil	Storage	
2004								
20.1	Algeria	Skikda	23	74		LNG	Refinery	845
2002								
12.4	UK	North Shields			100			
2001								
21.9	France	Toulouse	30	3,000		Ammonia	Petrochemicals	888
9.7	USA	Three Rivers, TX		4	200	Acid		
16.4	UK	Grimsby Humber				Hydrocarbons	Refinery	308
15.3	Brazil	Campos Basin	10		165	Natural Gas	Oil Platform	610
2000								
12.9	Mexico	Salamanca		300		Malathion	Petrochemicals	
27.3	USA	Pasadena, TX	1	69		Butadiene	Petrochemicals	81
25.1	Kuwait	Al-Ahmadi	5	50		Hydrocarbons	Refinery	512
1999								
2.12	Thailand	Sri Racha	8	13		Hydrocarbons	Refinery	44
8.6	Germany	Wuppertal		50			Petrochemicals	94
25.3	USA	Richmond, CA				Hydrocarbons	Refinery	94
27.1	Canada	Taylor, BC		15		Ethane	Natural Gas	28
1998								
28.9	Australia	Longford	2	8		Hydrocarbons		203
25.9	Australia	Victoria				Hydrocarbons	Natural Gas	203
9.7	Canada	St. John, NB	1			Hydrocarbons	Refinery	78
14.2	Cameroon	Yaoundi	220	130		Oil	Transport	

[a] D I E - Number of Deaths, number of Injuries, number of people Evacuated.

[b] M$: Material Damage in million $ (in 2005 prices).

Note: LNG, liquefied natural gas.

- Continued

Table C3.8 (cont.). Major Industrial Accidents Caused by Explosions.

Date	Country	City/State	D	I	E[a]	Chemical	Unit	M$[b]
1997								
25.12	Malaysia	Bintulu, Sarawak		12		Gas	Petrochemicals	348
25.11	India	Gujart	3	11		Hydrocarbons	Petrochemicals	
14.9	India	Visakhapatam	60	30		LPG	Refinery	76
22.6	USA	Deer Park, TX		30		Hydrocarbons	Petrochemicals	128
21.1	USA	Martinez, CA	1	46		Hydrocarbons	Refinery	26
1996								
26.7	Mexico	Cactus				LPG	Natural Gas	175
1995								
8.11	Jamaica	Kingston						
15.7	France	Annecy		4				
28.4	Korea	Taegu	101	104	>10,000	LPG		
12.3	India	Madras	100	23		Hydrocarbons	Transport	
1994								
28.12	Venezuela		50	10			Pipeline	
14.12	Mozambique	Palmeira	36			Natural Gas	Transport	
13.12	USA	Port Neal, IA			2,500	Ammonia	Petrochemicals	167
7.12	Korea	Seoul	7	50	10,000	LNG		
12.10	USA	Pasadena, TX				Hydrocarbons	Petrochemicals	71
27.5	USA	Belpre, OH				Hydrocarbons	Petrochemicals	140
7.5	Taiwan	Kaohsiung	1			Plastics	Chemicals	
1.1	India	Thane District	4	298		Chlorine	Transport	
1993								
26.11	China	Shuangpai	61					
4.11	Vietnam	Nam Khe	39	62			Pipeline	
6.8	China	Shenzhen	12	168		Natural Gas	Storage	
?.8	India	Panipat	3	25		Ammonia	Petrochemicals	
26.6	China	Zhengzhou	27	32				
25.3	Venezuela	Lake Maracaibo	11			Propane	Oil Platform	144
11.1	China	Baohe	70			Natural Gas		
1992								
9.11	France	La Mede				Hydrocarbons	Refinery	376
16.10	Japan	Sodegaura	10	7		Hydrocarbons	Refinery	232
8.10	USA	Wilmington, CA				Hydrogen	Refinery	114
1.9	Greece	Elefsina	14	21		LPG	Refinery	
8.8	Turkey	Corlu	32	64		Methane		
28.7	USA	Westlake, LA				Ammonia	Petrochemicals	35
29.4	India	New Delhi	43	20			Storage	
23.2	Korea	Kwangju		16	20,000	LPG		

[a] D I E - Number of Deaths, number of Injuries, number of people Evacuated.

[b] M$: Material Damage in million $ (in 2005 prices).

Note: LNG, liquefied natural gas; LPG, liquefied propane gas.

- Continued

Table C3.8 (cont.). Major Industrial Accidents Caused by Explosions.

Date	Country	City/State	D	I	E[a]	Chemical	Unit	M$[b]
1991								
20.6	Bangladesh	Dhaka				Ammonia	Petrochemicals	104
10.12	Germany	North Rhine				Hydrocarbons	Refinery	74
14.7	USA	Kensington, GA				Butadiene		
20.6	Bangladesh	Dhaka						92
1.5	USA	Sterlington, VA	>8	123	500	Nitromethane		
13.4	USA	Sweeney, TX				Hydrocarbons	Refinery	53
10.4	Italy	Livorno	141			Naphtha	Transport	
12.3	USA	Seadrift, TX	1			Ethylene oxide	Petrochemicals	116
11.3	Mexico	Pajaritos	3			Propane	Chemicals	133
3.3	USA	Lake Charles, LA				Hydrocarbons	Refinery	33
14.2	Korea	Daesan		2		Hydrogen		
1990								
6.11	India	Nagothane	31			Propane	Petrochemicals	33
5.11	India	Maharashtra	35			Natural Gas	Refinery	
3.11	USA	Chalmette, LA				Hydrocarbons	Refinery	30
24.9	Thailand	Bangong	68	100		LPG	Transport	
16.9	USA	Bay City, TX	1			Natural Gas	Transport	
22.7	Korea	Ulsan			10,000	Butane		
19.7	USA	Cincinnati, OH				Hydrocarbons	Petrochemicals	30
5.7	USA	Channei View, TX	5	13		Oil	Storage	
14.5	USSR	Tomsk				Ethylene	Chemicals	
1.4	USA	Warren, PA				LPG	Refinery	30
20.3	UK	Ellesmere Port	1	5			Petrochemicals	13
3.3	USA	N. Blenheim, NY	2	7		Propane	Pipeline	
1989								
24.12	USA	Baton Rouge, LA	4	12		Propane	Refinery	105
23.10	USA	Pasadena, TX	23	314	>1,300	Butane	Petrochemicals	1,030
2.7	USSR	Minnebeavo	4			Propane	Chemicals	
7.6	USA	Morris, IL				Ethylene	Petrochemicals	39
4.6	USSR	Siberia	645	706	>500	LPG	Pipeline	
20.3	Lithuania	Jonova	7	57	30,000	Ammonia	Storage	
19.3	USA	Gulf of Mexico,TX				Hydrocarbons	Oil Platform	79
7.3	Belgium	Antwerp	32	11		Ethylene oxide	Petrochemicals	117
14.2	Germany	Urdingen						55
1988								
21.11	Germany	Worms	3	25			Petrochemicals	
18.9	Norway	Bamble				VCM	Petrochemicals	17
8.9	Norway	Rafnes				VCM	Chemicals	16
6.7	UK	Piper Alpha	165			Natural Gas	Oil Platform	1,503
23.6	Mexico	Monterrey	4	15	10,000	Natural Gas		
15.6	Italy	Genoa	3	2	1,500	Hydrogen		
15.5	Mexico	Chihuahua		7	15,000	Oil	Storage	

[a] D I E - Number of Deaths, number of Injuries, number of people Evacuated.

[b] M$: Material Damage in million $ (in 2005 prices).

Note: LPG, liquefied propane gas; VCM, vinyl chloride monomer.

- *Continued*

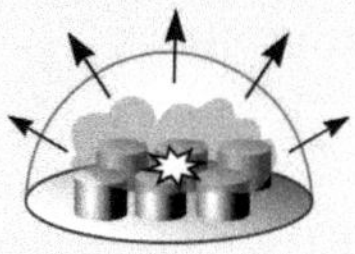

Table C3.8 (cont.). Major Industrial Accidents Caused by Explosions.

Date	Country	City/State	D	I	E[a]	Chemical	Unit	M$[b]
1988								
5.5	USA	Norco, LA	7	48		LPG	Refinery	398
7.4	Netherlands	Beek				Ethylene	Chemicals	
22.1	China	Shanghai	25	17		Hydrocarbons	Refinery	
1987								
14.11	USA	Pampa, TX	3	43		Butane	Petrochemicals	340
15.8	Saudi Arabia	Ras Tanura				Propane	Natural Gas	
17.7	Germany	Herborn	6	24		Natural Gas	Transport	
3.7	Belgium	Antwerp	5	20		Ethylene oxide	Petrochemicals	91
22.3	UK	Grangemouth	2	2		Hydrogen	Refinery	127
1985								
21.11	USA	Tioga, ND				Hydrocarbons	Natural Gas	15
5.11	USA	Mont Belieu, TX	4	13		Propane	Distribution	72
23.7	USA	Kaycee, WY	1	6		Hydrocarbons	Pipeline	
26.5	Spain	Algerais	18	56		Naphtha		
9.3	USA	Lake Charles, LA				Propane		
23.2	USA	Sharpsville, PA				Natural Gas	Pipeline	
19.2	Canada	Edmonton, AL				LNG	Pipeline	
18.1	Germany	Wesseling				Ethylene	Petrochemicals	65
1984								
13.12	Venezuela	Las Piedras				Oil		
1.12	Pakistan	Gahri Dhoda	60			Natural Gas	Pipeline	
19.11	Mexico	Mexico City	550	6,400		LPG	Distribution	34
30.9	USA	Basile, LA				Petrol	Natural Gas	51
25.9	USA	Phoenix, AZ				Natural Gas	Pipeline	
15.8	Canada	Ft. McMurray, AL				Hydrocarbons		
16.8	Brazil	Campos Basin	36	17		Petrol	Oil Platform	
23.7	USA	Romeoville, IL	15	76		LPG	Refinery	325
23.5	UK	Abbeystead	16	28		Methane		
20.4	Canada	Beaumont, ON	2			Hydrogen	Chemicals	
1983								
2.11	India	Dhurabar	47			Hydrocarbons	Train Wagon	
29.9	India	Dhulwari	41	100		Natural Gas		
26.6	USA	Bloomfield, NM		2		Natural Gas	Compressors	
1.5	Egypt	Nile River	317	44		LPG	Transport	
14.4	India	Bontang				LNG		
15.3	USA	West Odessa, TX	6			LPG	Pipeline	
8.3	India	Kerala				Hydrocarbons		18
10.1	Nicaragua	Corinto		17	25,000	Oil	Transport	18
7.1	USA	Port Newark, NJ	1			Natural Gas	Storage	53
29.12	Italy	Florence	5	30		Propane	Transport	
11.12	USA	Taft, CA			20,000	Acrolein		

[a] D I E - Number of Deaths, number of Injuries, number of people Evacuated.
[b] M$: Material Damage in million $ (in 2005 prices).
Note: LNG, liquefied natural gas; LPG, liquefied propane gas.

- Continued

Table C3.8 (cont.). Major Industrial Accidents Caused by Explosions.

Date	Country	City/State	D	I	E[a]	Chemical	Unit	M$[b]
1982								
4.11	USA	Hudson, IA	5			Natural Gas	Pipeline	
1.10	USA	Pine Bluff, AR				Natural Gas	Pipeline	
28.9	USA	Livingston, LA			3,000		Train Wagon	
28.6	USA	Portales, NM	6			Natural Gas	Pipeline	
6.5	USA	Duluth, MN				Acid	Petrochemicals	26
25.4	Italy	Todi	34	140		Natural Gas		
9.3	USA	Philadelphia, PA				Hydrocarbons	Petrochemicals	46
1981								
6.9	UK	Stalybridge	1	1		Hydrocarbons		
19.7	USA	Greens Bayou, TX					Reactor	17
15.5	Venezuela	San Rafael	18	35		LPG	Pipeline	
8.5	Sweden	Gothernburg	1	2		Propane	Pipeline	
7.4	USA	Corpus Christi, TX	9	30		Corn		72
7.4	USA	Bellwood, NE	2	1		Corn		8
13.2	USA	Louisville, KY		4	>100	Hydrocarbons		
11.2	USA	Chicago Height, IL					Reactor	23
1980								
29.11	Spain	Ortuella	51			Propane	Storage	
12.11	USA	Omaha, NE				Corn		4
21.10	USA	New Castle, DE	5			Hydrocarbons	Petrochemicals	131
18.8	Iran	Gach Saran	80	45		Nitroglycerin	Storage	
16.8	Japan	Shizuoka	15	222		Methane		
23.7	USA	Seadrift, TX				Ethylene oxide	Reactor	22
26.6	Australia	Sydney				Oil	Refinery	33
21.4	USA	St. Joseph, MO	1	4		Corn		5
26.3	Netherlands	Enschede				Propane		
3.3	USA	Los Angeles, CA	2	2		Natural Gas	Transport	
26.2	Canada	Brooks, AL				Natural Gas	Compressors	73
30.1	Puerto Rico	Bayamon	1			Hydrocarbons	Pipeline	
20.1	USA	Borger, TX				Hydrocarbons	Refinery	77
1.1	Italy	Naples		8		Corn	Storage	
1979								
15.11	Turkey	Istanbul	52	>2		Oil	Transport	
10.11	Canada	Mississauga, ON		8	21,500	Chlorine	Train Wagon	
17.10	Spain	Lerida	7			Corn	Storage	
6.10	USA	Cove Point, MD	1	1		LNG	Pipeline	
18.9	USA	Torrance, CA				Hydrocarbons		
8.9	USA	Paxton, TX		8			Train Wagon	
4.9	USA	Pierre Port, LA				LNG	Pipeline	
1.9	USA	Deer Park, TX				Hydrocarbons	Refinery	163
28.7	USA	Sauget, IL					Reactor	16
21.7	USA	Texas City, TX				Hydrocarbons		56

[a] D I E - Number of Deaths, number of Injuries, number of people Evacuated.

[b] M$: Material Damage in million $ (in 2005 prices).

Note: LNG, liquefied natural gas; LPG, liquefied propane gas.

- Continued

Table C3.8 (cont.). Major Industrial Accidents Caused by Explosions.

Date	Country	City/State	D	I	E[a]	Chemical	Unit	M$[b]
1979								
26.6	USA	Ypsilanti, MI				Propane	Storage	
3.6	Thailand	Phangnaga	50	15		Oil		
19.4	USA	Port Neches, TX				Oil	Transport	65
20.3	USA	Linden, NJ		1		LPG		35
6.2	Germany	Bremen	14	17		Flour		
1.1	Greece	Suda Bay	7	140		Propane	Transport	
1978								
2.11	Mexico	Sanch Magal	41	32		Natural Gas	Pipeline	
30.10	Romania	Pitesti				Propylene	Natural Gas	33
27.9	Spain	Oviedo	7			Natural Gas	Train Wagon	
16.9	UK	Immingham				Natural Gas	Chemicals	
4.8	USA	Donnellson, IA	3	2		LPG	Pipeline	
16.7	Mexico	Tula	12			Butane	Transport	
15.7	Mexico	Xilatopic	100	220		Butane	Transport	
7.7	Tunisia	Manouba	3	150		Ammonia		
12.6	Japan	Sendai	21	350		Oil	Storage	
11.6	Spain	San Carlos	211	200		Propylene	Transport	
15.4	Saudi Arabia	Abqaiq				Natural Gas	Natural Gas	139
24.2	USA	Waverly, TN	12	21		Propane		
11.2	Mexico	Poblado Tres	40			Natural Gas	Pipeline	
12.1	USA	Conway, KS				LPG		
1977								
28.12	India	Gujarat	5	35		Hydrogen		
23.12	USA	Westwego, LA	35	5		Corn	Storage	
10.12	Colombia	Pasacabolo	30	22		Ammonia	Fertilizers	
8.12	Italy	Brindisi	3			Ethylene	Petrochemicals	78
17.10	USA	Baton Rouge, LA				Oil		22
20.7	USA	Ruff Creek, PA	2			Propane	Pipeline	
8.7	Italy	Gela	1	25		Ethylene oxide		
19.6	Mexico	Puebla				VCM	Storage	
11.5	Saudi Arabia	Abqaiq	1	15		Oil	Pipeline	152
3.4	Qatar	Umm Said	7	87		Propane	Natural Gas	212
18.3	USA	Port Arthur, TX	4			Propane		
20.2	USA	Dallas, TX		1		Iso-Butane	Train Wagon	
27.1	USA	Baytown, TX	3			Natural Gas	Transport	
4.1	UK	Breahead		13			Storage	
1976								
17.12	USA	Los Angeles, CA				Oil	Distribution	30
7.12	USA	Robstown, TX				Natural Gas	Compressors	
26.11	USA	Belt, TX		22		LPG	Train Wagon	
15.10	USA	Longview, TX	1			Ethylene	Refinery	39
11.9	UK	Westoning		3		Hydrocarbons	Transport	

[a] D I E - Number of Deaths, number of Injuries, number of people Evacuated.

[b] M$: Material Damage in million $ (in 2005 prices).

Note: LPG, liquefied propane gas; VCM, vinyl chloride monomer.

- *Continued*

Table C3.8 (cont.). Major Industrial Accidents Caused by Explosions.

Date	Country	City/State	D	I	E[a]	Chemical	Unit	M$[b]
1976								
30.8	USA	Plaquemine, LA				Oil	Transport	
12.8	USA	Chalmette, LA	13			Ethylbenzene		
6.8	USA	Lake Charles, LA	7			Iso-Butane	Refinery	
27.6	UK	Kings Lynn	1	9			Reactor	
16.6	USA	Los Angeles, CA	6	35		Natural Gas	Pipeline	
4.6	Saudi Arabia	Abqaiq				Natural Gas	Chemicals	
1975								
2.12	USA	Watson, CA				Hydrogen	Chemicals	
21.11	Germany	Cologne				Hydrogen		
7.11	Netherlands	Beek	14	108		Propylene	Petrochemicals	72
5.9	Netherlands	Rosendaal				Natural Gas		
1.9	USA	Des Moines, IA		3		LPG	Train Wagon	
31.8	USA	Gadsden, AL	4	28		Natural Gas	Transport	
17.8	USA	Philadelphia, PA	8	20		Hydrocarbons	Refinery	40
5.4	UK	Ilford	1	3		Hydrogen	Electrolysis	
10.2	Belgium	Antwerp	6	13		Ethylene	Petrochemicals	109
1974								
29.11	USA	Beaumont, TX	2	10		Hydrocarbons	Petrochemicals	56
3.11	Japan	Tokyo Bay	33			Naphtha		
21.9	USA	Houston, TX	1	235		Butadiene	Train Wagon	
5.9	Spain	Barcelona				VCM	Chemicals	
30.8	UK	Fawley				Ethylene	Chemicals	
25.8	USA	Petal City, MO		24		Butane	Storage	
18.7	USA	Plaquemine, LA				Propylene	Chemicals	
17.8	USA	Los Angeles, CA				Hydrocarbons	Distribution	
6.8	USA	Wenatchee, WA	2	113			Train Wagon	
19.7	USA	Decatur, IL	7	152		Iso-Butane	Transport	56
7.7	Germany	Cologne				VCM	Chemicals	
29.6	USA	Climax, TX	7			VCM	Train Wagon	
1.6	UK	Flixborough	28	76		Cyclohexane	Petrochemicals	215
29.4	USA	Eagle Pass, TX	17	34		LPG	Transport	
2.3	USA	Munroe, LA				Natural Gas	Pipeline	
12.2	USA	Oneonta, NY		25		LPG	Train Wagon	
11.1	USA	West St. Paul, MN	4	6		LPG	Storage	
4.1	USA	Holly Hill, FL				Propane	Transport	
1973								
27.12	USA	Freeport, TX		29		Ethylene oxide	Transport	
28.10	Japan	Shinetsu	1	23		VCM	Chemicals	
24.10	UK	Sheffield	4	24		Natural Gas		
8.10	Japan	Ichihara City	4			Propylene	Petrochemicals	26
24.5	USA	Benson, AR				Explosives	Train Wagon	

[a] D I E - Number of Deaths, number of Injuries, number of people Evacuated.
[b] M$: Material Damage in million $ (in 2005 prices).
Note: LPG, liquefied propane gas; VCM, vinyl chloride monomer.

- Continued

Table C3.8 (cont.). Major Industrial Accidents Caused by Explosions.

Date	Country	City/State	D	I	E[a]	Chemical	Unit	M$[b]
1973								
23.5	Germany	Cologne	4	2		VCM	Chemicals	
22.2	USA	Austin, TX	6			NGL	Pipeline	
10.2	USA	Staten Island, NY	40	2		Natural Gas		
1.2	France	St. Amand	5	45		Propane	Transport	
1972								
21.9	Brazil	Duque Caxias	33	57		Butane		
14.8	USA	Billings, MT	1	4		Butane		18
4.8	USA	Weirton, WV	21	20		Natural Gas		
1.7	Mexico	Chihuahua	>8	800		Butane	Train Wagon	
14.5	USA	Hearn, TX	1	2		Petrol	Pipeline	
30.3	Brazil	Rio de Janeiro	37	53		Butane	Storage	
22.1	USA	East St. Louis, IL	1			Propylene	Transport	32
1971								
23.12	USA	Lake Charles, LA	4	3			Chemicals	
7.11	USA	Morris, IL		4		Ethylene oxide		21
19.10	USA	Houston, TX	1	50		VCM	Train Wagon	
15.9	USA	Houston, TX	1	6		Butadiene	Chemicals	
26.6	Poland	Czechowice	33			Petrol		
10.7	Netherlands	Amsterdam	8	21		Butadiene	Refinery	
25.2	USA	Longview, TX	4	60		Ethylene	Petrochemicals	26
19.1	USA	Baton Rouge, LA	0	21		Ethylene	Train Wagon	13
1970								
9.12	USA	Port Hudson, MO		10		Propane	Pipeline	
5.12	USA	Linden, NJ		40		Hydrocarbons	Refinery	164
30.5	USA	Brooklyn, NY	2	30		Oxygen	Transport	
1969								
28.12	UK	Fawley				Naphtha	Refinery	19
21.12	Switzerland	Basle	3	31			Reactor	
23.10	USA	Texas City, TX	3			Butadiene	Petrochemicals	39
1.10	Spain	Escombreras	4			Propane	Refinery	58
11.9	USA	Glendora, MS		1	30,000	VCM	Train Wagon	
9.9	USA	Houston, TX		9		Natural Gas	Pipeline	
12.8	USA	Flemington, NJ				VCM	Refinery	18
17.7	UK	Dudgeon Wharf	6			Oil	Storage	
25.1	USA	Laurel, MS	2	976		LPG	Transport	32
21.1	UK	Wilton	4			Ethylene		
1968								
21.1	Netherlands	Pernis	2	85		Hydrocarbons	Refinery	
1.1	USA	Dunreith, IN		5		VCM	Train Wagon	

[a] D I E - Number of Deaths, number of Injuries, number of people Evacuated.
[b] M$: Material Damage in million $ (in 2005 prices).
Note: LPG, liquefied propane gas; VCM, vinyl chloride monomer.

- Continued

Table C3.8 (cont.). Major Industrial Accidents Caused by Explosions.

Date	Country	City/State	D	I	E[a]	Chemical	Unit	M$[b]
1968								
1.1	Germany	East Germany	24			VCM		
1.1	UK	Hull	2	13		Acid		
1967								
8.8	USA	Lake Charles, LA	7	14		Iso-Butane		97
20.1	USA	Sacramento, CA				Hydrogen		
1966								
16.10	Canada	LaSalle	11	10		Styrene	Petrochemicals	16
23.5	USA	Philadelphia, PA				Hydrocarbons	Refinery	
6.2	USA	Scotts Bluff, LA	3			Butadiene	Reactor	
19.1	Germany	Raumheim	3	83		Methane	Petrochemicals	39
1965								
23.12	USA	Baltimore, MD				Hydrocarbons	Petrochemicals	
24.10	USA	Escambia, AL				Hydrogen	Chemicals	
7.8	UK	Bow	5	32				
31.7	USA	Baton Rouge, LA				Ethylchloride	Reactor	
13.7	USA	Lake Charles, LA				Ethylene	Petrochemicals	26
4.3	USA	Natchitoches, LA	17	56		Natural Gas	Pipeline	
1964								
25.10	USA	Texas City, TX	2	34		Ethylene	Petrochemicals	26
25.10	USA	Liberal, KS				Propane	Compressos	
17.7	Germany					Oxygen	Chemicals	
4.6	Belgium	Antwerp	4	20		Ethylene oxide	Reactor	26
12.1	USA	Attleboro, MA	7	40		VCM	Reactor	26
9.1	USA	Jackass Flats, NY				Hydrogen		
1963								
3.5	USA	Plaquemine, LA		7		Ethylene	Petrochemicals	26
3.4	USA	Norwick, CT	4	4			Transport	
1962								
25.7	USA	New Berlin, NY	10	75		LPG	Transport	5
10.5	USA	Toledo, OH	10	46				
27.4	USA	Marietta, OH	1	3		Hydrocarbons	Chemicals	26
17.4	USA	Doe Run, KY	2	19		Ethylene oxide	Petrochemicals	45
1961								
20.8	Japan	Minimata	4	10		VCM	Reactor	
1960								
17.12	USA	Freeport, TX	6	14				
14.10	USA	Kingsport, TN	15	55		Aniline		

[a] D I E - Number of Deaths, number of Injuries, Number of people Evacuated.
[b] M$: Material Damage in million $ (in 2005 prices).
Note: LPG, liquefied propane gas; VCM, vinyl chloride monomer.

- *Continued*

Table C3.8 (cont.). Major Industrial Accidents Caused by Explosions.

Date	Country	City/State	D	I	E[a]	Chemical	Unit	M$[b]
1959								
28.6	USA	Phillipsburg, NJ	6	6		Oil		
28.6	USA	Meldrin, GA	23	78		LPG	Train Wagon	
11.6	Japan	Ube	11	40		Ammonia	Fertilizers	
28.5	USA	McKittrick, CA		2		LPG	Storage	
27.2	USA	Roseberg, OR	13	74		Ammonia		
27.2	USA	Portland, OR				LPG	Transport	
1958								
15.4	USA	Ardmore, OK				Propane	Distribution	
22.1	USA	Niagara Falls, NY		>200		Nitromethane	Train Wagon	
3.1	Germany	Celle					Train Wagon	
1957								
24.10	USA	Sacramento, CA	1			LPG	Distribution	
8.1	Canada	Quebec, MO	1			Butane	Storage	
1956								
19.12	USA	Tonawanda, NY				Ethylene	Chemicals	
22.10	USA	Cottage Grove, OR	12	12		LPG	Storage	
7.8	Colombia	Cali	1,200	2,000		Explosives	Transport	
29.7	USA	Dumas, TX	19	32		Hydrocarbons	Storage	
26.7	USA	Baton Rouge, LA				Butylene	Alkylation unit	
1955								
27.8	USA	Whiting, IN	2	30		Naphtha		
22.7	USA	Wilmington, CA				Butane	Chemicals	
14.7	USA	Freeport, TX				Ethylene		
1954								
18.10	USA	Portland, OR				LPG	Train Wagon	
4.6	USA	Institute, WV				Acrolein	Train Wagon	16
1953								
23.9	USA	Tonawanda, NY	11	27				
6.8	Argentina	Campana	2			Hydrocarbons	Refinery	
1952								
22.12	USA	Bound Brook,NJ	5	21			Petrochemicals	
24.7	USA	Kansas City, KS				LPG	Distribution	
21.7	USA	Bakersfield, CA				Butane	Storage	
1951								
7.9	UK	Avonmouth	2			Oil	Storage	
16.8	USA	Baton Rouge, LA	2			Hydrocarbons		

[a] D I E - Number of Deaths, number of Injuries, number of people Evacuated.
[b] M$: Material Damage in million $ (in 2005 prices).
Note: LPG, liquefied propane gas.

- Continued

Table C3.8 (cont.). Major Industrial Accidents Caused by Explosions.

Date	Country	City/State	D	I	E[a]	Chemical	Unit	M$[b]
1951								
7.7	USA	Port Newark, NJ		14		Propane		
8.2	USA	St Paul, MN	14			LPG	Distribution	
1950								
15.8	USA	Wray, CO	2			Propane	Transport	
16.2	USA	Midland, MI				Butadiene	Chemicals	
1949								
30.12	USA	Detroit, IL	5			Propane	Cat cracker	
1948								
28.7	Germany	Ludwigshafen	245	2,500		Methyl ether	Transport	39
1947								
28.4	France	Brest	21			Ammonia		
20.2	USA	Los Angeles, CA	17	130			Electrolysis	
1945								
25.4	USA	Los Angeles, CA				Butane	Storage	
1944								
27.11	UK	Fauld	68	22		Explosives	Storage	
17.7	USA	Port Chicago, CA	321			TNT	Transport	
1943								
29.7	Germany	Ludwigshafen	57	37		Butadiene	Train Wagon	78
1942								
21.7	Belgium	Tessenderloo	>100			Ammonia	Chemicals	
1939								
12.12	USA	Witchita Falls, TX	1			Oil	Pipeline	
2.1	USA	Newark, NJ				Butane		
1934								
14.5	Hong Kong		65			Natural Gas		
1933								
10.2	Germany	Neunkirchen	63			Natural Gas	Natural Gas	
1927								
14.11	USA	Pittsburgh, PA	28	100		LNG	Refinery	

[a] D I E - Number of Deaths, number of Injuries, number of people Evacuated.

[b] M$: Material Damage in million $ (in 2005 prices).

Note: LNG, liquefied natural gas; LPG, liquefied propane gas.

- Continued

Table C3.8 (cont.). Major Industrial Accidents Caused by Explosions.

Date	Country	City/State	D	I	E[a]	Chemical	Unit	M$[b]
1921								
21.9	Germany	Oppau	430			Ammonia	Petrochemicals	
1917								
6.12	Canada	Halifax, NS	1,963	8,000		Explosives	Transport	

[a] D I E - Number of Deaths, number of Injuries, Number of people Evacuated.
[b] M$: Material Damage in million $ (in 2005 prices).

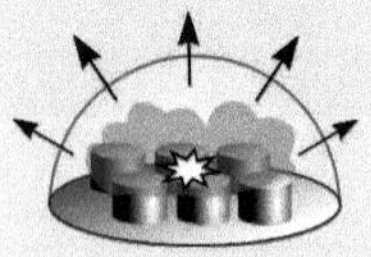

C4
BLEVE

The Boiling Liquid Expanding Vapor Explosion, known as BLEVE, is a consequence of the failure, because of an external cause (i.e., fire), of a pressurized vessel containing a gas or liquid stored in a higher-than-ambient pressure. The absorbed heat causes boiling and increase of the internal pressure, which in connection with the metal surface fatigue because of the increased temperature results in failure of the vessel and explosion. The released energy produces an intense shock wave, heat radiation and the rocketing of fragments and even whole vessels.

A BLEVE takes place in a vessel where the fluid stored is a gas at atmospheric pressure, but liquid at higher pressure (e.g., liquefied propane). The fluid is stored as a liquefied gas under pressure with its vapor filling the vessel's remaining space.

A series of explosions, VCE, fires and BLEVE completely destroyed the PEMEX LPG (Liquefied Propane Gas) Terminal, of 16,000 m^3 capacity in the outskirts of Mexico City on November 19, 1984. Final countdown was 550 deaths and over 6,400 wounded. Material damages were estimated at $34,000,000 in 2005 prices (see Example C3.1) [Lerdo de 1985].

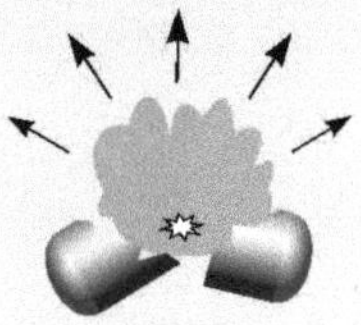

BLEVE

introduction

In the case of vessel failure with appearance of a crack (because of inside wall oxidation or other cause, like fire or wall weakening), the vapors that exist in the upper part of the vessel leak immediately result in a shock wave and a simultaneous drop of the pressure inside the vessel. This sudden drop of pressure causes intense boiling of the liquid inside the vessel resulting in the release of large quantities of vapors. The pressure of the released vapors can be quite large, resulting in turn, in a secondary large shock wave that is able to cause the explosion of the vessel and the rocketing of fragments in the surrounding area.

A BLEVE does not require the presence of a flammable fluid, and thus it is not considered a "chemical" explosion. In the case where a fluid is flammable, however, it is most probable that the intense release of gases will result, in addition to the explosion, in a fire ball with corresponding disastrous consequences.

The most common cause of a BLEVE is the submission of liquefied-gas storage vessels to fires that occurred in the plant. That is, BLEVEs usually take place in plants as a consequence of another incident where fire has started.

We can distinguish two types of BLEVEs, depending on their developments:

- 1-stage BLEVE

Usually encountered when the pressure increase of the enclosed gas (because of the external fire) is sufficient to create the initial crack and destroy the vessel. In such cases, the vessel's wall thickness is of the order of 4 mm and the outflow gas velocity from the crack is about 15 m/s.

- 2-stage BLEVE

In this case, gas is released from an initial small crack in the vessel's wall. The resulting pressure drop causes direct and intense boiling, release of large quantities of vapors, and a consequent pressure increase and the explosion of the vessel. It is encountered in vessels with thicker walls and the outflow gas velocity from the crack is about 1 m/s.

Figure C4.1 shows the resulting BLEVE in a tanker vehicle on June 22, 2002, at 13:30 on the C-44 road near the city of Tivisa in Catalonia, Spain [Planas-Cuchi *et al.* 2004]. The tanker, carrying natural gas, went off the road, probably because of speed. It capsized, slid onto its left side and finally stopped at a sand dune. Flames appeared immediately and 20 min later the tank exploded followed by a fire ball. The driver was killed while two persons (at a distance of 200 m) suffered 1st and 2nd degree burns. The tank was cylindrical, 2.33 m diameter, 13.5 m length and had been tested to a pressure of 9 bars. The temperature of the liquefied natural gas was -160°C and the pressure about 1 bar. The pressure developed during BLEVE was estimated to be about 10 bar.

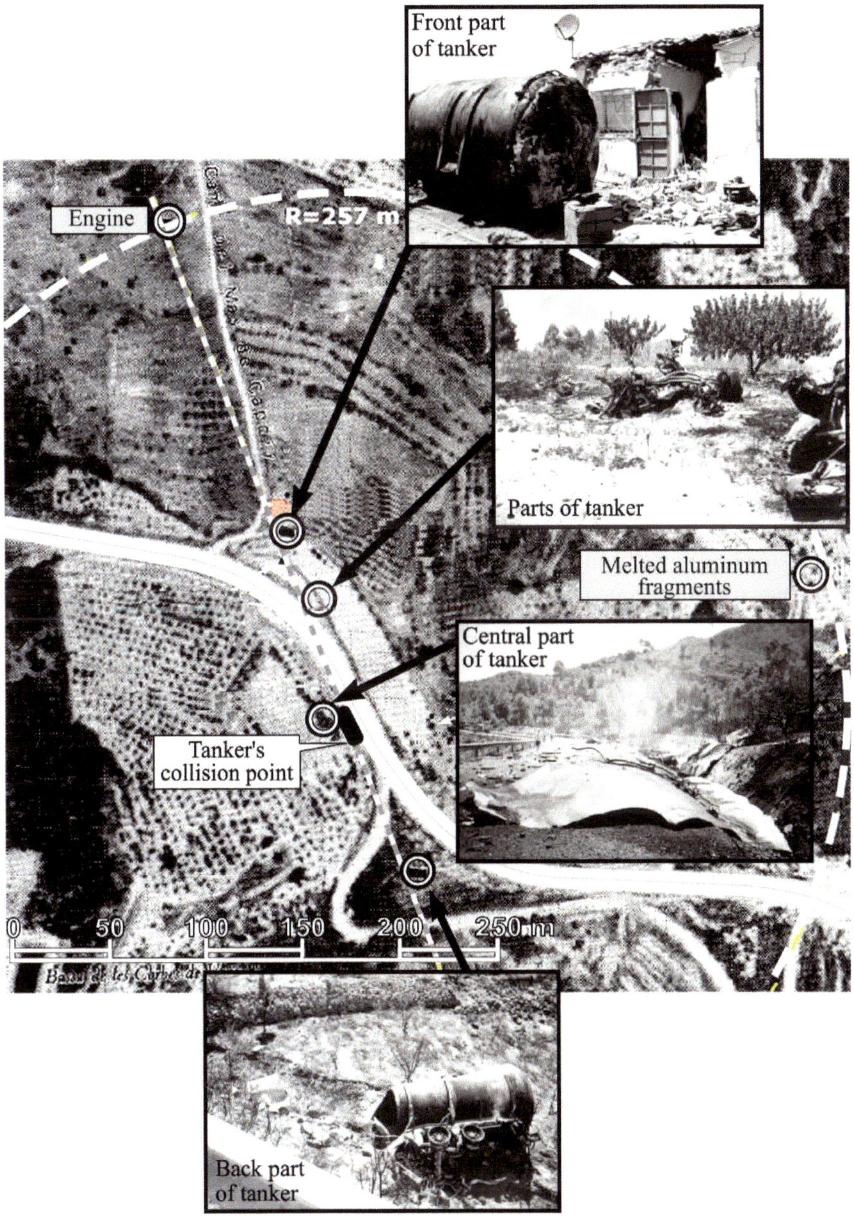

Figure C4.1. BLEVE in tanker vehicle [Planas-Cuchi *et al.* 2004].

BLEVE

introduction

In another case, a train wagon of 113 m^3 volume, carrying LPG (Liquefied Propane Gas) derailed and its wall was weakened in the collision. Forty hours later, because of a temperature increase, a BLEVE followed.

Most BLEVE cases are encountered when vessels are filled from one-half to two-thirds capacity with liquid. In these cases, the energy required for evaporation and expansion in relation to the ratio of part to the whole vessel can result in the rocketing of fragments up to 800 m, while the fire balls usually have diameters between 100 and 150 m. Death from heat radiation has been recorded at a distance of up to 75 m. The time between fire and BLEVE can range from minutes to hours.

C4.1. Estimation

Today, many investigators [Berg van den *et al.* 2004; Birk *et al.* 2006; Papazoglou & Aneziris 1999] work on the prediction and simulation of BLEVE in vessels. The most important consequences appear when the fluid is flammable, as in this case a fire ball is formed. The estimation of fire ball has already been discussed in Section C2.2, while the consequences from the heat radiated from a fire ball have been presented in Section C2.5.

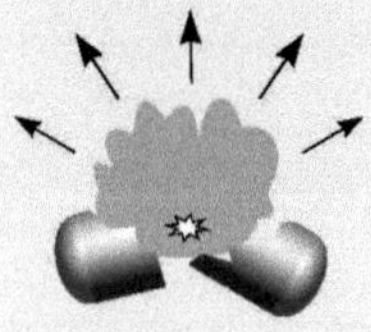

C4.2. Examples

C4.2.1. Major Industrial Accidents Caused by a BLEVE

Table C4.1 gives an indicative list of major industrial accidents caused by BLEVE. This list is only indicative, as a BLEVE can take place during any fire or series of explosions. The list is part of the database MinA - Major Industrial Accidents due to Fires, Explosions and Toxic-Gas Releases [MinA 2006].

Table C4.1. Major Industrial Accidents Caused by BLEVE.

Date	Country	City/State	D	I	E[a]	Chemical	Unit	M$[b]
1990								
1.4	Australia	Sydney			10,000		Storage	
1984								
19.11	Mexico	Ixhuatepec	550	23		LPG	Distribution	34
1982								
28.9	USA	Livingston, LA			3,000		Train Wagon	
1980								
21.4	USA	St. Joseph, MO	1	4		Grain		5
3.3	USA	Los Angeles, CA	2	2		Oil	Transport	
1979								
10.11	Canada	Mississauga, ON		8	215,000	Chlorine	Train Wagon	
8.9	USA	Paxton, TX		8			Train Wagon	
1977								
19.6	Mexico	Puebla				VCM	Storage	
1976								
26.11	USA	Belt, MT		22		LPG	Train Wagon	
1975								
1.9	USA	Des Moines, IA		3		LPG	Train Wagon	
31.8	USA	Gadsden, AL	4	28		Oil	Transport	
1974								
17.4	Germany	Bielefeld						
12.2	USA	Oneonta, NY		25		LPG	Train Wagon	
29.6	USA	Climax, TX	7			VCM	Train Wagon	

[a] D I E - Number of Deaths, number of Injuries, number of people Evacuated.

[b] M$: Material Damage in million $ (in 2005 prices).

Note: LPG, liquefied propane gas; VCM, vinyl chloride monomer.

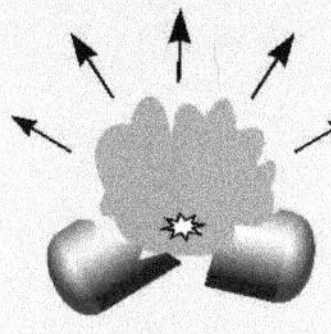

- *Continued*

Table C4.1 (cont.). Major Industrial Accidents Caused by BLEVE.

Date	Country	City/State	D	I	E[a]	Chemical	Unit	M$[b]
1974								
11.1	USA	West St. Paul, MN	4	6		LPG	Storage	
1972								
21.9	Brazil	Duque Caxias	37	53		Butane		
30.3	Brazil	Rio de Janeiro	37	53		Butane	Storage	
1971								
19.10	USA	Houston, TX	1	50		VCM	Train Wagon	
1968								
1.1	USA	Dunreith, IN		5		VCM	Train Wagon	
1966								
4.1	France	Feyzin	18	83		Propane	Refinery	114
1959								
2.6	USA	Deer Lake, PA	11	10		LPG	Train Wagon	
28.5	USA	McKittrick, CA		2		LPG	Storage	
1957								
8.1	Canada	Montreal, QC	1			Butane	Storage	
1956								
22.10	USA	Cottage Grove, OR	12	12		LPG	Storage	
29.7	USA	Dumas, TX	19	32		Hydrocarbons	Storage	
1951								
7.7	USA	Port Newark, NJ		14		Propane		

[a] D I E - Number of Deaths, number of Injuries, number of people Evacuated.

[b] M$: Material Damage in million $ (in 2005 prices).

Note: LPG, liquefied propane gas; VCM, vinyl chloride monomer.

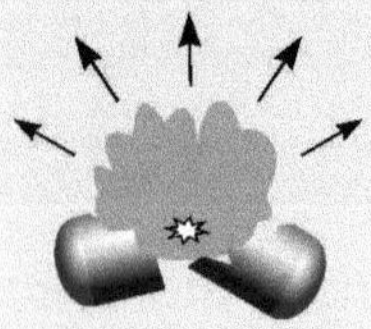

C5
Toxic Gas Dispersion

In this section the dispersion of a cloud of toxic gases in the atmosphere will be discussed. Emphasis will be given to the consequences of toxic dispersion either as the result of an industrial accident or a terrorist attack. Initially, major industrial accidents are presented, in addition to the most common toxic gases and their classification. Following that, the meteorological conditions that enter in the dispersion study are given. Finally, the methodology for calculating the dispersion of light and heavy gases from a continuous or instantaneous source is discussed.

Although the expressions that will be discussed can in general be also applied to atmospheric pollution, i.e., from vehicles, such cases will not be covered as there is a large amount of literature available on this topic.

C5.1. Types of Toxic Gases

As already mentioned, this work refers to the modeling of the dispersion of toxic gases encountered mostly in industrial accidents or terrorist attacks. Examples of such toxic gases are presented in brief in the following two sections.

C5.1.1. Toxic Gases from Industrial Accidents

The cause of the appearance of toxic gases following an industrial accident can be direct or indirect.

- Direct causes include leaks from a vessel, a pipeline, etc.
- Indirect causes can be another accident (i.e., fire) which had as a consequence the combustion of various materials and chemicals that resulted in the creation and release of toxic gases (i.e., dioxins).

In the next two subsections two very important recent industrial accidents which resulted in the release of toxic gases are presented in brief. Both accidents had a profound effect on the development of recent legislation for the prevention of toxic releases.

a) Seveso, Italy, 1976

On the morning of Saturday July 10, 1976, a safety valve vented on a reactor at the Industrie Chimiche Meda Societa Azionaria at Seveso, a town of 17,000 inhabitants situated about 15 miles from Milan, Italy. A white cloud of 50 m height, drifted over part of the town. Some 3,000 kg of chemicals were released, containing 2,4,5 trichlorophenoxyacetic acid and an undefined amount (between 100 g and 20 kg) of dioxin (2,3,7,8-tetrachloro dibenzo paradioxin, TCDD).

The release was not immediately addressed until the fourth day following the accident, when a child fell ill; on the fifth day, civil authorities declared a state of emergency in Seveso. An area of about two square miles was declared contaminated and people were asked to avoid contact with the vegetation or eating anything from this area. Small animals, like rabbits, began dying on a large scale. It took two weeks to realize that the cause of death was the dioxins. It is believed that about 200,000 people came in contact with the dioxin cloud. In addition to skin diseases, cancer cases increased significantly. About 4% of the animals in the area died, and those that did not - some 80,000 - were killed in order to avoid dioxins entering the food chain. Some 30,000 blood samples with dioxins were collected and these formed the basis of a very large on-going study on the dioxin consequences.

b) Bhopal, India, 1984

The pesticide unit of Union Carbide Corporation in Bhopal, India was commissioned by the Government of India in an effort to develop agricultural production and improve the trade deficit. Methyl isocyanate (MIC) is an intermediate product of their pesticide production and a very dangerous chemical. On the night of December 3, 1984, a large quantity of water accidentally entered the MIC storage tank, producing a catastrophic reaction. The contents of the storage tank became overheated and boiled, causing the relief valves to lift. The leak was initially noted at 11:30 p.m. from workers who got severe optical irritation. The foreman in charge took two hours to realize what was happening. During those two hours 40,000 kg of MIC (heavier than air) leaked into the atmosphere, and winds carried the toxic cloud to the nearby town of 90,000 inhabitants. Between 2,000 to 4,000 people died in their sleep, while the health of thousands of others - an estimated number between 20,000 and 400,000 - was affected to varying degrees. The survey on the consequences of MIC showed destruction of eyes and lungs, chronic bronchitis, gastritis, acute neurological problems, muscular malfunctions, and acute gynecological problems.

Union Carbide initially refused to pay the $220,000,000 compensation requested by the Indian government. After many court cases that lasted over 11 years, Union Carbide came to a compromise with the Indian government, in February 1999 on a final settlement of $470,000,000 (http://www.bhopal.net/).

C5.1.2. Toxic Gases in Terrorist Actions

The release of toxic gases as part of terrorist actions has, unfortunately, become quite important in recent times. On March 20, 1995, a terrorist group released a quantity of Sarin in three places on the Tokyo (Japan) underground, in the middle of rush hour, resulting in the death of 12 people and the injury of 980. Liquid Sarin was placed in metal beverage canisters, which the terrorists punctured with umbrellas as they got out of the train. This event and other similar ones, resulted in an increased demand for models that can simulate the dispersion of toxic gases in closed spaces.

The most common toxic gases that have been used so far or have been manufactured as chemical weapons are shown in Table C5.1. These can be classified in the following categories, which will only be briefly discussed here.

a) Nerve Agents

Nerve Agents, since World War II, are considered lethal chemical weapons. They are stable phosphorous-containing compounds, easily produced and disseminated, and extremely toxic with direct, fast consequences, as they intervene in the transmission of the neural impulses through the acetylcholine cycle. When disseminated as spray or aerosols, they form droplets that get absorbed by the skin, the eyes or by respiration, while as a gas, they are absorbed through respiration.

The transmission of neuromuscular impulses from the neural cells to the muscles takes place through acetylcholine (Figure C5.1). When this transmission is concluded, acetylcholine reacts with water in the presence of the cholinesterase enzyme, and is converted to oxalic acid and choline. Following this, another enzyme, acetylase, reconverts choline and oxalic acid to acetylcholine and the cycle is repeated. Nerve agents interfere in this cycle by neutralizing the cholinesterase. Hence acetylcholine remains active and continuously transmits nerve impulses. Muscle contractions do not stop, the heart gets overstressed and eventually fails or death is achieved by asphyxiation as control over respiratory muscles is also lost. Atropine injection is usually employed as an antidote for nerve agents, together with a reactivating agent. Atropine reduces the increased action of acetylcholine while the reactivating agent reactivates the cholinesterase enzyme.

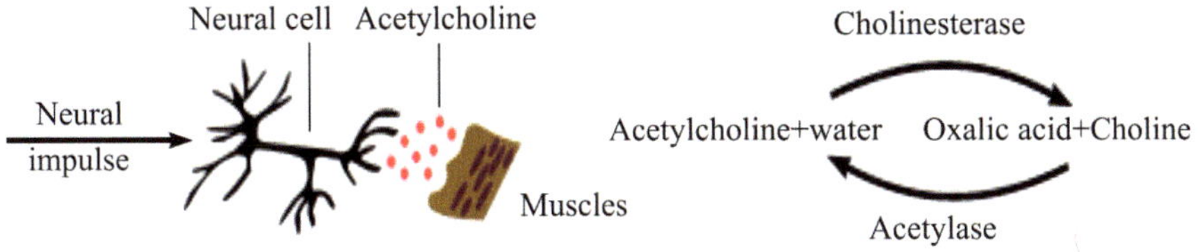

Figure C5.1. Effect of nerve agent on neural cells.

Table C5.1. Chemical Compounds Used or Designed as Chemical Weapons.

Code	Chemical Substance	LCt_{50} [a]	State [b]	Chemical Formula
	Nerve Agents			
GA	Tabun	70	L	$C_5H_{11}N_2O_2P$
GB	Sarin	35	L	$C_4H_{10}FO_2P$
GD	Soman	35	L	$C_7H_{16}FO_2P$
GF	Cyclo-Sarin	35	L	$C_7H_{14}FO_2P$
VX	Methylphosphonothiolate acid	15	L	$C_{13}H_{28}NO_3PS$
	Choking Agents			
CG	Phosgene	3,200	G	$COCl_2$
DP	Diphosgene	3,200	L	$C_2O_2Cl_3$
	Blister Agents			
HD	Sulfur Mustard gas	900	L	$C_4H_8Cl_2S$
HN-1	Nitrogen Mustard Gas	1,500	L	$(ClCH_2CH_8)_2NC_2H_5$
HN-2	Nitrogen Mustard Gas	3,000	L	$(ClCH_2CH_2)_2NCH_3$
HN-3	Nitrogen Mustard Gas	1,500	L	$(ClCH_2CH_2)_3N$
CX	Phosgene Oxime	3,200	L	CCl_2NOH
L	Lewisite	1,200-1,500	L	$C_2H_2AsCl_3$
HL	Mustard-Lewisite mixture	1,500	L	$C_4H_8Cl_2S + C_2H_2AsCl_3$
PD	Phenyldichloroarsine	2,600	L	$C_6H_5AsCl_2$
ED	Ethyldichloroarsine	4,000	L	$C_2H_3AsCl_2$
	Blood Agents			
AC	Hydrogen Cyanide	2,000-20,000	G	HCN
CK	Chlorocyanide	11,000	G	CNCl
SA	Arsine	5,000	G	AsH_3
	Tear Agents			
CN	Chloroacetophenone	7,000-14,000	S	C_8H_8ClO
CS	2-Chlorobenzalmalononitrile	61,000	S	$C_{10}H_5ClN_2$
CR	Dibenzoxazepine	> 61,000	S	$C_{13}H_9NO$
PS	Chloropicrin	2,000	L	Cl_3CNO_2
	Vomiting Agents			
DA	Diphenylchloroarsine	15,000	S	$(C_6H_5)_2AsCl$
DM	Adamsite	11,000	S	$C1_2H_9AsClN$
	Incapacitating Agents			
BZ	3-Quinuclidinyl benzilate	200,000	S	$C_{21}H_{23}NO_3$

[a] LCt_{50} (mg.min/m^3 - Median Lethal Concentration).
Represents the product of concentration (mg/m^3) of the chemical substance and the total exposure time necessary to cause death in 50% of the population exposed.

[b] State - State at 25°C: G - Gas, S - Solid, L - Liquid.

b) Choking Agents

Choking agents inflict injury mainly on the respiratory tract, including the nose, throat, and especially the lungs. They can all lead to pulmonary edema, respiratory and heart failure. Choking agents are in a liquid phase, initially held in a cell. Following explosion, they vaporize in a low-height cloud with a characteristic smell of freshly cut grass.

The symptoms, following a small exposure time, include initial failure of the respiratory system, which progressively can lead to death. Higher concentrations of these substances cause spasmodic coughing, suffocation and eventually death. The exact mechanism of the behavior of chocking agents is not exactly known, but it has been suggested that they block the action of enzymes, or produce HCl in the lungs' alveolus (fill or empty the air during respiration). Most probably, these substances (as they are highly active molecules), react directly with the alveolar-capillary membranes, leading to leakage of fluid from those capillaries into the interstitial portions of the lung. This disrupts the exchange of air, reduces the oxygen intake, and results in heart failure.

c) Blister Agents

Blister agents are named for their ability to cause severe chemical burns, resulting in large, painful water blisters on the bodies of those affected. Sulfur Mustard (known also as mustard gas) was used more than any other blister agent, because of its ease of production and its direct, very harmful and long lasting symptoms. It causes severe skin, eye and mucosal pain and irritation. Nitrogen mustards attack the skin directly. They were first produced after 1930, but there are no proofs up to today of their use in war actions.

d) Blood Agents

A blood agent or cyanogen agent is a chemical compound, carried by the blood for distribution through the body. Blood agents may contain the cyanide group, which can inactivate the energy-producing cytochrome oxidase enzymes of cells in the body. The term "blood agent" is a misnomer, because these agents do not typically affect the blood, but exert their toxic effect at the cellular level, by interrupting the electron transport chain in the inner membranes of mitochondria. The transfer of oxygen in the cells is disrupted and the cells die. Although they affect all tissues in the body, the most sensitive organs in the effect of cyanide are the central neural system and the heart.

Within seconds of exposure to high concentrations of cyanide gas an initial hyperpnea is followed by a loss of consciousness (within 30 s). This progresses to apnea (3-5 min), cessation of cardiac activity (5-8 min), and death. After exposure to lower concentrations, or exposure to lethal amounts via the oral or percutaneous routes, the effects develop more slowly.

e) Tear Agents

Tear agents (known also as Tear Gases or Lachrymatory Agents) are chemical compounds that stimulate the corneal nerves in the eyes to cause tearing, pain, and even temporary blindness, thus restricting the movement of people. They also irritate mucous membranes in the nose, mouth and lungs, and cause sneezing, coughing, etc. The effects persist as long as the concentration is high and diminish slowly as concentrations drop. They are mostly employed for crowd control.

The symptoms of tear agents are direct and fast. Normally they are found as solids of white color. They are stable compounds, can withstand heating and have very low vapor pressure. For this reason, they are usually disseminated as aerosols.

f) Vomiting Agents

Vomiting Agents are solids, primarily disseminated as aerosols. They have been produced for two purposes, as riot-control agents and as emesis-inducing agents to promote removal of personal protective gear during chemical warfare. The primary route of absorption is through the respiratory system. Exposure can also occur by ingestion, dermal absorption, or eye impact.

The effects of the vomiting agents by any route of exposure are slower in onset and longer in duration than typical riot control agents. On initial exposure, vomiting agents are slow irritants. By the time symptoms of irritation occur and personnel consider donning their protective equipment, significant contamination already may have occurred. Systemic symptoms (subsequently following) consist of headache, nausea, vomiting, diarrhea, abdominal cramps, and mental status changes. Symptoms typically persist for several hours after exposure. Death has been reported with excessive exposure. The protruding chlorine molecule can also disrupt the activity of the enzyme of cholinesterase and result in similar consequences to those of nerve agents.

g) Incapacitating Agents

Incapacitating agents produce temporary physiological or mental effects, or both, which will render individuals incapable of concerted effort in the performance of their assigned duties. The most common such agent is BZ (3-Quinuclidinyl benzilate) also known as Agent 15.

Incapacitation may result from physiological changes such as mucous membrane irritation, diarrhea, or hyperthermia, but also effects such as hallucinations or deep sleep are not unusual.

C5.2. Introduction to Cloud Dispersion

In this section, the effects of the various meteorological conditions will be discussed, as these affect the four types of dispersion that will be presented in this work. Following this, various dispersion models will be examined, and the selection of Gauss models for light gases and empirical relations for heavy gases will be justified.

C5.2.1. Meteorological Conditions

Irrespective of the model selected to describe the dispersion of the toxic gas in the atmosphere, meteorological conditions directly affect this dispersion because of air movement (winds at high altitudes strongly affect winds near the earth's surface) and atmospheric stability. An important role can also be attributed to temperature inversion. In the following subsections the above conditions will be examined.

a) Air Circulation

The circular shape of the earth is responsible for the uneven absorption of solar energy from the atmosphere and its surface. Without the transfer of heat, through the atmosphere and oceans, from the equator to the poles, the observed temperatures in the poles would have been much lower while those in the equator much higher.

Probably one of the first scientists to observe the movement of air was the Greek philosopher Thales of Miletus (624-546 BC), who in 586 BC reported on the hydrologic cycle. The hydrologic cycle or cycle of water consists of water evaporation into the air, the creation of clouds and the return of water to earth with rain. Two thousand years later, in 1686, the astronomer Edmond Halley (1656-1742, better known for the comet he discovered) suggested that near the Equator where solar heat is more intense, air rises towards places of lower temperature, while other colder air takes its place. He further proposed that the risen air followed the westward movement of the sun, while cold air had to follow east winds. The Halley theory of wind movement from east to west constituted a valid theory until 1735 when it was questioned by George Hadley (1685-1768), an English lawyer and amateur meteorologist.

Hadley agreed with Halley on the fact that hot air rises in the Equator and moves towards the Poles, while it gradually cools. As it cools, it descends and returns towards the Equator, because of the aforementioned rising movement on the Equator. These closed circulation loops of air behavior (see Figure C5.2) exist in both hemispheres. In contrast to Halley's hypothesis, Hadley proposed that air returns to the Equator from the east. His explanation is based on the fact that the earth rotates with high speed towards the east, and hence the atmosphere should

Gas Dispersion

Meteorological Conditions
↓
Air Circulation
↓
Atmospheric Stability
↓
Wind Speed

Temperature Inversion

Dispersion Models
↓
Light Gases Model Selection
↓
Heavy Gases Model Selection

follow this rotation. As, however, the earth's surface rotates with higher speed at the places where its diameter is larger, i.e., the Equator, the wind coming from the Poles will be delayed and hence it will appear to come from the east.

Hadley also gave an explanation for the west winds outside the tropical zones. If the wind in the Equator moved from west to east with a velocity higher than that near the Poles, then the wind moving towards the Poles would move faster than the earth's surface. Hence the wind in middle geographical latitudes (see Figure C5.2) will seem to come from the west.

Hadley's theories were mathematically proven and developed further in 1855, from William Ferrel (1817-1891) an American meteorologist who employed the theories of French mathematician Gaspard Gustave de Coriolis (1792-1843), published only 20 years before studying in detail the behavior of bodies in movement over a rotating surface.

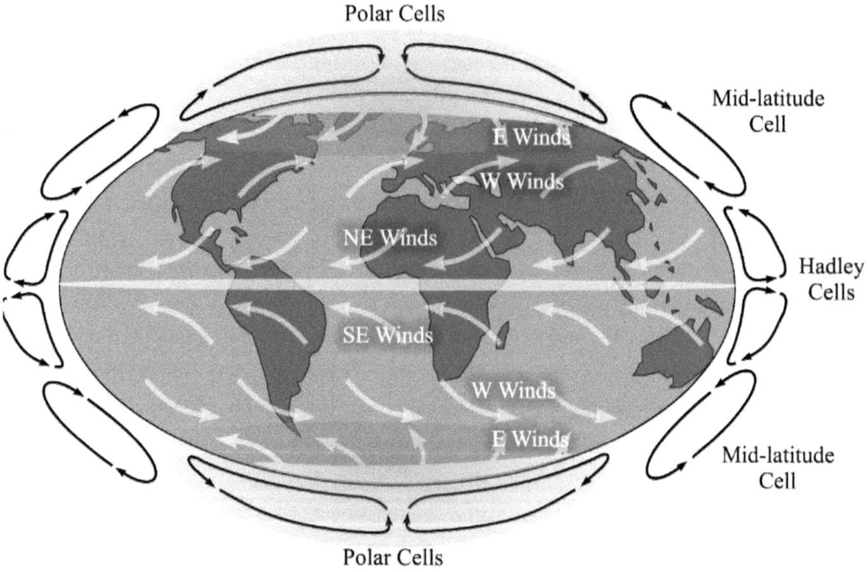

Figure C5.2. Earth's atmosphere global circulation.

b) Atmospheric Stability

In meteorology three main atmospheric conditions of the surface atmospheric layer are distinguished:

- unstable,
- neutral, and
- stable.

In Figure C5.3 these three atmospheric stability conditions are shown for a mass of air, which starts from the altitude marked by a circle, and moves vertically. In this particular altitude, the temperature of the air mass is equal to that of the environment.

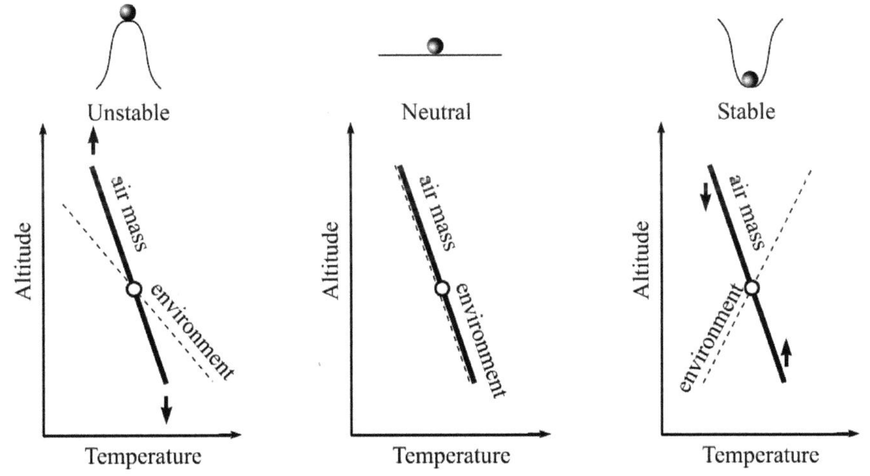

Figure C5.3. Atmospheric stability.

If the temperature of the moving air mass is equal to that of the environment (which means that densities are equal) then the air mass continues with its initial velocity (neutral stability) and no force is exerted on the air mass from the surrounding environment. During the unstable case, movement is directly related to the relation of the temperature of the air mass and the surrounding environment. If the temperature of the air mass is higher than that of the environment, it will move upwards, otherwise (in the opposite case) downwards. Under the stable conditions, the temperature relation works against any movement of the air mass.

In 1961 Pasquill [Turner 1994] presented a method of calculation of atmospheric stability (Table C5.2), which takes into consideration the buoyancy forces (due to solar radiation), and the relation of sunlight and clouds in conjunction with existing wind speed.

Gas Dispersion

Meteorological Conditions
↓
Air Circulation
↓
Atmospheric Stability
↓
Wind Speed

Temperature Inversion

Dispersion Models
↓
Light Gases Model Selection
↓
Heavy Gases Model Selection

Table C5.2. Pasquill Atmospheric Stability Classes.

Surface Wind Speed[a] (m/s)	Relative Cloud Coverage				
	Day			Night	
	0/8 - 2/8	3/8 - 5/8	6/8 - 8/8	< 3/8	> 4/8
< 2	A	A - B	B	F	F
2 - 3	A - B	B	C	E	F
3 - 5	B	B - C	D	D	E
5 - 6	C	C - D	D	D	D
> 6	C	D	D	D	D

[a] At a height of 10 m.

Neutral conditions, denoted by atmospheric stability class D, exist when the velocity of the moving air mass (wind speed) is large or there are clouds.

Unstable conditions can be categorized in three classes:
1) Very unstable, stability class A.
2) Unstable, stability class B.
3) Slightly unstable, stability class C.

Stable conditions can be separated into the following classes:
1) Slightly stable, stability class E.
2) Stable, stability class F.
3) Very stable, stability class G (also denoted as "-").

Based upon the above classification, in Table C5.2 the atmospheric stability is shown as a function of the wind speed and the relative cloud coverage. Table C5.2 is based upon the following assumptions:

- The case of strong solar radiation (relative cloud coverage 0), corresponds to the sun's radiation in the middle of the summer in England.
- Similarly, the case of low sun radiation (relative cloud coverage 8/8), corresponds to the sun's radiation in the middle of the winter in England.
- Night is the time period that starts one hour before sunset and ends one hour after sunrise.

Neutral conditions of atmospheric stability class D can in general be employed independent of wind speed, of cloud coverage during day or night, and for any meteorological condition during the night.

c) Wind Speed

The wind speed discussed so far referred to a particular reference height (usually 10 m). The power law is used to adjust the observed wind speed, u_{ref} (m/s), in a particular reference height, z_{ref} (m), to the wind speed, u_s (m/s), in another height, h_s (m). The wind power law is of the form [Turner 1994]

$$u_s = u_{ref}\left(\frac{h_s}{z_{ref}}\right)^p, \qquad \text{(C5.1)}$$

where p (-) denotes the wind profile exponent, depending upon the atmospheric stability. Values of exponent p are given in Table C5.3 for rural and urban areas.

Table C5.3. Wind Profile Exponent p, Eq. (C5.1).

Stability class	Rural area	Urban area
A	0.07	0.15
B	0.07	0.15
C	0.10	0.20
D	0.15	0.25
E	0.35	0.30
F	0.55	0.30

d) Temperature Inversion

The change of temperature in the atmosphere, as a function of altitude from the earth's surface, follows the shape shown in Figure C5.4. The dispersion of toxic gases or pollutants takes place in the troposphere, where the temperature is reduced with altitude. There are, however, cases where this temperature gradient is interrupted by small changes (i.e., increase of temperature with altitude), of thickness ranging from a few meters to a few kilometers, a phenomenon known as temperature inversion. This phenomenon usually appears in calm atmospheric conditions and in some cases at extremely low altitudes from the ground (e.g., in 100 m).

It is obvious that a cloud cannot pass through the temperature inversion zone, unless it has a speed from initial momentum. Hence it gets trapped either over or under the temperature inversion zone, depending on its release height. When the cloud is trapped under the temperature inversion zone, the entrapment is known as "fumigation."

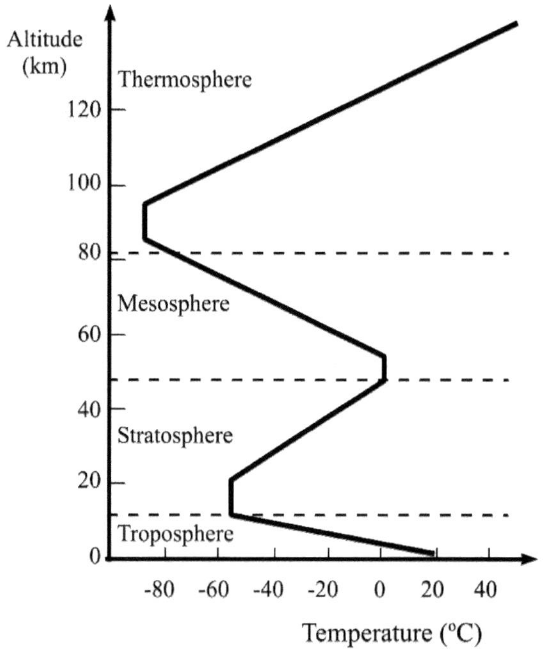

Figure C5.4. Change of temperature with altitude.

The opposite phenomenon is called "lofting." More analytically (see Figure C5.5),

- During fumigation, the temperature inversion takes place in a height larger than that of the stack. Hence, it forces the cloud to remain in low attitudes. This case is very dangerous for the areas around the cloud source.
- During lofting, the cloud cannot descend lower than the height where the temperature inversion takes place.

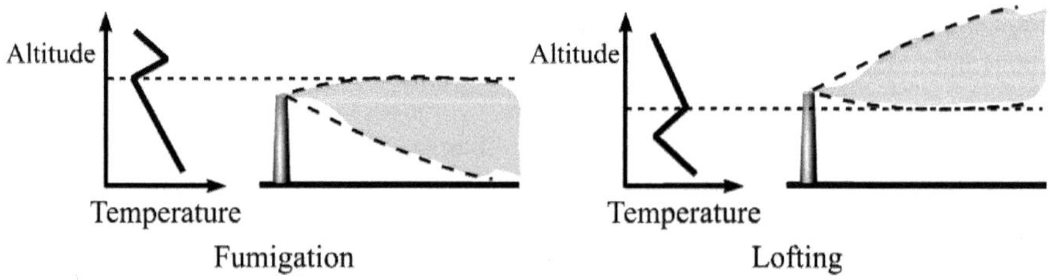

Figure C5.5. Fumigation and lofting phenomenon.

One of the reasons why industrial stacks are so tall is to avoid the possibility of pollutants encountering such an inversion layer and not dispersing properly.

C5.2.2. Dispersion Models

There are many possible categories of dispersion models. One can classify them according to the way the pollutant is produced (instantaneous, continuous source), or the spatial type of the source (point, line, area, volume source), the ground morphology and the atmospheric conditions, the composition of the pollutants (chemical, radioactive, etc.), their state (solid, liquid, gas) or the scale of their consequences (local, middle or large scale). The most scientific way of classification, however, is according to their mathematical approach. Hence we distinguish three categories:

- empirical models,
- Lagrangian models,
- Eulerian models.

Empirical models are not based on a full mathematical analysis but are clearly empirical. Lagrangian and Eulerian models are based upon the mathematical approach of transport phenomena, but they differ in relation to their reference system. More specifically, in Eulerian models, pollutants or pollution plume parcels are tracked in relation to a fixed spatial reference point fixed by the user, while in Lagrangian models pollutants or pollution plume parcels are tracked as they move, that is the reference point considered to move with the pollutants. Some more information for each type of models is given in the following subsections.

a) Empirical Models

This class of models includes most of the models employed today for environmental control reasons. They can be further classified as:

- Gaussian models.
- Box models.

The mathematical form of the Gaussian models starts from the mathematical solution of the theoretical problem of an emission from a continuous point source in an infinite ideal medium where steady dispersion takes place. The various parameters are changed to empirical parameters to accommodate phenomena such as deposition, various losses, etc. There is an abundance of models of this type. The large advantage of these models [Turner 1994] is their simplicity and the very small computing times required for the calculations.

Accumulated experience makes these models particularly valuable and irreplaceable in simple atmospheric conditions with sufficient wind, normal topography and uniform use of land. On the other hand, it is easy for someone to realize the "abuse" of the use of Gaussian models in literature and especially on the

Meteorological Conditions
↓
Air Circulation
↓
Atmospheric Stability
↓
Wind Speed

Temperature Inversion

Dispersion Models
↓
Light Gases Model Selection
↓
Heavy Gases Model Selection

Internet, where they are applied over conditions for which they are not designed; much more in cases where the study of multiple scenarios is required coupled with the need for fast computing times.

The box model is the simplest of the model types. It assumes that the airshed (i.e., a given volume of atmospheric air in a geographical region) is in the shape of a box. It also assumes that the air pollutants inside the box are homogeneously distributed and uses that assumption to estimate the average pollutant concentrations anywhere within the box area. Although useful, this model is very limited in its ability to accurately predict dispersion of air pollutants over an airshed because the assumption of homogeneous pollutant distribution is much too simple. These models have been employed primarily for the dispersion of heavy gases from particular industries that needed to conform to the European Union's SEVESO directive.

b) Lagrangian Models

In this class of models, the algorithm mathematically follows pollution plume parcels (also called particles) as the parcels move in the atmosphere and they model the motion of the parcels as a random walk process. The Lagrangian model then calculates the air pollution dispersion by computing the statistics of the trajectories of a large number of the pollution plume parcels. A Lagrangian model uses a moving frame of reference as the parcels move from their initial location. It is said that an observer of a Lagrangian model follows along with the plume.

c) Eulerian Models

In this class of models, the mathematical approach is based on the usual differential equations of continuity, integrated over the turbulent time scale. Eulerian models have the flexibility to deal with problems where atmospheric phenomena and pollutant distribution are complex (three dimensional flows, buoyancy phenomena, phase changes) and the topography is complicated. However, their complexity demands high computing times, making them not attractive in environmental controlling situations.

Increased computing capability is also required for point sources where a high spatial resolution is required near the source in order to obtain valid results. Eulerian models are invaluable for area or spatial sources and for large-scale spatial calculations.

C5.2.3. Selection of Gaussian Model for Light Gas Dispersion

The model that will be employed in the study of the dispersion of light gas pollutants is an empirical Gaussian model. The reasons for selecting this model are the following:

- it produces results that agree with experimental data as well as any other similarity function,
- mathematical calculations on the Gaussian equation are relatively easy,
- it is consistent with the random nature of turbulence,
- other "theoretically based" similarity functions include a much larger degree of empiricism in their final stages.

According to the aforementioned reasons, Gaussian models are adopted by most of the national environmental agencies [EPA–454/B–95–003a, 1995].

One can easily understand that the validity of the aforementioned models is a function of many parameters. Some of these parameters are discussed in the following section.

The presence of large pollutant sources near the sea shore can lead to high and persistent concentrations in the area. The phenomenon is attributed to the formation of the atmospheric boundary layer* which prevents the vertical mixing of pollutants. The height of the atmospheric boundary layer over the sea is typically of the order of 100 m while over land is typically of the order of 1,500 to 2,000 m. In the seashore area, the movement of cold air from the sea restricts the height of the boundary layer which is very low near the seashore, and thus restricts the vertical dispersion of pollutants. Hence the presence of a large point source near the seashore produces the risk of fumigation.

Urban territories differ from rural ones in relation to their thermal characteristics and the surface roughness. More important are the thermal differences, which are attributed to the anthropogenic sources of heat as well as to the fact that road pavement and concrete walls heat faster and accumulate more heat than grass or ground due to their high thermal diffusivity. As a consequence of this, urban areas are characterized by higher temperatures than rural ones, and especially during the night. It is also worth noting that in many cases, during the night, unstable meteorological conditions are observed in urban territories because of these phenomena.

* The atmospheric boundary layer (ABL), also known as the planetary boundary layer (PBL), is the lowest part of the atmosphere and its behavior is directly influenced by its contact with a planetary surface. On Earth it usually responds to changes in surface forcing in an hour or less. In this layer physical quantities such as flow velocity, temperature, moisture, etc. display rapid fluctuations (turbulence) and vertical mixing is strong.

Gas Dispersion

Meteorological Conditions
↓
Air Circulation
↓
Atmospheric Stability
↓
Wind Speed

Temperature Inversion

Dispersion Models
↓
Light Gases Model Selection
↓
Heavy Gases Model Selection

C5.2.4. Selection of Empirical Models for Heavy Gas Dispersion

The dispersion of heavy gases cannot be described by a simple Gaussian model as in this case many complex phenomena take place. A more analytic presentation of these phenomena will be carried out in Section C5.5.

For the description of the dispersion of heavy gases, we will use a series of empirical diagrams resulting from an extensive number of large-scale experiments. The main reasons for this choice are similar to those discussed for Gaussian models in the case of light gases:

- The model must be suitable for a personal computer (both ease of use and speed of results).
- The results must be validated with corresponding measurements and similar cases.

For the dispersion of heavy gases, one could also use some of the box models that are freely distributed by the Environmental Protection Agency, EPA (http://epa.gov/scram001). Unfortunately, however, these algorithms include complex mathematical expressions that are beyond the scope of this book.

Toxic plume from a fire in Moldes Barcelona plastic plant, in Polinya, Catalonia Spain, on March 14, 2008. (Reproduced by kind permission of S. Busquets.)

C5.3. Light Gas Dispersion from a Continuous Source

Continuous Source (Light Gas)

calculation procedure

Rise due to Buoyancy or Momentum
↓
Distance of Maximum Plume Rise
↓
Gradual and Final Rise
↓
Dispersion Equation

A source producing a continuous steady release of pollutants, known as a plume, is classified as a continuous source. In other words, in a continuous plume the release and the sampling time are longer compared with the travel time. In the next subsections we will discuss the calculation of the plume rise of gases lighter than air, and subsequently the equations describing its dispersion will be presented.

C5.3.1. Plume Rise

An important variable that ought to be taken into consideration in calculating the plume dispersion is the actual height of the source from the ground. This height could be the actual stack height, or the actual height at which the release takes place. The plume rise is directly related to the momentum of the release that is the product of the mass of the gas released times its exit velocity.

Because of the difference in densities between the gas that is being released and the surrounding air, it is probable that a further rise of the plume just after its release will occur, because of buoyancy forces. If the gas released is warmer than the surrounding air (case of gases produced in a fire), the effect of buoyancy forces will be greater because of its lower density. In general, if the exit temperature of the gas is higher by 10 to 15°C from the surrounding air, then the plume rise attributed to buoyancy forces will be greater than that attributed to momentum forces. It should also be mentioned that the effect of momentum forces does not last more than 30 to 40 s after the release, while the effect of buoyancy remains stable until enough air has been mixed with the released gas to drop its temperature to that of the surrounding air. Depending on the degree of turbulence, the effect of buoyancy lasts between 3 to 4 min.

In order to calculate the plume rise, one must:

a) examine whether the predominate phenomenon in plume rise is the buoyancy or momentum, and
b) calculate the distance of maximum plume rise (measured downwind from the plume source) and the plume's rise before and after that point.

In the special case of stacks, an additional parameter that must be taken into consideration in the plume rise is the phenomena associated with the low pressure that exists in the stack tip. A vertical vortex appears near the stack tip, and as a result of this, the effective stack height is lower than the actual one. Hence a correction of the stack height is given by the Briggs equations [Briggs 1975, Turner 1994]. This correction is accomplished by the empirical subtraction of a portion of the actual stack height. The effective stack height, h_s' (m), is calculated from the actual stack height, h_s (m), as a function of the gas exit speed, v_s (m/s),

Continuous Source (Light Gas)

calculation procedure

Rise due to Buoyancy or Momentum
↓
Distance of Maximum Plume Rise
↓
Gradual and Final Rise
↓
Dispersion Equation

the wind speed at the stack height, u_s (m/s), and the internal diameter, d_s (m), of the stack. This correction is only applied in the case when $(v_s/u_s) < 1.5$.

$$h_s' = h_s + 2\, d_s \left(\frac{v_s}{u_s} - 1.5 \right). \tag{C5.2}$$

As it can be seen in the above equation, the maximum correction applied is equal to $-3d_s$.

a) Plume Rise Due to Buoyancy or Momentum

Plume rise occurs because of buoyancy (instead of momentum), when the following empirical criterion [Briggs 1975, Bowers *et al.* 1979] is satisfied

$$\left(T_s - T_a\right) \geq \Delta T_c, \tag{C5.3}$$

where, T_s (K) denotes the temperature of the released gas and T_a (K) the temperature of the surrounding air. The calculation of the critical difference ΔT_c (K) is accomplished by the empirical equations of Table C5.4. In the equations of Table C5.4, g (m/s^2) denotes the acceleration due to gravity (= 9.81 m/s^2) while $(\partial\theta / \partial z)$ (K/m) denotes the change in temperature with height (empirically it takes the value of 0.020 K/m for stability conditions E, and the value of 0.035 K/m for stability conditions F). Moreover, F_b (m^4/s^3) is the buoyancy flux parameter.

Table C5.4. Equations for Selecting Plume Rise Because of Buoyancy or Momentum[a].

Unstable (A,B,C) or Neutral (D) Conditions	$\Delta T_c = 0.0297\, T_s \left(\frac{v_s}{d_s^2}\right)^{1/3}$	where $F_b < 55$	(C5.4)
	$\Delta T_c = 0.00575\, T_s \left(\frac{v_s}{d_s^2}\right)^{1/3}$	where $F_b \geq 55$	(C5.5)
	where $F_b = g\, v_s\, d_s^2 \left(\frac{T_s - T_a}{4\, T_s}\right)$		(C5.6)
Stable (E,F) Conditions	$\Delta T_c = 0.019582\, T_s v_s \sqrt{s}$	and $s = \frac{g}{T_a}\frac{\partial\theta}{\partial z}$	(C5.7)

[a] Equations are empirical and values should be entered only at specified units.

Continuous Source (Light Gas)

calculation procedure

Rise due to Buoyancy or Momentum
↓
Distance of Maximum Plume Rise
↓
Gradual and Final Rise
↓
Dispersion Equation

b) Distance of Maximum Plume Rise

The maximum plume rise distance, x_f (m), is the downwind distance from the source where the plume has reached its maximum rise. Having established from the previous section the type of plume rise, i.e., whether the driving force is buoyancy or momentum, the calculation of the total plume rise will be separated into two parts:

- gradual rise until the distance x_f, and
- final plume rise (the rise after the distance x_f).

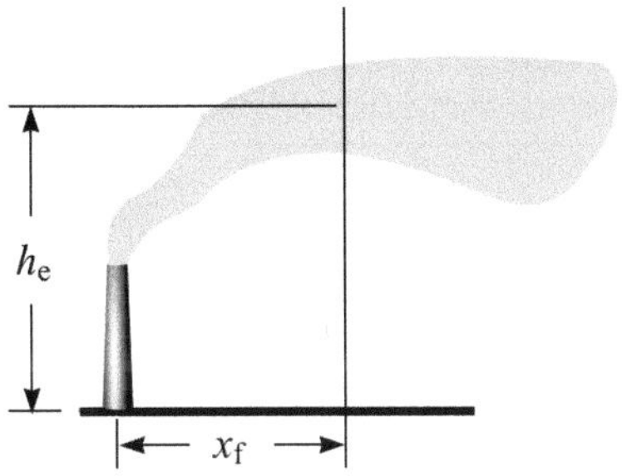

Figure C5.6. Distance of maximum plume rise.

Calculations are entirely empirical [Bowers *et al.* 1979] and are separated, as before, depending on the atmospheric conditions and as to whether the plume rise driving force is buoyancy or momentum. The equations are shown in Table C5.5.

Table C5.5. Equations for Calculating Distance, x_f, of Maximum Plume Rise[a].

	Unstable (A,B,C) or Neutral (D) Conditions			Stable (E,F) Conditions	
Buoyancy	$F_b < 55$	$x_f = 49\, F_b^{5/8}$	(C5.8)	$x_f = 2.0715 \dfrac{u_s}{\sqrt{s}}$	(C5.10)
	$F_b \geq 55$	$x_f = 119\, F_b^{2/5}$	(C5.9)	where $s = \dfrac{g}{T_a} \dfrac{\partial \theta}{\partial z}$	
Momentum	$F_b = 0$	$x_f = \dfrac{4 d_s (v_s + 3 u_s)^2}{v_s\, u_s}$	(C5.11)	$x_f = 0.50\, \pi \dfrac{u_s}{\sqrt{s}}$	(C5.14)
	$F_b < 55$	$x_f = 49\, F_b^{5/8}$	(C5.12)		
	$F_b \geq 55$	$x_f = 119\, F_b^{2/5}$	(C5.13)		

[a] Equations are empirical and values should be entered only at specified units.

Continuous Source (Light Gas)

calculation procedure

Rise due to Buoyancy or Momentum
↓
Distance of Maximum Plume Rise
↓
Gradual and Final Rise
↓
Dispersion Equation

c) Gradual and Final Plume Rise

Having calculated the distance x_f, the gradual plume rise before x_f and the final plume rise after x_f can be calculated. Table C5.6 shows the plume rise, h_e (m), for the two types of meteorological conditions, as a function of the distance, x (m), from the source [Bowers *et al.* 1979, Briggs 1975].

Table C5.6. Equations for Gradual and Final Plume Rise[a].

	Unstable (A,B,C) or Neutral (D) Conditions	Stable (E, F) Conditions
	Gradual plume rise (rise until distance x_f)	
Buoyancy	$h_e = h_s' + 1.60 \frac{(F_b x^2)^{1/3}}{u_s}$ (C5.15)	
Momentum	$h_e = h_s' + 1.60 \left(\frac{3F_m x}{\beta_j^2 u_s^2}\right)^{1/3}$ (C5.16)	$h_e = h_s' + \left(3F_m \frac{\sin(x\sqrt{s}/u_s)}{\beta_j^2 u_s \sqrt{s}}\right)^{1/3}$ (C5.17)
	where $\beta_j = \frac{1}{3} + \frac{u_s}{v_s}$, $s = \frac{g}{T_a}\frac{\partial \theta}{\partial z}$, $F_m = v_s^2 d_s^2 \left(\frac{T_\alpha}{4T_s}\right)$ (C5.18)	
	Final plume rise (rise after distance x_f)	
Buoyancy	$F_b < 55$: $h_e = h_s' + 21.425 \frac{F_b^{3/4}}{u_s}$ (C5.19) $F_b \geq 55$: $h_e = h_s' + 38.710 \frac{F_b^{3/5}}{u_s}$ (C5.20)	$h_e = h_s' + 2.6 \left(\frac{F_b}{u_s s}\right)^{1/3}$ (C5.21)
Momentum	$h_e = h_s' + 3d_s \frac{v_s}{u_s}$ (C5.22)	$h_e = h_s' + 3d_s \frac{v_s}{u_s}$ (C5.23) $h_e = h_s' + 1.5 \left(\frac{F_m}{u_s \sqrt{s}}\right)^{1/3}$ (C5.24) The smaller of the two values is selected[b]

[a] Equations are empirical and values should be entered only at specified units.

[b] The smaller of the two values is selected as the rise at stable conditions can not be higher than the rise at unstable or neutral conditions.

EXAMPLE C5.1. *Plume Rise Calculation*

Calculate the plume rise of pollutants released from a stack, as a function of the distance from the stack. The release took place in a rural area during the night. The following data are available:

-	Cloud coverage	: 5/8	
	Wind speed (10 m), u_{ref}	: 2	m/s
-	Ambient temperature, T_{a}	: 280	K
	Gas exit temperature, T_{s}	: 400	K
-	Stack diameter, d_s	: 1	m
	Stack height, h_{s}	: 25	m
	Exit gas speed, v_{s}	: 4	m/s

Since the release takes place during the night with a cloud coverage of 5/8 and a wind speed at 10 m equal to 2 m/s, from Table C5.2, the atmospheric conditions according to Pasquill are "stable type F."
The wind speed at the top of the stack (25 m) is calculated from Eq. (C5.1) with exponent $p = 0.55$ (Table C5.3, rural area, stability F),

$$u_{\text{s}} = u_{\text{ref}} \left(\frac{h_{\text{s}}}{z_{\text{ref}}} \right)^{p} = 3.31 \text{ m/s}.$$

As the ratio $(v_{\text{s}}/u_{\text{s}}) = 1.21 < 1.5$, a correction to the actual stack height is required. From Eq. (C5.2)

$$h_S' = h_S + 2\, d_{\text{S}} \left(\frac{v_{\text{S}}}{u_{\text{S}}} - 1.5 \right) = 24.4 \text{ m}.$$

a) Plume Rise Because of Buoyancy or Momentum
The criterion for the selection as to whether the plume rise occurs because of buoyancy or momentum forces is given by Eq. (C5.3) as

$$\left(T_{\text{s}} - T_{\text{a}}\right) \geq \Delta T_{\text{c}}.$$

For stable conditions, from Table C5.4 and Eq. (C5.7)

$$\Delta T_{\text{c}} = 0.019582\, T_{\text{S}} v_S \sqrt{s} = 1.097 \text{ K} \qquad \text{for} \quad s = \frac{g}{T_{\text{a}}} \frac{\partial \theta}{\partial z} = 0.00123 \text{ s}^{-2}$$

(where $(\partial\theta / \partial z) = 0.035$ K/m, for stability F).
Therefore, since $(T_{\text{s}} - T_{\text{a}}) = 120$ K >> $\Delta T_{\text{c}} = 1.097$ K, buoyancy dominates.

Continuous Source (Light Gas)

example

Rise due to Buoyancy or Momentum
↓
Distance of Maximum Plume Rise
↓
Gradual and Final Rise
↓
Dispersion Equation

Continuous Source (Light Gas)

example

Rise due to Buoyancy or Momentum
↓
Distance of Maximum Plume Rise
↓
Gradual and Final Rise
↓
Dispersion Equation

b) Distance of Maximum Plume Rise

The maximum plume rise distance, x_f (m), is obtained from Eq. (C5.10), as

$$x_f = 2.0715\frac{u_s}{\sqrt{s}} = 195.8 \text{ m}.$$

c) Gradual and Final Plume Rise

Having calculated the distance x_f, the gradual plume rise can be calculated from Eq. (C5.15)

$$h_e = h_s' + 1.60\frac{(F_b x^2)^{1/3}}{u_s},$$

$$\text{where} \quad F_b = g\, v_s\, d_s\left(\frac{T_s - T_a}{4T_s}\right) = 2.94 \text{ m}^4/\text{s}^3,$$

and the final plume rise from Eq. (C5.21)

$$h_e = h_s' + 2.6\left(\frac{F_b}{u_s\, s}\right)^{1/3} = 47.8 \text{ m}$$

Calculated results are shown in Figure C5.7.

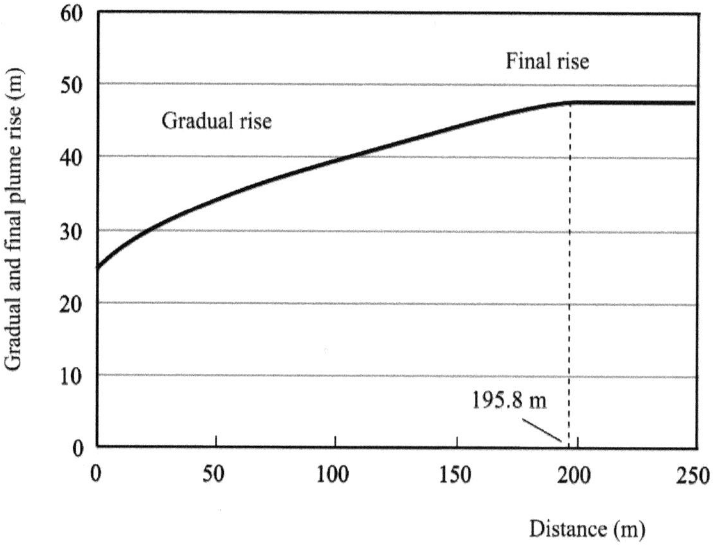

Figure C5.7. Plume rise as a function of distance from the source.

❐

C5.3.2. Dispersion Equation

The Gaussian model that will be presented is a steady-state model, applied to urban or rural areas according to user's needs, that refers to a continuous point source release. The model is based upon the following assumptions:

1) The mass release rate (in units of mass per time) is continuous and stable with time.
2) The "release time" (the time from the start of the release until the time examined) is equal or longer than the "travel time" (time required for the cloud to travel from its source to the point of interest) in a downwind direction, so that the dispersion in this direction can be neglected.
3) The material being dispersed is a stable gas or aerosol, with a particle diameter smaller than 20 μm, and remains airborne for a long time.
4) During the cloud's travel from the source to the point of interest, the mass released by the source remains stable and airborne. None of the gases reacts or remains on the ground because of reactions, gravity or turbulent collision. It is assumed that the part of the cloud dispersed close to the ground surface by turbulent eddies is again dispersed away from the ground by other subsequent eddies. This phenomenon is called eddy reflection.
5) Meteorological conditions are assumed stable with time, or at least during the cloud's travel from the source to the receptor, that is about 1 hour, on average. This assumption becomes more realistic in cases when the receptor (the person in the distance in question) is near and under normal conditions. Contrary to that, in cases of light wind or when the receptors are far away, this assumption may not be satisfied.
6) The wind speed and direction are considered stable with height.
7) The influence of the wind's shear stresses in the horizontal direction is not taken into consideration (it only becomes an influence after 10 km).
8) Dispersion parameters are assumed to be independent of the vertical direction z, but are a function of distance x (and wind speed).
9) Concentrations downwind in any crosswind (in the vertical or horizontal plane) distance are described by a Gaussian distribution.

Continuous Source (Light Gas)

calculation procedure

Rise due to Buoyancy or Momentum

↓

Distance of Maximum Plume Rise

↓

Gradual and Final Rise

↓

Dispersion Equation

Continuous Source (Light Gas)

calculation procedure

Rise due to Buoyancy or Momentum
↓
Distance of Maximum Plume Rise
↓
Gradual and Final Rise
↓
Dispersion Equation

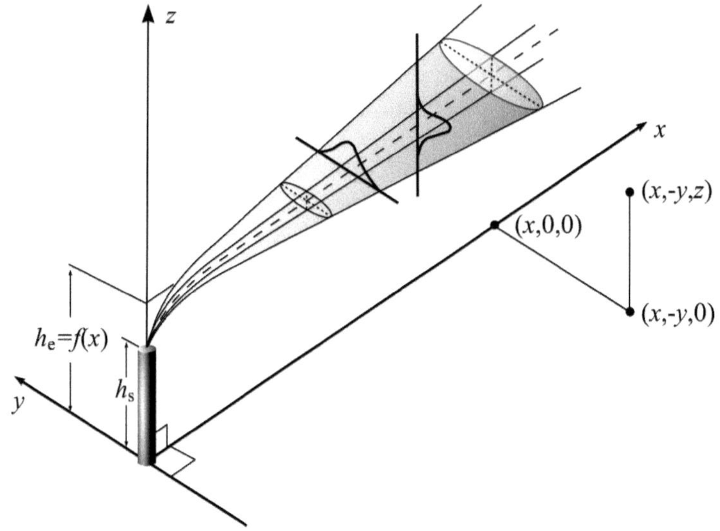

Figure C5.8. Gaussian dispersion coordinates system.

The center of coordinates for a continuous elevated source (Figure C5.8) is defined at the ground of the stack or at the ground projection of the point of the source. The x-axis is positive downwind from the emission source point, the y-axis is crosswind from the emission plume centerline, while the z-axis is vertical.

a) Concentration Equation

For a steady-state plume Gaussian dispersion, the concentration at a point with coordinates (x, y, z), where x (m) is the downwind distance from the emission source point, y (m), the crosswind distance from the emission plume centerline and z (m) the height above ground level, is given by the equation [EPA–454/B–95–003a 1995, Turner 1994]:

$$C = \frac{Q_c}{u_s}\frac{10^9}{2\pi\sigma_y}\exp\left[-\frac{y^2}{2\sigma_y^2}\right]\frac{1}{\sigma_z}\left\{\exp\left[-\frac{(h_e - z)^2}{2\sigma_z^2}\right] + \exp\left[-\frac{(h_e + z)^2}{2\sigma_z^2}\right]\right\}, \quad (C5.25)$$

where C (μg/m^3) is the concentration of the gas pollutant, Q_c (kg/s), the source pollutant emission rate, u_s (m/s), the horizontal wind speed along the plume centerline (constant in time and space), σ_z (m), the vertical dispersion coefficient and σ_y (m), the lateral dispersion coefficient. The plume rise, h_e (m), in the above equation is a function of the distance x (m) and represents the gradual and final rise

Table C5.7. Terms of Eq. (C5.25).

source emission rate	Q_c
downwind dispersion	$\dfrac{1}{u_s}$
lateral dispersion	$\dfrac{1}{\sqrt{2\pi}\,\sigma_y}\exp\left[-\dfrac{y^2}{2\sigma_y^2}\right]$
vertical dispersion	$\dfrac{1}{\sqrt{2\pi}\,\sigma_z}\left\{\exp\left[-\dfrac{(h_e-z)^2}{2\sigma_z^2}\right]+\exp\left[-\dfrac{(h_e+z)^2}{2\sigma_z^2}\right]\right\}$

of the plume respectively. The coefficient 10^9 enters so that final concentration units are μg/m^3. Every term of Eq. (C5.25) expresses a different phenomenon as can be seen in Table C5.7.

In Eq. (C5.25) the following observations can be made:

a) Concentrations at receptors are directly proportional to emission rate.
b) Downwind (x-axes) concentrations are inversely proportional to the wind speed.
c) Crosswind (lateral, y-axis) concentrations are inversely proportional to the lateral dispersion coefficient, σ_y. The larger the lateral distance from the plume centerline, the larger the lateral dispersion coefficient, σ_y, and hence the lower the concentration will be. The exponential term that incorporates the ratio (y/σ_y) corrects for the distance of the receptor from the dispersion centerline (y is equal to zero in the dispersion centerline, i.e., exactly on the x-axis).
d) Crosswind (vertical, z-axis) concentrations are inversely proportional to the vertical dispersion coefficient, σ_z. The larger the vertical distance from the plume centerline, the larger the vertical dispersion, and thus the lower the concentration will be. The addition of the two exponential terms in the vertical dispersion represent the plume's symmetry with reference to the xy-plane.
   The first term, (h_e - z), represents the direct distance of the receptor from the plume centerline (Figure C5.8). The second term represents the distance of the eddy reflectivity of the receptor from the plume centerline, plus the distance z to return to the receptor after the reflection (h_e is the plume rise, i.e., the distance between the ground and the plume centerline).

Continuous Source (Light Gas)

calculation procedure

Rise due to Buoyancy or Momentum
↓
Distance of Maximum Plume Rise
↓
Gradual and Final Rise
↓
Dispersion Equation

b) Dispersion Coefficients

Pasquill (1961) proposed the following empirical equations for the calculation of the dispersion coefficients σ_z (m) and σ_y (m), as a function of the distance x (m) for

urban areas $\sigma_y = e\,x(1+c_1\,x)^{-1/2}$ $\sigma_z = f\,x(1+g\,x)^h$ (C5.26)

rural areas $\sigma_y = c_2\,x\,\tan(TH)$ $\sigma_z = a(c_3\,x)^b$ (C5.27)

where $TH = c_4[c - d\ln(c_3\,x)]$. Values for parameters *a, b, c, d, e, f, g, h* are given in Table C5.8 [EPA–454/B–95–003a 1995, Turner 1994]. Also $c_1 = 0.0004$ m^{-1}, $c_2 = 0.4651$, $c_3 = 0.001$ m^{-1} and $c_4 = 0.01745$.

Table C5.8. Parameters of Eqs. (C5.26)-(C5.27) as a Function of Atmospheric Stability.

Atm. Stab.	*x* (m)	*a* (m)	*b* (-)	Atm. Stab.	*x* (m)	*a* (m)	*b* (-)
A	< 100	122.800	0.94470	E	< 100	24.260	0.83660
	100 - 150	158.080	1.05420		101 - 300	23.331	0.81956
	151 - 200	170.220	1.09320		301 - 1,000	21.628	0.75660
	201 - 250	179.520	1.12620		1,001 - 2,000	21.628	0.63077
	251 - 300	217.410	1.26440		2,001 - 4,000	22.540	0.57154
	301 - 400	258.890	1.40940		4,001 - 10,000	24.703	0.50527
	401 - 500	346.750	1.72830		10,001 - 20,000	26.970	0.46713
	501 - 3,110	453.850	2.11660		20,001 - 40,000	34.420	0.37615
	> 3,110	-	-		> 40,000	47.618	0.29592
B	< 210	90.673	0.93196	F	< 200	15.209	0.81558
	211 - 400	98.483	0.98332		201 - 700	14.457	0.78407
	> 400	109.300	1.09710		701 - 1,000	13.953	0.68465
					1,001 - 2,000	13.953	0.63227
C	Any *x*	61.141	0.91465		2,001 - 3,000	14.823	0.54503
					3,001 - 7,000	16.187	0.46490
D	< 310	34.459	0.86974		7,001 - 15,000	17.836	0.41507
	310 - 1,000	32.093	0.81066		15,001 - 30,000	22.651	0.32681
	1,001 - 3,000	32.093	0.64403		30,001 - 60,000	27.074	0.27436
	3,001 - 30,000	36.650	0.56589		> 60,000	34.219	0.21716
	> 30,000	44.053	0.51179				

Atm. Stab.	*c* (-)	*d* (-)	*e* (-)	*f* (-)	*g* (m^{-1})	*h* (-)
A	24.1670	2.5334	0.32	0.24	0.001	0.5
B	18.3330	1.8096	0.32	0.24	0.001	0.5
C	12.5000	1.0857	0.22	0.20	0	0
D	8.3330	0.7238	0.16	0.14	0.0003	-0.5
E	6.2500	0.5428	0.11	0.08	0.0015	-0.05
F	4.1667	0.3619	0.11	0.08	0.0015	-0.05

Continuous Source (Light Gas)

example

EXAMPLE C5.2. *Concentration Calculation*

Calculate the concentration of pollutants in the plume detailed in Example C5.1. The following data are available:

Source emission rate, Q_c	: 0.020	kg/s

The concentration will be calculated from Eq. (C5.25) as

$$C = \frac{Q_c}{u_s} \frac{10^9}{2\pi\sigma_y} \exp\left[-\frac{y^2}{2\sigma_y^2}\right] \frac{1}{\sigma_z} \left\{ \exp\left[-\frac{(h_e - z)^2}{2\sigma_z^2}\right] + \exp\left[-\frac{(h_e + z)^2}{2\sigma_z^2}\right] \right\}.$$

We note the following:

- The dispersion coefficients σ_z (m) and σ_y (m), as a function of the distance x (m), are calculated from Eq. (C5.27) for rural area as

$$\sigma_z = a(0.001x)^b \qquad \sigma_y = 0.4651 \tan(TH)$$

  where $TH = 0.01745[c - d \ln(0.001x)]$. Values for coefficients a, b, c and d, are given in Table C5.8.

- The gradual and final plume rise, h_e (m), has been calculated in Example C5.1.

- The source emission rate is given (Q_c = 0.020 kg/s) while the wind speed u_s = 3.31 m/s, has been calculated in Example C5.1.

- The height, z (m), is equal to 2 m, taken as the height of a person.

- For y = 0, concentration C (μg/m^3), can be calculated as a function of the distance x (m).

Results are plotted in Figure C5.9, while numerical values are given in Table C5.9 (in the same table values for the dispersion coefficients σ_y and σ_z are also included). The concentration profile is shown in the upper diagram, while two concentration contour lines are given immediately underneath. The typical shape of the concentration profile, and particularly the zero concentration values very near the source (point zero), are noted. The first 500 m where the concentration is

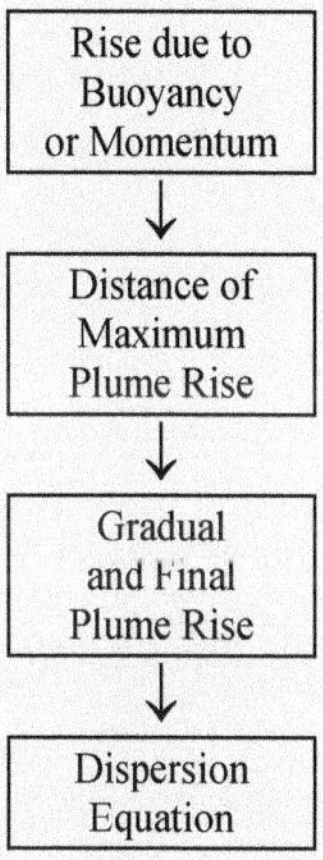

Continuous Source (Light Gas)

example

zero corresponds to the distance that the cloud must travel until it lowers to a height of 2 m (emissions started from a stack of 25 m height). This point is actually quite interesting, as measurements near the stack will show zero concentrations, while further away those values will increase.

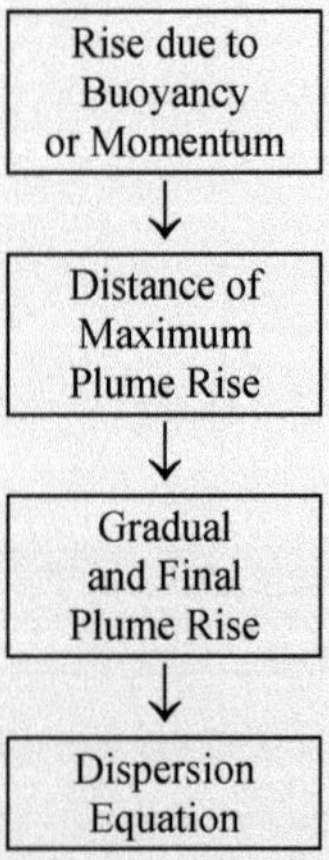

Table C5.9. Calculation of Concentration, C (μg/m^3), as a Function of the Distance x (m) for $y = 0$ and $z = 2$ m.

x (m)	h_e (m)	σ_y (m)	σ_z (m)	C (μg/m^3)
100	39.34	4.07	2.33	0.0
200	47.77	7.73	4.09	0.0
500	47.77	17.97	8.40	0.0
750	47.77	26.05	11.46	1.4
1,000	47.77	33.88	13.95	12.9
1,500	47.77	49.03	18.03	67.5
2,000	47.77	63.68	21.63	123.8
3,000	47.77	91.92	26.98	162.7
4,000	47.77	119.17	30.84	158.1
5,000	47.77	145.67	34.21	145.8
6,000	47.77	171.58	37.23	132.3
7,000	47.77	196.99	40.00	119.7
8,000	47.77	221.99	42.28	108.3
9,000	47.77	246.61	44.40	98.5
10,000	47.77	270.90	46.38	90.1
20,000	47.77	500.95	60.29	46.5
30,000	47.77	715.60	68.84	30.7
40,000	47.77	920.24	74.49	22.8
50,000	47.77	1,117.44	79.19	18.1
60,000	47.77	1,308.72	83.25	15.0

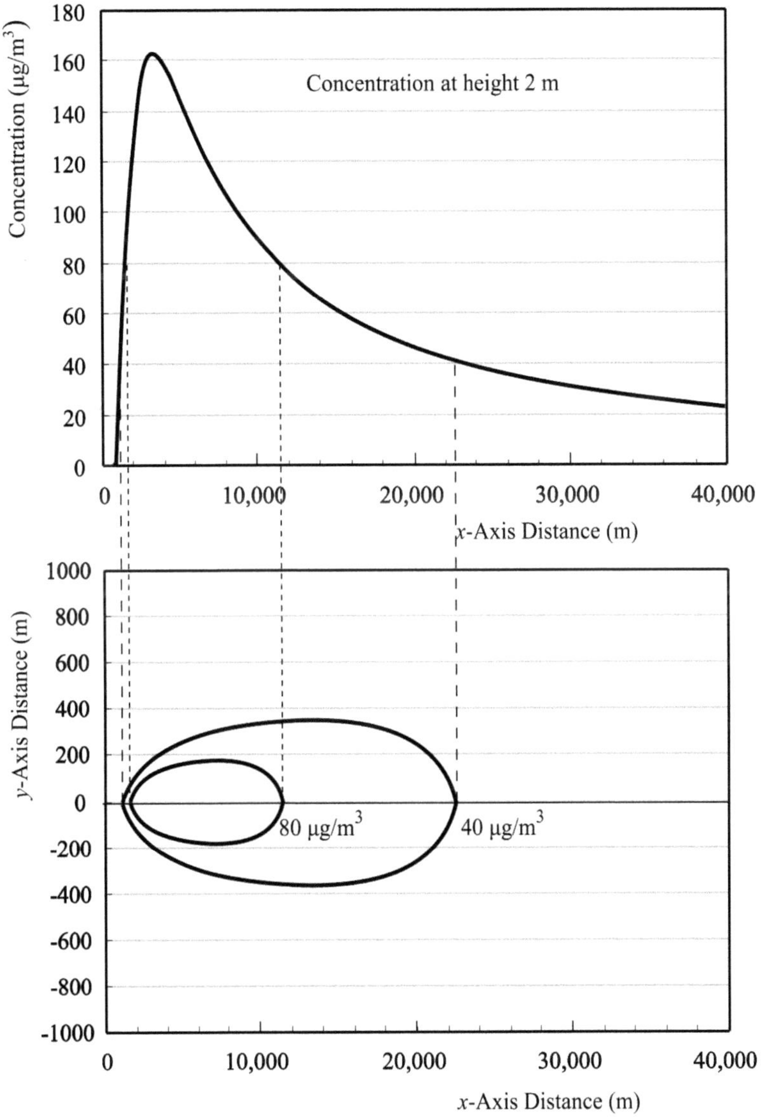

Figure C5.9. Concentration at various distances.

Continuous Source (Light Gas)

example

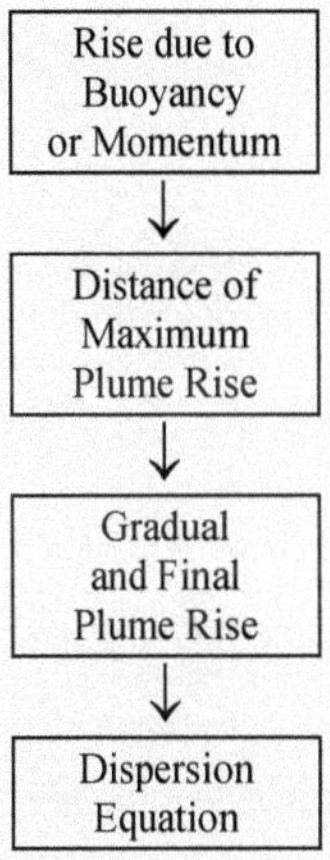

Continuous Source (Light Gas)

calculation procedure

Rise due to Buoyancy or Momentum
↓
Distance of Maximum Plume Rise
↓
Gradual and Final Rise
↓
Dispersion Equation

C5.3.3. Chemical Deposition Mechanism

In its full expansion, Eq. (C5.25) which describes the concentration, C (μg/m^3), of a pollutant in a point with coordinates (x, y, z) is given [EPA–454/B–95–003a 1995, Turner 1994] as

$$C = \frac{Q_c}{u_s}\frac{D\ 10^9}{2\pi\sigma_y}\exp\left[-\frac{y^2}{2\sigma_y^2}\right]\frac{1}{\sigma_z}\left\{\exp\left[-\frac{(h_e - z)^2}{2\sigma_z^2}\right] + \exp\left[-\frac{(h_e + z)^2}{2\sigma_z^2}\right]\right\} \quad \text{(C5.28)}$$

The coefficient, D (-), represents a decay term and it is a simple method of accounting for pollutant removal by chemical or physical processes. It is of the form

$$D = \exp\left(-\psi\frac{x}{u_s}\right) \qquad \text{for } \psi > 0, \qquad \text{(C5.29)}$$

$$D = 1 \qquad \text{for } \psi = 0. \qquad \text{(C5.30)}$$

The coefficient ψ (s^{-1}), is called the decay coefficient (a value of zero means decay is not considered), while x (m) is the downwind distance, and u_s (m/s) the mean wind speed at the height of the pollutant's emission.

In cases when the pollutant half-life, $T_{1/2}$ (s), is known, then the decay coefficient, ψ (s^{-1}), can be calculated from the expression

$$\psi = \frac{0.693}{T_{1/2}}. \qquad \text{(C5.31)}$$

As an example we note that during the modeling of the dispersion of sulfur dioxide, SO_2, in the atmosphere, measurements indicated $T_{1/2}$ = 4 h, and so ψ = 0.0000481 s^{-1}.

C5.4. Light Gas Dispersion from an Instantaneous Source

Instantaneous Source (Light Gas)

calculation procedure

The modeling of the dispersion of toxic gases from an instantaneous source will follow the methodology developed by Rocket Exhaust Effluent Diffusion Model (REEDM) [Bjorklund *et al.* 1998], which was based on the Briggs equations [Briggs 1975] for a continuous source (see Section C5.3). The model is known as OBODM (Open Burn/Opcn Detonation Dispersion Model). An instantaneous source is considered a source whose duration is shorter than 15 s.

As in the case of the continuous source, the equations for the puff rise will be first presented followed by the dispersion equations. Note that in the case of a continuous source emissions form a "plume" while in the case of instantaneous source emissions form a "puff" is released.

Gradual and Final Puff Rise

↓

Dispersion Equation

C5.4.1. Puff Rise

In the case of a continuous release, the rise of the plume is attributed either to buoyancy or to momentum forces. When one refers to puffs created by instantaneous sources, two types are generally distinguished depending upon the type of the release:

a) release at a temperature much higher than the ambient temperature, and
b) release following a detonation, or a burn of a duration less than 15 s.

The puff rises because of the energy ΔH_g (kJ/kg), that it has, and the corresponding instantaneous source buoyancy flux parameter, F_{bi} (m^4/s^2) - see corresponding Section C5.3.1 and Eq. (C5.6) - is given by the expression [Briggs, 1975]

$$F_{bi} = \frac{g Q_i}{\pi \rho_{air}} \frac{\Delta H_g}{C_{pair} T_a}, \qquad \text{(C5.32)}$$

where, g (m/s^2) is the acceleration due to gravity (9.81 m/s^2), Q_i (kg) denotes the total gas mass in the puff, ρ_{air} (kg/m^3) the air density, C_{pair} ($kJ \cdot kg^{-1} \cdot K^{-1}$) the air heat capacity and T_a (K) the ambient temperature. Usually, the energy of the gas in the puff is taken as equal to the difference in enthalpy as a result of the difference in temperatures (high temperature release), that is $\Delta H_g = C_{pg}(T_g - T_a)$, where subscript "g" refers to the gas and. Assuming that $C_{pg} \approx C_{pair}$, Eq. (C5.32) becomes

$$F_{bi} = \frac{g Q_i}{\pi \rho_{air}} \frac{(T_g - T_a)}{T_a}. \qquad \text{(C5.33)}$$

Instantaneous Source (Light Gas)

calculation procedure

In the case of gas released as a result of an explosion, the gas energy is considered to be equal to the total gas mass in the puff, Q_i (kg), times the heat of combustion, ΔH_c (kJ/kg). This value usually results in an underestimation of the puff rise and hence in an overestimation of the gas concentration near the ground. To avoid this, an empirical modification of Eq. (C5.33) is proposed [Bjorklund *et al.* 1998], as

$$F_{bi} = \frac{3}{4} \frac{g Q_i}{\pi \rho_{air}} \frac{\Delta H_c}{C_{pair} T_a}, \tag{C5.34}$$

Note that the instantaneous source buoyancy flux parameter , F_{bi} (m^4/s^2), has slightly different units than the continuous source buoyancy flux parameter , F_b (m^4/s^3).

Gradual and Final Puff Rise

↓

Dispersion Equation

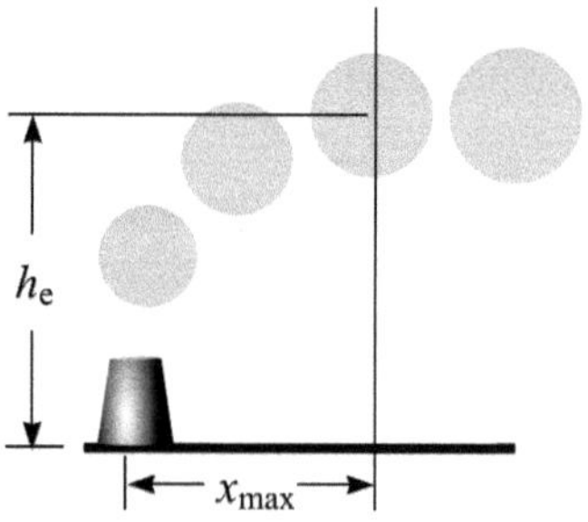

Figure C5.10. Puff rise from instantaneous source.

The puff rises until it reaches its final height, see Figure C5.10. The distance from the source to this point of final height is known as maximum distance, x_{max} (m) (similar to the distance x_f traveled by a plume until it reaches maximum height in the case of a continuous source). The equations for calculating the maximum distance, x_{max} (m), of the gradual and the final rise, h_e (m), of the puff are given in Table C5.10, as a function of the atmospheric conditions [Bjorklund *et al.* 1998].

Table C5.10. Gradual and Final Puff Rise Equations*.

Unstable (A,B,C) or Neutral (C) Conditions	Stable (E, F) Conditions
Gradual puff rise (rise until maximum distance x_{max})	
$h_e = h_s + \left(\dfrac{2F_{bi}\, x^2}{c^3 u_s^2} \right)^{1/4}$ (C5.35)	$h_e = h_s + \left\{ \dfrac{4F_{bi}}{c^3 s} \left[1 - \cos\left(\dfrac{x\sqrt{s}}{u_s} \right) \right] + \left(\dfrac{R}{c} \right)^4 \right\}^{1/4} - \left(\dfrac{R}{c} \right)$ (C5.36)
Calculation of maximum distance x_{max}	
$x_{max} = 12\, F_{bi}^{1/2}\, u_s^{1/3}$ $F_{bi} \le 300 u_s^{2/3}$ (C5.37) $x_{max} = 50\, F_{bi}^{1/4}\, u_s^{1/2}$ $F_{bi} > 300 u_s^{2/3}$ (C5.38)	$x_{max} = \left(\dfrac{\pi\, u_s}{\sqrt{s}} \right)$ (C5.39)
Final puff rise (rise after maximum distance x_{max})	
$h_e = h_s + \left(\dfrac{2F_{bi}\, x_{max}^2}{c^3 u_s^2} \right)^{1/4}$ (C5.40)	$h_e = h_s + \left\{ \dfrac{8F_{bi}}{c^3 s} + \left(\dfrac{R}{c} \right)^4 \right\}^{1/4} - \left(\dfrac{R}{c} \right)$ (C5.41)

where $$s = \frac{g}{T_a} \frac{\partial \theta}{\partial z}. \qquad \text{(C5.42)}$$

* Equations are empirical and values should be entered only at specified units.

In Table C5.10, h_s (m) denotes the height of the source from the ground, u_s (m/s), the wind speed at the height of the source, while c (m) is the instantaneous source entrainment coefficient, whose default value is usually 0.64 m. Finally R (m) is the initial radius of the puff, g (m/s^2), the acceleration due to gravity (= 9.81 m/s^2), while $(\partial\theta / \partial z)$ (K/m) denotes the change of temperature with height (empirically, it takes the value of 0.020 K/m for atmospheric stability class E, and the value of 0.035 K/m for atmospheric stability class F).

Instantaneous Source (Light Gas)

calculation procedure

Gradual and Final Puff Rise

↓

Dispersion Equation

Instantaneous Source (Light Gas)

calculation procedure

C5.4.2. Dispersion Equation

Having calculated the gradual and final rise of the puff, we can now proceed to calculate the concentration and the dispersion coefficients. A more accurate estimate for the calculation of the wind speed will also be presented.

Gradual and Final Puff Rise

↓

Dispersion Equation

a) Concentration Equation

According to Turner [Turner 1994], for an instantaneous dispersion of a puff following a Gaussian distribution, the concentration, C ($\mu g/m^3$), at a specific point with coordinates (x, y, z) and at time t (s) after the release, where x (m) is the downwind distance while y (m) and z (m), are the lateral and vertical distances, is

$$C = \frac{2Q_i\,10^9}{(2\pi)^{3/2}\sigma_x\sigma_y\sigma_z}\exp\left[-\frac{(x-u_s t)^2}{2\sigma_x^2}\right]\exp\left[-\frac{y^2}{2\sigma_y^2}\right] \times\left\{\exp\left[-\frac{(h_e-z)^2}{2\sigma_z^2}\right]+\exp\left[-\frac{(h_e+z)^2}{2\sigma_z^2}\right]\right\} . \quad \text{(C5.43)}$$

In the above equation, Q_i (kg), is the total gas mass, u_s (m/s), the average wind speed at the height of the release (constant in time and space), σ_z (m), the vertical dispersion coefficient, and σ_x (m) and σ_y (m), the downwind and lateral dispersion coefficients, respectively. The rise, h_e (m), is a function of the distance x (m) and represents the gradual and final rise of the puff, respectively. The coefficient 10^9 enters so that final concentration units are $\mu g/m^3$.

Similarly to the case of the continuous dispersion of light gases, in Eq. (C5.43) the following observations can be made:

a) Concentrations at receptors are directly proportional to the emission rate.

b) Downwind (x-axis) concentrations are inversely proportional to the downwind dispersion coefficient, σ_x. The larger the distance from the puff center, the larger the downwind dispersion coefficient, σ_x, and hence the lower the concentration will be. The exponential term that incorporates the ratio (x - $u_s t$) to σ_x corrects for the distance of the receptor from the dispersion center (x is equal to zero in the dispersion center).

c) Crosswind (lateral, y-axis) concentrations are inversely proportional to the lateral dispersion coefficient, σ_y. The larger the lateral distance from the puff center, the larger the lateral dispersion coefficient, σ_y, and hence the lower the concentration will be. The exponential term that incorporates the ratio y to σ_y corrects for the distance of the receptor from the dispersion center (y is equal to zero in the dispersion center, i.e., exactly on the x-axis).

Instantaneous Source (Light Gas)

calculation procedure

d) Crosswind (vertical, z-axis) concentrations are inversely proportional to the vertical dispersion coefficient, σ_z. The larger the vertical distance from the puff center, the larger the vertical dispersion, and thus the lower the concentration will be. The addition of the two exponential terms in the vertical dispersion represent the height of the receptor from the puff center of rise. The first term, $(h_e - z)$, represents the direct distance of the receptor from the puff center (Figure C5.10). The second term, $(h_e + z)$, represents the distance of the eddy reflectivity of the receptor from the plume center, plus the distance z to return to the receptor after the reflection (h_e is the plume rise, i.e., the distance between the ground and the plume center).

Gradual and Final Puff Rise

↓

Dispersion Equation

b) Dispersion Coefficients

Slade in 1965 [Turner 1994] proposed the following equations for the dispersion coefficients σ_x (m), σ_y (m) and σ_z (m),

$$\sigma_x = \sigma_y = a\left(\frac{x}{e}\right)^b \quad \text{and} \quad \sigma_z = c\left(\frac{x}{e}\right)^d . \qquad \text{(C5.44)}$$

Values for the parameters a (m), b (-), c (m) and d (-) are given in Table C5.11 as a function of the atmospheric stability, while $e = 1\ \text{m}^{-1}$.

Table C5.11. Parameters of Eq. (C5.44) as a Function of Atmospheric Stability.

Atmospheric Stability	a (m)	b (-)	c (m)	d (-)
A	0.18	0.92	0.72	0.76
B	0.14	0.92	0.53	0.73
C	0.10	0.92	0.34	0.72
D	0.06	0.92	0.15	0.70
E	0.045	0.91	0.12	0.67
F	0.030	0.90	0.08	0.64

c) Mean Wind Speed

The wind speed, u_s (m), used in the equations, was defined by Eq. (C5.1), as a function of the observed wind speed, u_{ref} (m/s), in a particular reference height, z_{ref} (m). In the case of an instantaneous dispersion, and when the puff has large dimensions, it is better to employ the mean wind speed, $\overline{u}$ (m), which can be calculated empirically as a function of the puff lower and upper part heights, z_b (m) and z_t (m), respectively, as

Instantaneous Source (Light Gas)

calculation procedure

Gradual and Final Puff Rise

↓

Dispersion Equation

$$\overline{u} = \frac{1}{z_t - z_b} \int_b^t u_s \, dz , \qquad \text{(C5.45)}$$

where

$$z_b = h_e - 2.15\sigma_z \qquad \text{when} \quad h_e - 2.15\sigma_z > 2 , \qquad \text{(C5.46)}$$

$$z_b = 2 \text{ m} \qquad \text{when} \quad h_e - 2.15\sigma_z \leq 2 , \qquad \text{(C5.47)}$$

and

$$z_t = h_e + 2.15\sigma_z \qquad \text{when} \quad h_e + 2.15\sigma_z < h_e^{max} , \qquad \text{(C5.48)}$$

$$z_t = h_e^{max} \qquad \text{when} \quad h_e + 2.15\sigma_z \geq h_e^{max} , \qquad \text{(C5.49)}$$

where h_e^{max} denotes the final height of the puff. Substituting Eq. (C5.1) in Eq. (C5.45) one obtains

$$\overline{u} = \frac{u_{ref}}{(z_t - z_b) z_{ref}^{p} (1+p)} \left(z_t^{(1+p)} - z_b^{(1+p)} \right) . \qquad \text{(C5.50)}$$

Values for the exponent p (-) depend upon the atmospheric stability and can be found in Table C5.3.

d) Puff Arrival and Departure Time

As already mentioned, in an instantaneous source, concentrations are a function not only of the Cartesian coordinates but also of time. Hence for a given point in space, there is a specific time, t_{in} (s), in which the front part of the puff arrives at this point, and another specific time, t_{out} (s), when the puff departs from this point. These two times, as well as the specific time, t_{peak} (s), in which the concentration has reached its maximum value, refer to the worst case (i.e., refers to the points located on the puff center with coordinates (x, 0, h_e)), and are calculated by the equations

$$t_{in} = \frac{x - 2.45\sigma_x}{\overline{u}} , \qquad t_{out} = \frac{x + 2.45\sigma_x}{\overline{u}} , \qquad t_{peak} = \frac{x}{\overline{u}} . \qquad \text{(C5.51)}$$

EXAMPLE C5.3. *Concentration Calculation*

Instantaneous Source (Light Gas)

example

An accident in a rural area results in the instantaneous release of 500 kg of a gas lighter than air, at a temperature of 420 K. Ambient temperature is 298 K and the sky is clear. Calculate at a height of 2 m

a) Concentration $C\,(x,0,z,t_{peak})$ as a function of the distance x.
b) Concentration $C\,(500,0,z,t)$ as a function of the time t.
c) Concentrations $C\,(x,0,z,60)$ and $C\,(x,0,z,240)$ as a function of the distance x.

The following data are available:

Wind speed (at 10 m), u_{ref}	:	4	m/s
Source height, h_s	:	2	m
Density of air, ρ_{air}	:	1.21	kg/m^3

Gradual and Final Puff Rise → Dispersion Equation

Since the sky is clear (cloud coverage 0/8) and the wind speed is 4 m/s, from Table C5.2 the atmospheric stability is B.

From Eq. (C5.1), for rural area and atmospheric stability B, $p = 0.07$, and wind speed, u_s (m/s) = 3.57 m/s.

a) Puff Rise

The instantaneous source buoyancy flux parameter, F_{bi} (m^4/s^2), is calculated from Eq. (C5.33), as

$$F_{bi} = \frac{gQ_i}{\pi \rho_{air}} \frac{(T_g - T_a)}{T_a} = 528.6\ \text{m}^4/\text{s}^2\,.$$

In order to calculate the rise, first x_{max} (m) must be obtained. From Eq. (C5.38),

and as $528.6 = F_{bi} < 300\ u_s^{2/3} = 701, \quad x_{max} = 12\,F_{bi}^{1/2}\,u_s^{1/3} = 421.8$ m.

The puff gradual and final rise is calculated from Eq. (C5.35) and Eq. (C5.40) in Table C5.10. In Table C5.12 indicative values of the gradual and final puff rise are shown, while they are all drawn in Figure C5.11.

b) Dispersion Coefficients and Mean Wind Speed

The dispersion coefficients σ_x, σ_y, and σ_z are calculated from Eq. (C5.44). Indicative values are shown in Table C5.12. Having calculated the dispersion coefficients and the puff rise as a function of the distance x, from Eqs. (C5.46)-(C5.50) the mean wind, $\bar{u}$ (m), can be calculated. Corresponding values are given in Table C5.12.

Instantaneous Source (Light Gas)

example

Gradual and Final Puff Rise

↓

Dispersion Equation

Table C5.12. Calculated Values at Different Distances.

x (m)	h_e (m)	σ_x, σ_y (m)	σ_z (m)	z_b (m)	z_t (m)	$\overline{u}$ (m/s)	t_{in} (s)	t_{out} (s)	t_{peak} (s)
1	6.2	0.1	0.5	5.1	7.4	3.87	0.2	0.3	0.3
50	31.8	5.1	9.2	12.0	51.6	4.32	8.7	14.5	11.6
100	44.2	9.7	15.3	11.3	77.0	4.41	17.3	28.1	22.7
200	61.6	18.3	25.4	7.1	88.6	4.42	35.1	55.5	45.3
300	75.0	26.6	34.1	2.0	88.6	4.38	53.6	83.4	68.5
421.8	**88.6**	**36.4**	**43.7**	**2.0**	**88.6**	**4.38**	**76.0**	**116.7**	**96.3**
500	88.6	42.6	49.5	2.0	88.6	4.38	90.4	138.0	114.2
600	88.6	50.4	56.5	2.0	88.6	4.38	108.9	165.2	137.0
700	88.6	58.0	63.3	2.0	88.6	4.38	127.4	192.3	159.9
800	88.6	65.6	69.7	2.0	88.6	4.38	146.0	219.4	182.7
900	88.6	73.1	76.0	2.0	88.6	4.38	164.6	246.5	205.6
1,000	88.6	80.6	82.1	2.0	88.6	4.38	183.3	273.5	228.4
1,200	88.6	95.3	93.8	2.0	88.6	4.38	220.8	327.4	274.1
1,400	88.6	109.8	104.9	2.0	88.6	4.38	258.3	381.2	319.7
1,600	88.6	124.1	115.7	2.0	88.6	4.38	296.0	434.9	365.4
1,800	88.6	138.3	126.1	2.0	88.6	4.38	333.7	488.5	411.1
2,000	88.6	152.4	136.2	2.0	88.6	4.38	371.5	542.1	456.8
3,000	88.6	221.4	183.1	2.0	88.6	4.38	561.3	809.0	685.2

Table C5.12 (cont.). Calculated Values at Different Distances.

x (m)	$C(x, 0, 2, t_{peak})$ (μg/m³)	$C(x, 0, 2, 60)$ (μg/m³)	$C(x, 0, 2, 240)$ (μg/m³)	t (s)	$C(500, 0, 2, t)$ (μg/m³)
1	0.1	0	0	0	0
50	1,721,621	0	0	15	0
100	1,452,286	0	0	30	0
200	790,289	1,469	0	45	0
300	470,300	176,163	0	60	0
421.8	**282,258**	**20**	**0**	**75**	**4.2**
500	285,960	0	0	90	12,939
600	259,992	0	0	100	121,359
700	223,936	0	0	110	212,987
800	188,956	0	126	120	175,301
900	158,549	0	18,891	130	101,320
1,000	133,221	0	109,182	140	49,831
1,200	95,527	0	28,039	150	24,673
1,400	70,323	0	447	160	13,260
1,600	53,144	0	3	170	7,661
1,800	41,124	0	0	180	4,702
2,000	32,492	0	0	190	3,036
3,000	12,597	0	0		

Instantaneous Source (Light Gas)

example

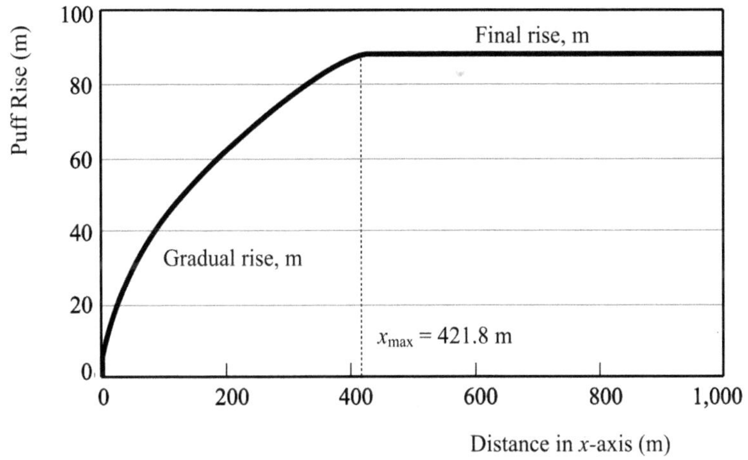

Figure C5.11. Puff rise as a function of distance x.

Gradual and Final Puff Rise

↓

Dispersion Equation

c) Puff Arrival and Departure Times

The times, t_{in} (s), t_{out} (s) and t_{peak} (s), are calculated as a function of the dispersion coefficient σ_x (x) and the mean wind speed, $\overline{u}$ (m), for every distance x, from Eq. (C5.51). Indicative values are given in Table C5.12.

d) Concentrations

The concentration, $C(x,0,2,t_{peak})$, is directly calculated from Eq. (C5.43). Indicative values are given in Table C5.12, while Figure C5.12 shows the corresponding plot.

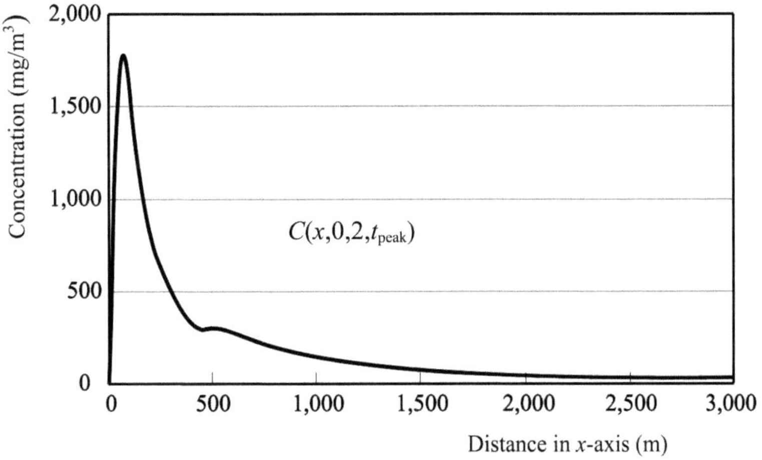

Figure C5.12. Concentration $C(x,0,2,t_{peak})$ as a function of distance x.

Instantaneous Source (Light Gas)

example

Gradual and Final Puff Rise

↓

Dispersion Equation

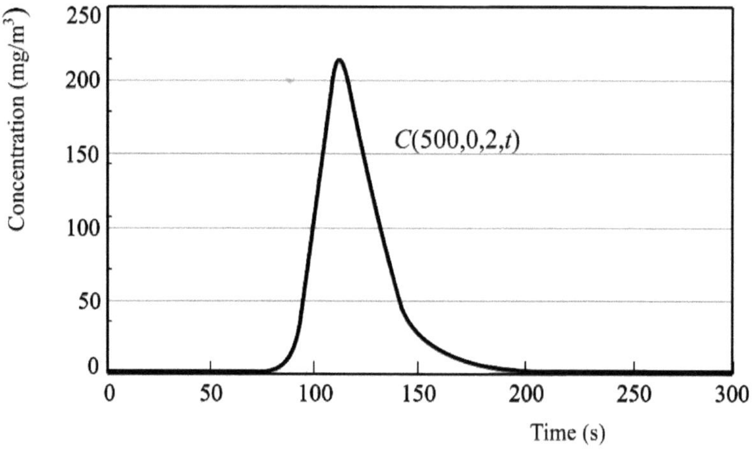

Figure C5.13. Concentration $C(500,0,2,t)$ as a function of time t.

We note that the puff rises until the maximum distance x_{max}= 421.8 m, while after this point concentration changes only because of dispersion in the air.

Concentrations $C(500,0,2,t)$, $C(x,0,2,60)$ and $C(x,0,2,240)$ are also directly calculated from Eq. (C5.43). Indicative values are shown in Table C5.12, while their plots are shown in Figures C5.13 and C5.14.

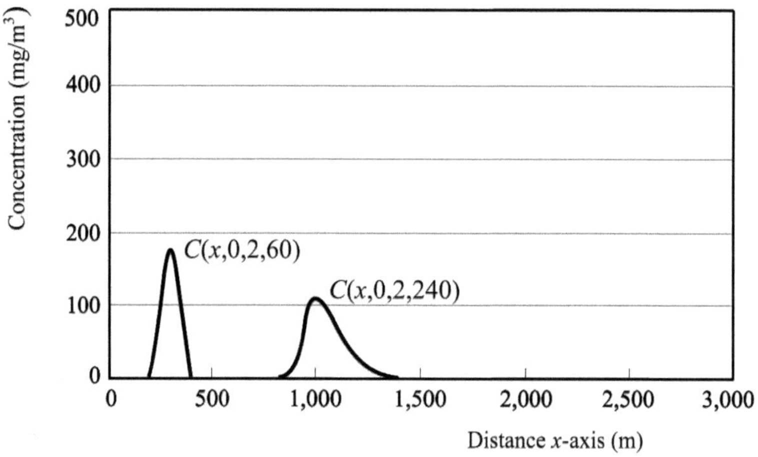

Figure C5.14. Concentrations $C(x,0,2,60)$ and $C(x,0,2,240)$ as a function of distance x.

❒

Heavy Gas Dispersion

introduction

C5.5. Heavy Gas Dispersion

The dispersion of a large number of toxic gases cannot be modeled with the schemes already presented in the previous sections, as their density is considerably higher than that of air. These gases are known as "heavy gases" and their release into the atmosphere introduces a series of phenomena that directly influence their dispersion in air.

Upon the release, heavy gas descends to the ground and spreads radially under the influence of gravitational forces. This self-induced flow produces a shallow cloud with increased horizontal extent. At the edges of this shallow cloud a front develops with a strong vorticity (see Figure C5.15). For the duration of its expansion, this deterministic field replaces random atmospheric turbulence, and increases the rate of mixing, especially at a point just behind the gravity front.

Once all the heavy gas is on the ground, the vertical variation of density in the cloud will cause a stable stratification inside the cloud which in turn reduces the dispersion in the vertical direction. At the end, the effects of density are dispersed and become negligible, and the dispersion becomes passive.

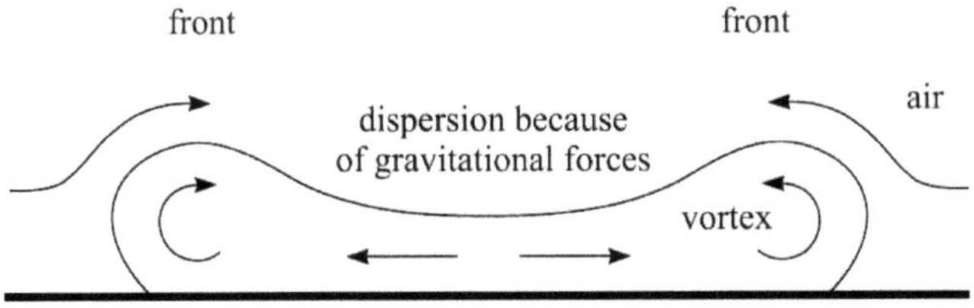

Figure C5.15. Heavy gas dispersion because of gravity.

Hence the modeling of the dispersion of a heavy toxic gas, in essence, consists of its spreading on the ground. Even if the release takes place at a certain height, because of gravity, this will descend to the ground. The modeling of this type of dispersion can be carried out in many ways, the two most usual ones are:

- Computational models such as DEGADIS [Collenbrander 1980], or HEGADAS [Havens 1985] and the SLAB code [Ermak 1990] - some of them are freely available from the Environmental Protection Agency (EPA). These models certainly require the use of adequate computers.
- The Britter and McQuaid method [Britter & McQuaid 1988] employs an empirical diagram for the calculation of the reduction of the concentration during a continuous or instantaneous dispersion, as a function of the downwind distance. The method is very simple and does not require the use of computers.

Heavy Gas Dispersion

introduction

In this work the Britter and McQuaid method [Britter & McQuaid 1988] will be employed, as it can easily and quickly produce relatively good results. The method is based upon a series of empirical expressions that refer to the average values of the dispersion variables. The method was presented for the first time in 1988, as a logical algorithm, that has the distinct advantage of allowing even non-experts to predict the dispersion of a heavy gas. The method allows the calculation of

- the average concentration levels along the plume axis, in the case of a continuous release, and
- the maximum concentration levels along the downwind puff path, in the case of an instantaneous release.

The difference in densities between the released heavy gas and atmospheric air causes the following phenomena:

1) The velocity field set up by the horizontal density difference, in a gravitational field, is an additional transport mechanism to that provided by the environmental flows. This "self-induced" flow produces a plume or puff that spreads horizontally not randomly but with a predetermined pattern that results in profiles of concentration in the lateral direction which are frequently quite uniform.
2) The velocity shear introduced by this velocity field may lead to more enhanced mixing of the gases and eventually generate more turbulence and consequent turbulent mixing and plume or puff dilution. This mechanism of dilution is of principal importance when the self-induced velocities are large compared with the mean environmental velocity.
3) The variation of density in the vertical direction, in a gravitational field, will be stably stratified and turbulence and turbulent mixing can be significantly reduced or entirely inhibited. This effect can extend to the atmospheric turbulence in the wind flow over the cloud, as well as to the cloud itself.
4) The inertia of the released gas depends directly upon its density. When the density difference is small compared with the density of the gas or of the air, the influence of the density difference on the inertia is small and may be neglected (Boussinesq approximation).

In general there exists an extensive literature on the dispersion of released light gases with density similar to that of the atmospheric air (see Sections C5.1 - C5.4). This kind of release is usually known as "passive," in contrast to those where the density difference influences the motion, and these are referred to as "dynamically active."

C5.5.1. Heavy Gas Dispersion from a Continuous Source

Continuous Source (Heavy Gas)

calculation procedure

In the case of a heavy gas dispersion from a continuous source, the principle of the Britter and McQuaid [Britter & McQuaid 1988] method is shown in Figure C5.16. Initially there is a continuous source of radius b_o (m). The dispersion takes place downwind and its width, b (m), is a function of the distance x. We observe that, although normally the dispersion front should have been curved (dotted line), the method considers it as flat. Behind the source some dispersion also takes place up to a distance x_u (m). Here also, the dispersion "front" is considered flat and not curved, as we are only interested in the distance x_u.

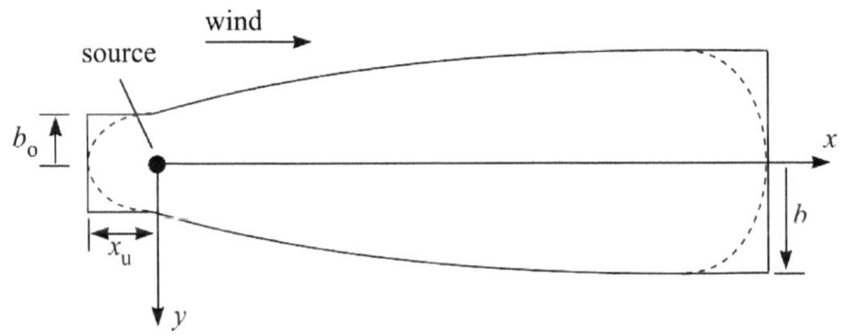

Figure C5.16. Line of constant concentration.

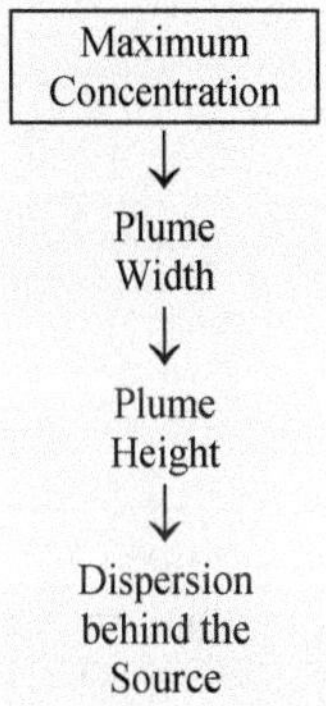

The method is based on the empirical diagram of Figure C5.17. V_c (m^3/s) denotes the dispersion volume, while u_{ref} (m/s) is the observed wind speed at a height of 10 m (see Eq. (C5.1)). Moreover, C_o (kg/m^3), is the initial concentration at the source, while C_{max} (kg/m^3) is the maximum concentration of the gas at a specific distance, x (m). Finally the acceleration g_o (m/s^2) is a "correction" to the acceleration due to gravity, $g = 9.81$ m/s^2, given by

$$g_o = g\frac{(\rho - \rho_{air})}{\rho_{air}}, \qquad \text{(C5.52)}$$

where ρ (kg/m^3) and ρ_{air} (kg/m^3) denote the density of the gas and of the atmospheric air respectively.

Continuous Source (Heavy Gas)

calculation procedure

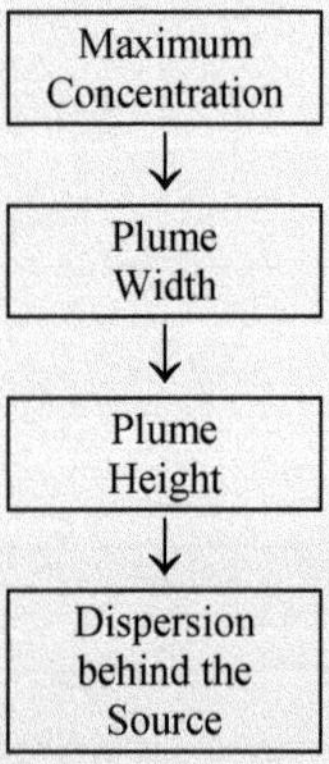

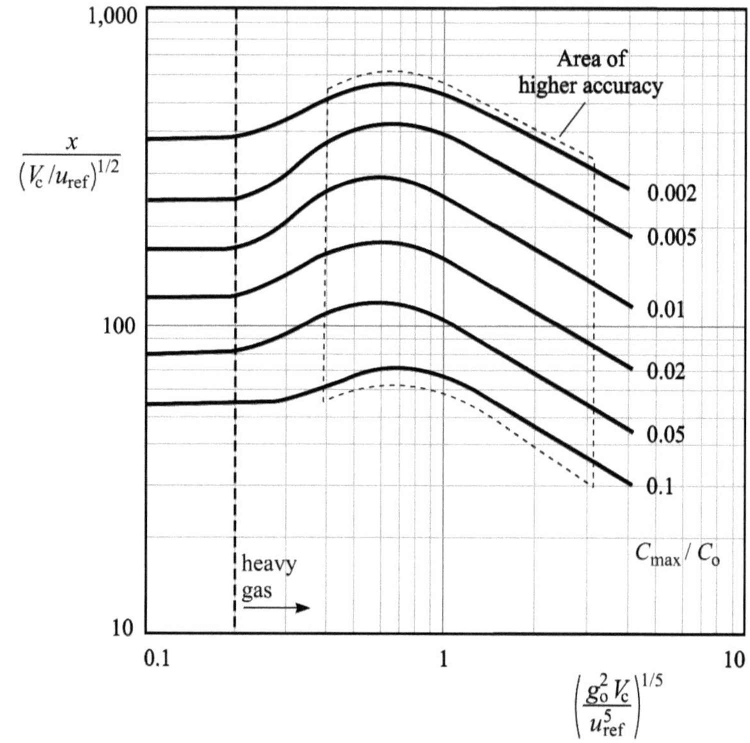

Figure C5.17. Britter and McQuaid diagram [Britter & McQuaid 1988].

The algorithm of Britter and McQuaid [Britter & McQuaid 1988] consists of the following four steps:

1) The ratio $(g_o^2 V_c / u_{ref}^5)^{1/5}$ is calculated, and from the diagram in Figure C5.17, for every ratio of concentrations (C_{max}/C_o), the ratio $x/(V_c/u_{ref})^{1/2}$ is obtained. Since the variables C_o, V_c, and u_{ref}, are known, a table $C_{max} = f(x)$ is created. This way the maximum concentration in every distance from the source is calculated.

2) The width, b (m), of the dispersion downwind is calculated as a function of the distance, x (m), from the expressions [Fay & Zemba 1986]

$$b = 2b_o + 8L_b + 2.5L_b^{1/3}x^{2/3}, \tag{C5.53}$$

where parameter L_b is

$$L_b = \frac{g_o V_c}{u_{ref}^3}. \tag{C5.54}$$

3) The height, b_z (m), of the dispersion from the ground, is calculated as a function of the dispersion width (which in turn, is a function of the distance, see Eq. (C5.53)), from the empirical expression:

$$b_z = \frac{V_c}{2 u_{ref} b}. \quad \text{(C5.55)}$$

4) Finally, the spread, x_u (m), of the dispersion behind the source is given by the expression

$$x_u = b_o + 2L_b. \quad \text{(C5.56)}$$

This algorithm is successfully applied when in a distance, x, the duration, t_r (s), of the dispersion is greater than $2.5\ x/u_{ref}$, so that steady-state conditions have been reached.

Continuous Source (Heavy Gas)

example

Maximum Concentration → Plume Width → Plume Height → Dispersion behind the Source

EXAMPLE C5.4. *Calculation of Heavy Gas Dispersion*

Calculate the dispersion of a toxic heavy gas, released from a continuous source of 1 m^3/s and initial concentration of 6 kg/m^3. The following data are available:

Release radius, b_o	:	2	m
Wind speed (at 10 m), u_{ref}	:	4	m/s
Density of gas, ρ	:	6	kg/m^3
Density of air, ρ_{air}	:	1.21	kg/m^3

To calculate the abscissa of the diagram in Figure C5.17, first the acceleration, g_o (m/s^2), must be obtained from Eq. (C5.52), as

$$g_o = g\frac{(\rho - \rho_{air})}{\rho_{air}} = 38.83 \text{ m/s}^2 \quad \text{and thus} \quad \left(\frac{g_o^2 V_c}{u_{ref}^5}\right)^{1/5} = 1.081.$$

From the diagram of Figure C5.17, for every concentration ratio (C_{max}/C_o), the ratio $x/(V_c/u_{ref})^{1/2}$ is calculated. Since the initial concentration C_o is known (= 6 kg/m^3), a table of $C_{max} = f(x)$, can be constructed (see Table C5.13).

The width, b (m), of the dispersion plume, downwind, is calculated as a function of the distance, x (m), from the empirical Eqs. (C5.53) and (C5.54). The results are shown in Table C5.13, while L_b = 0.607 m. The height, b_z (m), of the dispersion plume centerline from the ground can be calculated from Eq. (C5.55). Finally, the distance, x_u (m) of the dispersion behind the source, can be obtained from Eq. (C5.56), as equal to 3.21 m.

Continuous Source (Heavy Gas)

example

Maximum Concentration → Plume Width → Plume Height → Dispersion behind the Source

Table C5.13. Concentration and Size of Plume.

C_{max}/C_o (-)	$x/(V_c/u_{ref})^{1/2}$ (-)	C_{max} (kg/m^3)	x (m)	b (m)	b_z (m)
0.1	64.3	0.600	32.15	30.25	0.0041
0.05	97.6	0.300	48.80	37.12	0.0034
0.02	152.2	0.120	76.10	46.86	0.0027
0.01	241.2	0.060	120.60	60.52	0.0021
0.005	383.3	0.030	191.65	79.21	0.0016
0.002	518.0	0.012	259.00	94.85	0.0013

The results of Table C5.13 are shown in Figure C5.18.

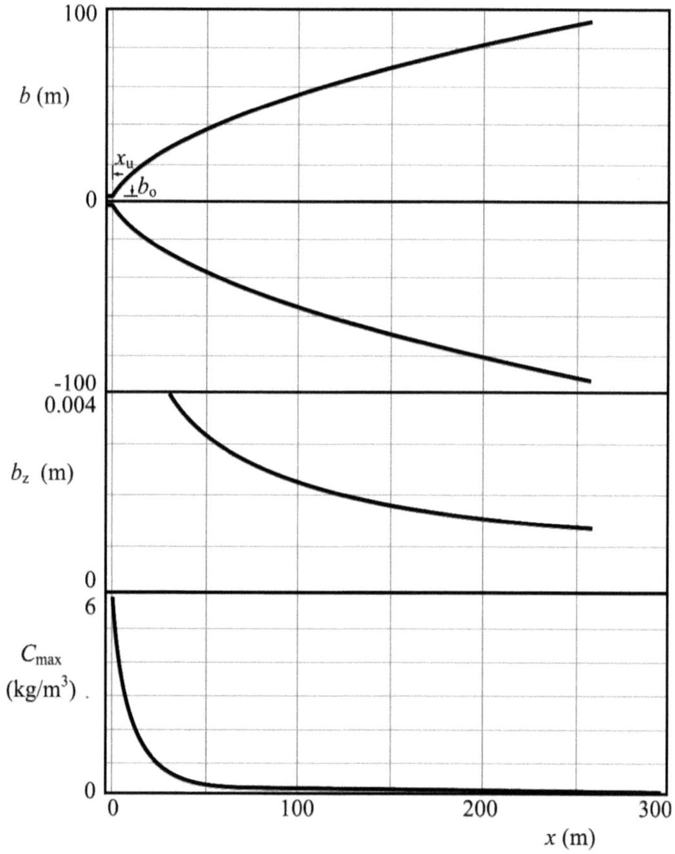

Figure C5.18. Concentration and size of plume.

❐

C5.5.2. Heavy Gas Dispersion from an Instantaneous Source

In the case of an instantaneous source, the principle of the Britter and McQuaid method [Britter & McQuaid 1988] is shown in Figure C5.19. There is an instantaneous release of a gas of volume, V_i (m^3), that results in a puff of initial radius b_o (m). The center of the puff moves downwind (x-direction) while its radius, b (m), increases as a function of the distance x.

Instantaneous Source (Heavy Gas)

calculation procedure

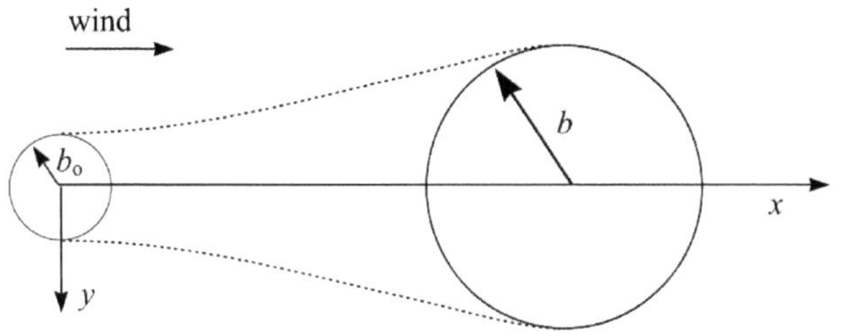

Figure C5.19. Instantaneous source.

Maximum Concentration
↓
Puff Width
↓
Puff Height

The method can be explained easier using the empirical diagram of Figure C5.20. The symbols are the same with those employed in the case of the heavy gas continuous dispersion. In addition, u_{ref} (m/s) denotes the observed wind speed at a height of 10 m (see Eq. (C5.1)), C_o (kg/m^3) is the initial concentration at the source, while C_{max} (kg/m^3) is the maximum concentration of the gas at a specified distance, x (m). Finally the acceleration g_o (m/s^2) is the "correction" to the acceleration due to gravity, $g = 9.81$ m/s^2, as it was given in Eq. (C5.52).

The Britter and McQuaid algorithm [Britter & McQuaid 1988] consists of the following three steps:

1) The ratio $(g_o V_i^{1/3} / u_{ref}^2)^{1/2}$ is calculated, and from the diagram in Figure C5.20, for every concentration ratio (C_{max}/C_o), the ratio $(x / V_i^{1/3})$ is obtained. Since the variables C_o, V_i, and u_{ref}, are known, a table of $C_{max} = f(x)$ is constructed. Hence the maximum concentration at every distance from the source is obtained.

2) To calculate the time, t (s), and the width of the puff, b (m), of the dispersion downwind, as a function of the distance x, the following two empirical expressions must be simultaneously solved [Brighton *et al.* 1985, Wheatly & Prince 1987].

Instantaneous Source (Heavy Gas)

calculation procedure

Maximum Concentration
↓
Puff width
↓
Puff Height

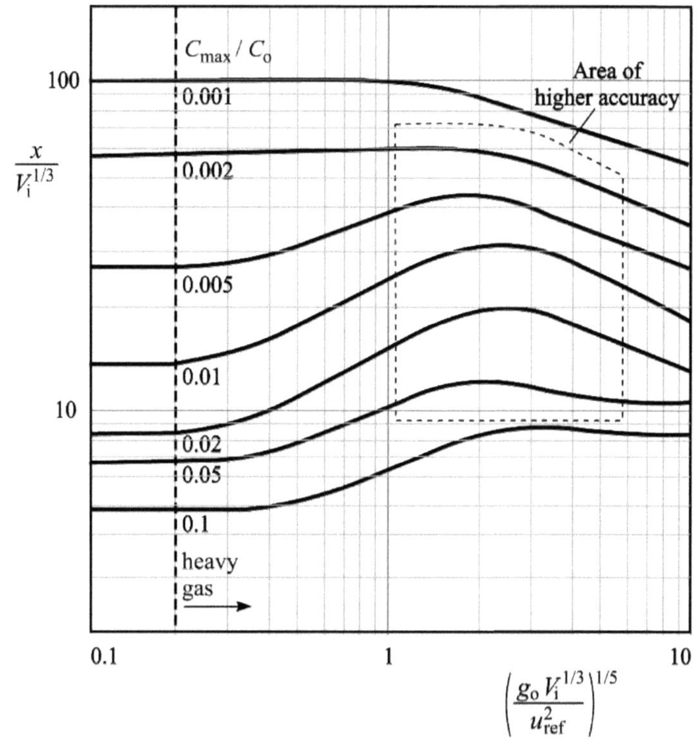

Figure C5.20. Britter and McQuaid diagram [Britter & McQuaid 1988].

$$b = \sqrt{b_o^2 + 1.2\, t \sqrt{g_o V_i}}\,, \tag{C5.57}$$

and

$$x = 0.4 u_{ref} t + b\,. \tag{C5.58}$$

3) Finally, the average height, b_z (m), of the puff from the ground, at every distance x, can be found from the expression

$$b_z = \frac{C_o V_i}{\pi b^2 C_{max}}\,. \tag{C5.59}$$

The aforementioned procedure is better illustrated in the following example.

Instantaneous Source (Heavy Gas)

example

EXAMPLE C5.5. *Calculation of Heavy Gas Dispersion*

Calculate the dispersion of a toxic heavy gas that resulted from the instantaneous release of 10 m^3 of initial concentration of 6 kg/m^3. The following data are available:

Release radius, b_o	:	2	m
Wind speed (at 10 m), u_{ref}	:	4	m/s
Density of gas, ρ	:	6	kg/m^3
Density of air, ρ_{air}	:	1.21	kg/m^3

To calculate the abscissa of the diagram in Figure C5.20, first the acceleration, g_o (m/s^2), must be obtained from Eq. (C5.52), as

$$g_o = g\frac{(\rho - \rho_{air})}{\rho_{air}} = 38.83 \text{ m/s}^2 \quad \text{and hence} \quad \left(\frac{g_o V_i^{1/3}}{u_{ref}^2}\right)^{1/2} = 2.287.$$

Maximum Concentration → Puff Width → Puff Height

From the diagram of Figure C5.20, for every concentration ratio (C_{max}/C_o), the ratio $x/V_i^{1/3}$ is calculated. Since the initial concentration C_o is known (= 6 kg/m^3), a table of $C_{max} = f(x)$ can be constructed (see Table C5.14).

The time, t (s), and the width of the puff, b (m), of the dispersion downwind, as a function of the distance x, are calculated from the simultaneous solution of the empirical Eqs. (C5.57) and (C5.58).

$$b = \sqrt{b_o^2 + 1.2t\sqrt{g_o V_i}},$$

and

$$x = 0.4u_{ref}t + b.$$

Results are shown in Table C5.14.

Having obtained the width, b (m), as a function of the distance x, the average height, b_z (m), of the puff from the ground is obtained from Eq. (C5.59), as

$$b_z = \frac{C_o V_i}{\pi b^2 C_{max}}.$$

Instantaneous Source (Heavy Gas)

example

Table C5.14. Calculated Concentration and Width Values at Various Distances.

C_{max}/C_o (-)	$x/(V_c/u_{ref})^{1/2}$ (-)	C_{max} (kg/m^3)	x (m)	t (s)	b (m)	b_z (m)
0.1	7.2	0.600	15.5	3.71	9.47	0.35
0.05	11.0	0.300	23.8	6.79	12.75	0.39
0.02	17.5	0.120	37.8	12.67	17.37	0.53
0.01	28.3	0.060	60.9	23.36	23.55	0.57
0.005	42.9	0.030	92.4	38.79	30.32	0.69
0.002	62.5	0.012	134.6	60.49	37.85	1.11
0.001	96.7	0.006	208.3	99.82	48.61	1.35

Maximum Concentration → Puff Width → Puff Height

In Figure C5.21, the width b of the puff for every distance x of Table C5.14 is shown, while Figure C5.22 shows the distance-radius where the maximum concentration is met.

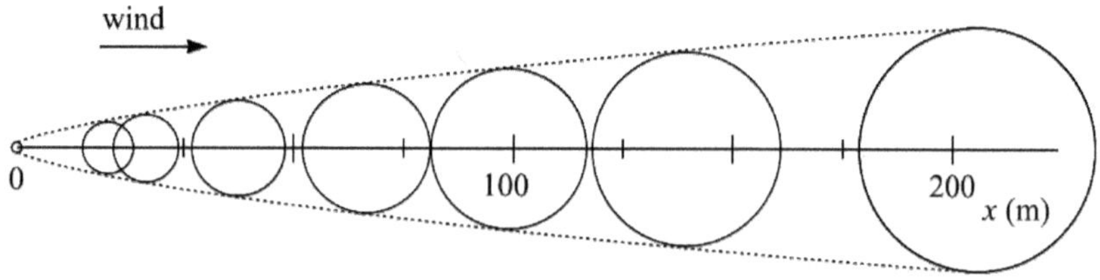

Figure C5.21. Width of puff at various distances.

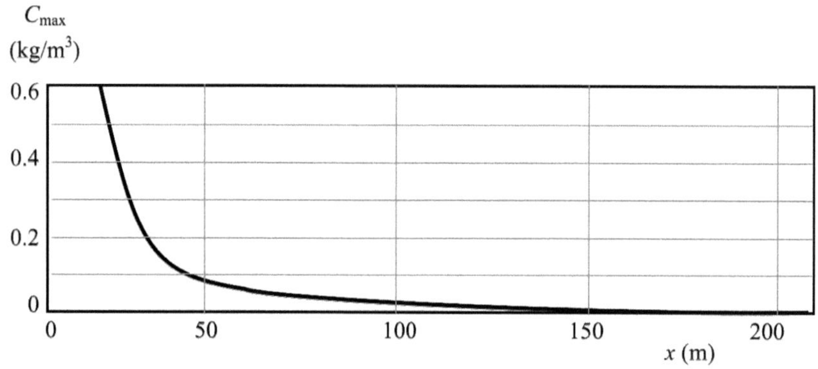

Figure C5.22. Maximum concentration at various distances.

❐

Solid Particles Dispersion

C5.6. Dispersion of Solid Particles (e.g., PM_{10})

In recent years airborne particles with diameters of less than 100 μm have attracted particular interest in various research areas. More specifically particulate matter with diameter smaller than about 10 μm, referred to as PM_{10}, are not filtered by the nose and throat and can settle in the bronchi and lungs and can cause health problems. Increased levels of such coarse particles in the air are now linked to health hazards such as heart disease, altered lung function, lung cancer, alzheimer, atherosclerosis, etc. In urban areas, such microparticles are usually of carbon and various other cancerous or harmful chemicals absorbed in them. They usually originate from industries or oil combustion (vehicles or heating). Particulate matter with aerodynamic diameters smaller that 10 μm are known, as already mentioned, as PM_{10}, while particulate matter with diameters smaller than 2.5 μm are known as $PM_{2.5}$.

In Europe, the observed microparticle concentration in cities is particularly high and this is attributed mostly to petroleum driven vehicles, like taxis, buses or trucks. It should further be mentioned that an old petroleum-driven vehicle produces a thousand times more harmful emissions than a new environment-friendly technology vehicle.

The dispersion of particles with aerodynamic diameters in the region of microns (<100 μm) is considered [Lees 2003] to be adequately described by the various light gas models, if deposition on the ground is taken into consideration. During the emission of solid particles in the atmosphere, the two predominant phenomena which influence their dispersion are the momentum (wind motion) and the gravitational settling. In the case of PM_{10}, however, the predominant phenomenon is gravitational settling.

The deposition speed, u_d (m/s), of solid particles with diameter, d_p (m) and density, ρ_p (kg/m^3), can be obtained from Stokes law, as:

$$u_{d} = \frac{d_{p}^{2}\, g\, \rho_{p}}{18\eta} \quad . \tag{C5.60}$$

In this expression, g (m/s^2) is the acceleration due to gravity (= 9.81 m/s^2), and η (Pa s) the kinetic viscosity of air.

The concentration of microparticles is obtained, if the concentration, C (μg/m^3), calculated by Eq. (C5.43), is multiplied by the dimensionless empirical factor, DP (-), calculated by the following expression [Lees 2003]:

$$DP = DP_{0}^{(u_{so} u_{d} / u_{s} u_{do})} \quad . \tag{C5.61}$$

In the above expression, u_s (m/s), denotes the mean wind speed at the point of the release. The correction factor DP_o is given in literature [Lees 2003] as a function of the height of the source, its distance from the source, and the atmospheric stability class (A, B, C, D, E, and F), for a deposition speed $u_{do} = 0.01$ m/s and a wind speed $u_{so} = 1$ m/s (see also Table C5.15).

Table C5.15. Correction Factor DP_o [Lees 2003].

Atmospheric stability	Source height (m)	Distance from source (m)			
		10	100	1,000	10,000
A	0	0.90	0.82	0.75	
	10	1.00	0.96	0.90	
	100	1.00	1.00	1.00	
B	0	0.90	0.80	0.65	
	10	0.98	0.95	0.80	
	100	1.00	1.00	1.00	
C	0	0.90	0.70	0.60	0.45
	10	1.00	1.00	0.80	0.60
	100	1.00	1.00	1.00	0.80
E	0	0.90	0.60	0.15	0.012
	10	1.00	1.00	0.80	0.10
	100	1.00	1.00	1.00	0.80
F	0	0.90	0.65	0.40	0.07
	10	1.00	1.00	0.80	0.20
	100	1.00	1.00	1.00	0.80

On September 26, 2006 [European Parliament Press release, 2006], the European Parliament discussed the proposal of H. Krahmer (rapporteur on the directive) which called for, among other factors, the reduction of the maximum mean concentration of PM_{10} particles, to 33 μg/m^3 for the year 2010, replacing the existing limit of 40 μg/m^3. The imposition of a new limit value for the $PM_{2.5}$ concentration was also discussed. Regardless of these discussions, the European Council published the directive 2008/50/EC [Directive Council 2008/50/EC] where the 2010 limit value for PM_{10} concentration, was kept as in 2005, at 40 μg/m^3 [Directive Council 1999/30/EC]. Furthermore, the yearly average limit value for $PM_{2.5}$ was set to 25 μg/m^3 from 2010, and to 20 μg/m^3 from 2020.

C5.7. Effects of Toxic Gas Dispersion

Effects of Dispersion

calculation procedure

Toxic gases can appear in the atmosphere as a consequence of an industrial accident or a terrorist attack. Their appearance can be traced to three prime sources: (a) direct dispersion of toxic gases, (b) indirect dispersion following an explosion, and (c) indirect appearance as combustion products following a fire of harmful chemicals. Their effects on people and on the environment are very important and spread in a large distance from the point of the initial source.

In this section the toxic exposure limits, which are the toxic concentrations over which specific harmful effects are noted to people exposed to them, are presented. The limiting values imposed by international organizations are also discussed. Following this, and based on these limiting values, safety zones are determined, that is the safe distance from the release point of a harmful chemical.

Toxic Exposure Limits
↓
Atmospheric Pollutants
↓
Risk Zones

C5.7.1. Toxic Exposure Limits

As toxic exposure limits for the people, we consider those limits over which specific harmful effects are noted to people exposed to them.

The toxic exposure limits are determined according to the intake dose, which is directly related to the extent and the duration of the exposure to the harmful chemical. In the case of toxic gas dispersion, the exposure can be:

a) accidental, directly following the gas release, and/or
b) longterm, i.e., ground or water pollution, or effects from polluted agricultural products, etc.

Certainly more intense are the effects that appear immediately following the release of the toxic gas, as at that time the highest concentrations appear in the environment.

During the procedure of calculating the toxic exposure limits of a longterm exposure, the exposure time is considered as constant, while for the exposure intensity, the mean concentration of the harmful gas is taken. In the case, however, of an accidental exposure, both the exposure time and the exposure intensity are variables and thus the relative toxic exposure dose is calculated by mathematical expressions. Hence, the "dose" that a person is subjected to, for a toxic chemical for a specific exposure time, is a critical parameter in calculating effects.

An "exposure dose" is how much of a substance is encountered in the environment. An "absorbed dose" is the amount of a substance that actually got into the body through the eyes, skin, stomach, intestines, or lungs.

Effects of Dispersion

calculation procedure

The absorbed dose, D (mg/kg), by inhalation for a person, is calculated from the empirical expression

$$D = C^n t \tag{C5.62}$$

where C (mg/m^3) denotes the concentration of the chemical at a particular point in space, n is an empirical constant and t (min) denotes the exposure time. The above expression is valid for a continuous constant-rate release of a toxic substance and it assumes that the person stays in that point for the whole duration of the release (stationary observer), and then moves to a point of zero concentration.

Toxic Exposure Limits
↓
Atmospheric Pollutants
↓
Risk Zones

The calculation of the consequences of a harmful event is carried out according to the personal hazard technique. The personal hazard technique denotes the percentage of people in a population, which is expected to show an injury, or the probability that a person has to be subjected to an injury for a specified level of dose. Usually, personal hazard is expressed in limiting values that correspond to cases where an injury appears to 50% or 1% of the people in a population. All necessary parameters for their determination are usually obtained from corresponding experiments in different types of animals or actual observations after large industrial accidents. It should be noted that in the corresponding indices that will be discussed here, although the exposure time is not mentioned, it is always taken as constant (usually 30 min).

The most usual indices are the following:

- LC_{50} (Lethal Concentration 50),
- LC_1 (Lethal Concentration 1),
- LD_{50} (Lethal Dose 50),
- IDLH (Immediately Dangerous to Life and Health),
- LOC (Level of Concern),
- PEL (Permissible Exposure Level).

These indices limits will be discussed in the following subsections.

a) LC_{50} (Lethal Concentration 50)

LC_{50} (mg/m^3 air) represents the concentration of a toxic substance in the air, which is likely to cause death to 50% of the population, through inhalation for 30 min exposure time.

In order to calculate the concentration LC_{50} for an exposure time of 30 min when you know its value at a different exposure time, one can use the following expression [TNO 1989]

$$\mathrm{LC}_{50}(30\,\mathrm{min}) = \mathrm{LC}_{50}(t)\left(\frac{t}{t_o}\right)^{1/n}, \qquad \text{(C5.63)}$$

where the time t is expressed in hours and $t_o = 0.5$ h. The value of the constant n can be calculated through this equation, employing experimental data. Otherwise, it can approximately be taken as equal to 3.

Concentration LC_{50} can also be obtained from observations, in particular laboratory animals (guinea pigs, rats, etc.), of equivalent 30 min exposure time, from the expression

$$\mathrm{LC}_{50}(\mathrm{people}) = F_d\, \mathrm{LC}_{50}(\mathrm{laboratory\ animals})\,. \qquad \text{(C5.64)}$$

Values for the parameter F_d (-) are shown in Table C5.16.

LCt_{50} (mg.min/m^3 air), the Median Lethal Concentration, represents the concentration of a toxic substance in the air, multiplied by the total exposure time, which is likely to cause death to 50% of the population through its inhalation.

Hence,

$$\mathrm{LCt}_{50} = 30\, \mathrm{LC}_{50}\,. \qquad \text{(C5.65)}$$

Concentration values LCt_{50} for various chemical substances, used or designed to be used as chemical weapons, were shown in Table C5.1.

Table C5.16. Parameter F_d of Eq. (C5.64) [TNO 1989].

Laboratory animal	F_d (-)
Rat	0.25
Mouse	0.5
Guinea Pigs	0.2
Hamster	0.3

b) LC_1 (Lethal Concentration 1)

LC_1 (mg/m^3 air) represents the concentration of a toxic substance in the air, which is likely to cause death to 1% of the population, through inhalation for 30 min exposure time.

Concentration LC_1 can be approximately calculated from concentration LC_{50}, through the equation [TNO 1989]

$$LC_1 = \exp\left[\ln(LC_{50}) - \frac{2.33}{1.5}\right]. \quad (C5.66)$$

c) LD_{50} (Lethal Dose 50)

LD_{50} (mg/kg animal) represents the lowest concentration of a toxic substance, which is likely to cause death to 50% of the laboratory animals. Toxicologists usually use rats or mice for their experiments. One very important factor in determining LD_{50} is the method used to administer the toxic substance to the animal, usually either through skin absorption or oral intake. So, although LC_{50} refers to the concentration in air of the toxic substance, which is likely to cause death to 50% of the human population through inhalation for 30 min exposure time, LD_{50} refers to the lethal dose that is likely to cause death to 50% of them. The limit LD_{50} expresses more directly the toxic capability of a substance.

In literature, one can still find the following toxic exposure limits:

- LD_1 (Lethal Dose 1).
  Minimum dose to cause fatality in 1% of laboratory animals.
- LD_{100} (Lethal Dose 100).
  Minimum dose to cause fatality in 100% of laboratory animals.
- LDLO (Lowest Dose causing Lethality).
- TDLO (Lowest Dose causing Toxic effect).

Since small animals are employed in the laboratory to determine LD_{50}, the obvious question is how to carry over the results from animal tests to humans. To do exactly that, we employ today the empirical Toxicology Classes Scale of Hodge & Sterner [Hodge & Sterner 1956]. This scale (see Table C5.17], although created in 1956, is still valid. To carry over the dose results from laboratory animals to humans, large safety factors were employed, and certainly the scale produces only empirical values.

Effects of Dispersion

calculation procedure

Toxic Exposure Limits
↓
Atmospheric Pollutants
↓
Risk Zones

Table C5.17. Toxicity Classes Scale [Hodge & Sterner 1956].

Toxicity Classes	LD_{50} Oral (single dose to rats) (mg/kg)	LD_{50} Inhalation (exposure of rats for 4 h) (ppm)	LD_{50} Dermal (single application to skin of rabbits) (mg/kg)	Probable Lethal Dose for Man
1 Extremely Toxic	<1	<10	<5	1 sip
2 Highly Toxic	1 - 50	10 - 100	5 - 43	4 ml
3 Moderately Toxic	50 - 500	100 - 1,000	44 - 340	30 ml
4 Slightly Toxic	500 - 5,000	1,000 - 10,000	350 - 2,810	600 ml
5 Practically Non-Toxic	5,000 - 15,000	10,000 - 100,00	2,820 - 22,590	1 l
6 Relatively Harmless	>15,000	>100,000	>22,600	1 l

According to the Toxicity Classes Scale of Hodge & Sterner shown in Table C5.17, a chemical of class "2 Highly Toxic" is indeed highly toxic. Even 2-3 ml in a 1-kg rat would result in its death, while 4 ml of the same substance would result in the death of a human being.

Unfortunately, the Toxicity Classes Scale of Hodge & Sterner is not the only one. Table C5.18, presents the Toxicity Classes Scale of Gosselin, Smith and Hodge [Gosselin *et al.* 1984]. This scale unfortunately has a reverse numbering of classes in comparison with the aforementioned scale and the user must be particularly careful (an Extremely Toxic substance belongs to class 1 according to the Hodge & Sterner Scale, but to class 5 according to the scale of Gosselin, Smith and Hodge).

Table C5.18. Toxicity Classes Scale [Gosselin *et al.* 1984].

Toxicity Classes	LD_{50} (Oral) (mg/kg person)	Dose For 70-kg Person
1 Practically Non-Toxic	>15,000	>1,200 ml
2 Slightly Toxic	5,000 - 15,000	600 - 1,200 ml
3 Moderately Toxic	500 - 5,000	30 - 600 ml
4 Very Toxic	50 - 500	30 ml
5 Extremely Toxic	5 - 50	4 ml
6 Super Toxic	<5	1 sip (< 7 drops)

d) IDLH (Immediately Dangerous to Life and Health)

IDLH (mg/m^3 air) represents the maximum concentration of a toxic substance in the air, to which a healthy employee can be exposed for 30 min, and still be able to escape without suffering non-reversible health damages or injuries that would prevent him from escaping (mainly eye or lung irritation). IDLH limits refer exclusively to non-reversible health damages resulting from inhaling a toxic substance. It is of interest that the IDLH limits proposed in the mid-1970s by NIOSH (National Institute of Occupational Safety and Health, USA) propose exposure levels only for healthy working people and not sensitive population groups. They are also based upon a 30-min exposure time and not on longterm exposure.

If there are no experimental data for the direct determination of the IDLH limits, then an estimation can be done based upon the values of the other limits that have been obtained from experiments in laboratory animals, at the LD_{50} limit. The IDLH limits that have been developed for serious injuries to people are usually one or two orders of magnitude lower than the lethality limits for 50% of the animal population. More specifically the estimated IDLH limits can be calculated from the following expressions

$$\mathrm{IDLH} = 0.10\ \mathrm{LC}_{50}\,, \tag{C5.67}$$

$$\mathrm{IDLH} = 0.01\ \mathrm{LD}_{50}\,, \tag{C5.68}$$

$$\mathrm{IDLH} = 0.10\ \mathrm{LDLO}\,. \tag{C5.69}$$

The IDLH limits* have been determined and employed by the NIOSH (National Institute for Occupational Safety and Health, USA) and the OSHA (Occupational Safety and Health Administration, USA), in the mid-1970s for 387 chemical substances, aiming to determine the protective respiration equipment that workers in industry must wear according to the USA legislation. In 1994, NIOSH proposed new values for the IDLH limits (these new IDLH values are given in Appendix B).

* IDLH limits are often quoted in ppm. To change the units of mg/m^3 to ppm, divide by the gas density, e.g., if the gas density is 4 kg/m^3, then 2 mg/m^3 correspond to 2/4 = 0.5 ppm.

e) LOC (Level of Concern)

LOC (mg/m^3 air) represents the concentration in air of an extremely hazardous substance above which there may be serious immediate health effects to anyone exposed to it for short periods (usually 30 min). LOC limits are defined in "Technical Guidance for Hazard Analysis - Emergency Planning for Extremely Hazardous Substances, U.S. Environmental Protection Agency, December 1987."

The LOC limit is directly determined from the IDLH limit as

$$LOC = 0.10\ IDLH\,, \tag{C5.70}$$

and is considered as a realistic approximate value for the protection of the whole population. LOC limits refer to the whole population, and include sensitive population groups like older people, children, and those with health problems, and is valid for short exposure periods (usually 30 min).

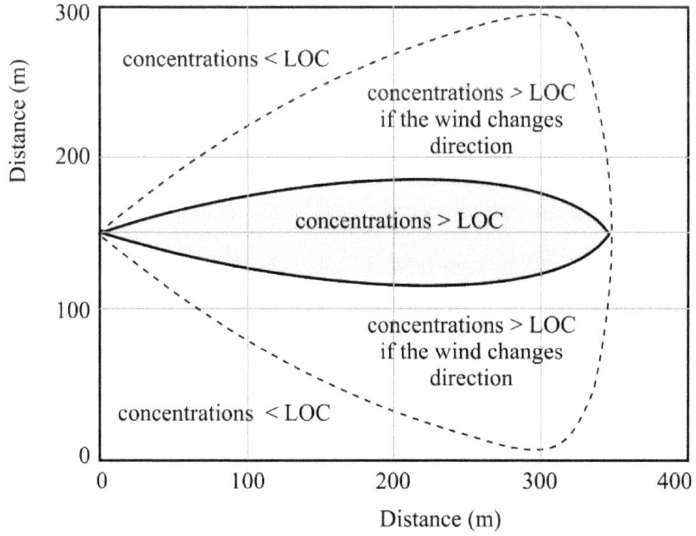

Figure C5.23. LOC concentration footprint.

f) PEL (Permissible Exposure Level)

PEL (mg/m^3 air) represents the minimum permissible exposure limits in toxic substance concentrations in air in areas where workers exist. They are mostly employed by OSHA (Occupational Safety and Health Administration, USA) and refer to averages over 8 hours exposure time. More information on PEL limits as well as full catalogues of limits can be easily found at http://www.osha.gov/.

Effects of Dispersion

example

Toxic Exposure Limits

↓

Atmospheric Pollutants

↓

Risk Zones

EXAMPLE C5.7. *Calculation of Toxic Exposure Limits*

When a group of rats was exposed to 420 mg/m^3 of phosgene, half of them died in 30 min. Calculate the toxic exposure limits for phosgene. The following data are available:

Density of phosgene, ρ	:	4	kg/m^3

Since half of the rats died in 30 min, then

$$LC_{50}\text{ (rats)} = 420\text{ mg/m}^3.$$

Employing Eq. (C5.64) and $F_d = 0.25$ (Table C5.16), we obtain

$$LC_{50} = 0.25\times420 = 105\text{ mg/m}^3.$$

From Eq. (C5.65),

$$LCt_{50} = 30\times105 = 3{,}150\text{ mg min/m}^3.$$

We observe a very good agreement with the value of 3,200 mg min/m^3 given in Table C5.1.

Also from Eq. (C5.66), we can calculate

$$LC_1 = \exp\left[\ln(LC_{50})\text{-}(2.33/1.5)\right] = 22\text{ mg/m}^3.$$

From Eqs. (C5.67)-(C5.69), we obtain

$$IDLH = 0.1\ LC_{50} = 10\text{ mg/m}^3\ \text{ (or } 10/4 = 2\text{ ppm)}$$

$$LD_{50} = 10\ LC_{50} = 1{,}050\text{ mg/m}^3$$

$$LOC = 0.1\ IDLH = 1\text{ mg/m}^3\ \text{ (or } 1/4 = 0.2\text{ ppm)}$$

We also observe that the IDLH value is in full agreement with that in Appendix B.

❐

Effects of Dispersion

calculation procedure

Toxic Exposure Limits
↓
Atmospheric Pollutants
↓
Risk Zones

C5.7.2. Special Atmospheric Pollutants Limits

For the common atmospheric pollutants NO_2, NO_x, CO and SO_2, the European Union is trying to reduce their emissions as shown in Table C5.19. These limits, valid at the time of print, are quite low because they refer to large exposure times.

Table C5.19. Special Limits for NO_2, NO_x, CO and SO_2.

Reason	Measurement duration	Concentration limit	Attained until
		NO_x, NO_2 (IDLH 20 ppm or 36 mg/m^3)	
Protection of human health	1 hour	200 μg/m^3, NO_2 not to be exceeded by more than 18 times/year	1.1.2010
Protection of human health	1 year	40 μg/m^3, NO_2	1.1.2010
Hazardous concentration	3 hours	400 μg/m^3, NO_2	
		CO (IDLH 1,200 ppm or 1,350 mg/m^3)	
Protection of human health	8 hours	10 mg/m^3	1.1.2005
		SO_2 (IDLH 100 ppm or 250 mg/m^3)	
Protection of human health	1 hour	350 μg/m^3, not to be exceeded by more than 24 times/year	1.1.2005
Protection of human health	24 hours	125 μg/m^3, not to be exceeded by more than 3 times/year	1.1.2005
Hazardous concentration	3 hours	500 μg/m^3	1.1.2005

Source: Directive Council 1999/30/EC, Official Journal of the European Communities, Directive Council Proposal 2005/0183 COD, COM 2005 447 final.

C5.7.3. Risk Zones

A Risk Zone is a specified geographical region which can be submitted to the effects of a toxic cloud or puff dispersion in such concentration levels that could place the public health in danger (by death, serious injuries or health problems). The geographical region covers an area that is defined from the point of source of the emission of the toxic substance, or the end of the previous zone, until the last point (or line) where the exposure limit that defines this safety zone is valid.

The definition of the risk zones is based on the evaluation of the exposure levels, with reference to the event (toxic substance concentration dose, or heat flux dose in case of fires, overpressure dose in case of explosions), and in parallel with the application of a specified limit (see also Table C2.9). To better determine each risk zone, one of the aforementioned limits (LC_{50}, LC_1 or IDLH) is employed.

Zone I - Very Serious Consequences (LC_{50} region)

This corresponds to the zone with probable deaths to 50% of the exposed population, through inhalation of a toxic substance. Zone I directly surrounds the point of emissions/explosion. The definition of this zone is the first duty not only for the protection of people, but also for undertaking control and safety actions to deal with the event.

Zone II - Serious Consequences (LC_1 region)

This corresponds to the zone of serious injuries, assuming a probable 1% death rate of the exposed population, by inhaling toxic substance. It follows immediately after Zone I, and people in it are mostly in danger of serious, non-reversible injuries (especially sensitive population groups, e.g., youngsters, sick people, the elderly, etc.). Usual response measures are the immediate evacuation of the population that is in danger.

Zone III - Moderate Consequences (IDLH region)

This corresponds to the final zone, where the probability exists for reversible injuries following the inhalation of toxic substance. It is the third and last zone. People in this zone are not threatened by serious consequences to their health, but mostly threatened by light injuries and other disturbance (dizziness, temporary eyesight restriction, loss of consciousness), which could however affect their safety in the case of more sensitive people. Outside Zone III is the "safe zone" where people are not in any danger and no measures are necessary.

C5.8. Examples

In this section a case study will demonstrate the full significance of dispersion modeling. Following that, other industrial accidents that resulted in the emission of toxic gases will be listed.

C5.8.1. Case Study: Toxic Gas Emissions at the City of Kamini

The inhabitants of the city of Kamini (hypothetical case - see Figure C5.24) are strongly protesting due to their respiratory problems, naming as the cause of their health problems the nearby fertilizer plant. However, the plant manager and the environmental engineer refuse to accept that the NO_2 emissions from the plant stacks could affect the city, as the concentrations measured at the plant's meteorological station (which is at the far end of the plant towards the city) are below the allowable limits.

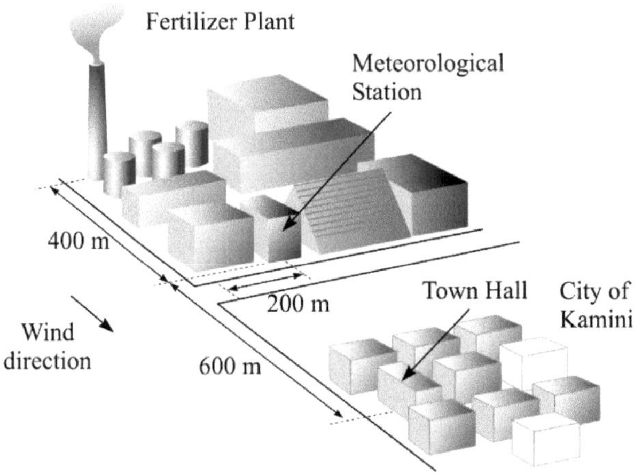

Figure C5.24. The city of Kamini and the nearby fertilizer plant.

In order to find a solution to this conflict and as both sides are insistent in their views, it was decided that independent and in-parallel concentration measurements would be carried out in the meteorological station and in the town hall.

Based on the data shown below and assuming that the NO_2 does not take part in photochemical reactions, estimate the toxic concentration levels, and decide as an independent expert, which side is right.

The following data are available:

Case Study

Kamini City

Plume Rise

Rise due to Buoyancy or Momentum
↓
Distance of Maximum Plume Rise
↓
Gradual and Final Plume Rise
↓
Dispersion Equation
↓
Concentrations Comparison

Case Study

Kamini City

Cloud coverage	:	7/8	-
Wind speed (at 10m height), u_{ref}	:	4.5	m/s
Stack height, h_s	:	50	m
Stack diameter, d_s	:	1.7	m
Exit speed of gases, v_s	:	12	m/s
Ambient temperature, T_a	:	298	K
Stack exit temperature, T_s	:	673	K
Emission mass rate, Q	:	150	g/s

Plume Rise

Rise due to Buoyancy or Momentum
↓
Distance of Maximum Plume Rise
↓
Gradual and Final Plume Rise
↓
Dispersion Equation
↓
Concentrations Comparison

In order to decide upon the level of the toxic concentrations and the validity of the measurements, the best way is to calculate the concentration for both places and to examine these calculated values in relation to the measured ones. The experimental procedure for the calculation of the concentrations consists of the following steps:

a) calculation of wind speed at the stack exit,
b) determination as to whether the predominant effect is buoyancy forces or momentum forces,
c) calculation of gradual and final height,
d) calculation of dispersion coefficients,
e) calculation of dispersion concentrations.

The wind speed at the stack exit, is given as a function of the wind speed, u_{ref} (m/s), at 10 m, from Eq. (C5.1). The parameter p (-), in the equation, is determined as a function of the atmospheric stability class and the type of the region. In this particular case, we refer to an urban region, and from the cloud coverage and the wind speed, from Table C5.2, the atmospheric stability class is D.

For atmospheric stability class D, parameter p is equal to 0.25 and thus from Eq. (C5.1), the wind speed at the stack height is calculated as, $u_s = 6.73$ m/s.

To decide if the predominant effect is the buoyancy or momentum forces, the criterion is Eq. (C5.3),

$$\left(T_s - T_a\right) \geq \Delta T_c ,$$

where, T_s (K) is the temperature of the released gases (673 K) and T_a (K) the ambient temperature (298 K). The calculation of the critical difference, ΔT_c (K), is carried out from the empirical equations of Table C5.4, as a function of the buoyancy flux parameter, F_b (m^4/s^3), which is obtained from Eq. (C5.6), as

$$F_b = g\, v_s\, d_s^2 \left(\frac{T_s - T_a}{4\,T_s}\right) = 47.4\ \mathrm{m^4/s^3}.$$

As the value of the buoyancy flux parameter is smaller than 55, the critical difference, ΔT_c (K), is calculated from Eq. (C5.4), as equal to 32.13 K. This result is obviously smaller than the difference (673-298), a fact that determines that the predominant effect is that of the buoyancy.

Before calculating the gradual and the final height we must first find the maximum plume rise, x_f (m). For neutral atmospheric conditions, where the predominant effect is buoyancy and for buoyancy flux parameter, $F_b = 47.4\ m^4/s^3 < 55$, from Eq. (C5.8), the maximum plume rise is calculated as

$$x_f = 49 F_b^{5/8} = 546.5 \text{ m}.$$

The gradual and final rise, h_e (m), of the plume is calculated from Eqs. (C5.15) and (C5.19) respectively. Results are shown in Table C5.20 and in Figure C5.25.

Table C5.20. Calculated Values of Gradual, Final Rise and Concentration as a Function of Distance.

x (m)	h_e (m)	σ_y (m)	σ_z (m)	$C(x, 0, 2)$ ($\mu g/m^3$)	$C(x, 200, 2)$ ($\mu g/m^3$)
50	61.68	7.9	6.9	0.00	0.00
100	68.54	15.7	13.8	0.24	0.00
150	74.30	23.3	20.5	30.19	0.00
200	79.43	30.8	27.2	162.0	0.00
250	84.15	38.1	33.8	331.7	0.00
300	88.57	45.4	40.2	461.8	0.03
350	92.74	52.4	46.6	536.3	0.37
400	96.72	59.4	52.9	567.2	1.97
450	100.54	66.3	59.1	569.6	6.00
500	104.21	73.0	65.3	555.4	13.1
546.5	**107.53**	**79.2**	**70.9**	**533.9**	**22.0**
550	107.53	79.7	71.3	534.9	22.9
600	107.53	86.2	77.3	540.0	36.6
700	107.53	99.0	89.1	518.0	67.3
800	107.53	111.4	100.6	476.9	95.2
900	107.53	123.5	111.8	431.6	116.3
1,000	107.53	135.2	122.8	388.4	130.1

Case Study

Kamini City

Plume Rise

Rise due to Buoyancy or Momentum

↓

Distance of Maximum Plume Rise

↓

Gradual and Final Plume Rise

↓

Dispersion Equation

↓

Concentrations Comparison

Case Study

Kamini City

Plume Rise

Rise due to Buoyancy or Momentum

↓

Distance of Maximum Plume Rise

↓

Gradual and Final Plume Rise

↓

Dispersion Equation

↓

Concentrations Comparison

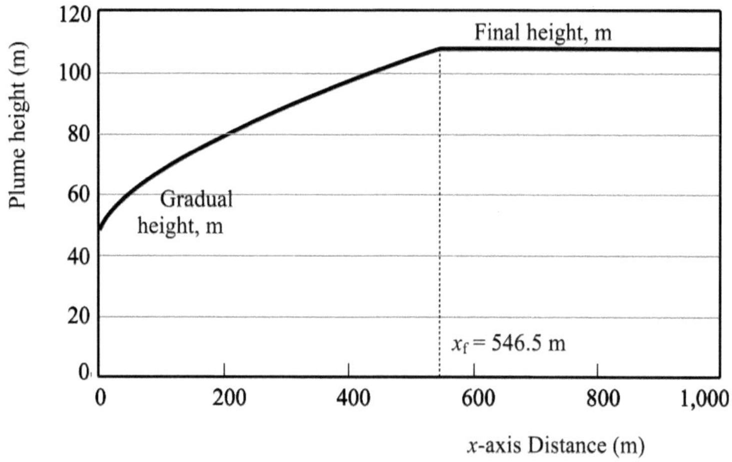

Figure C5.25. Plume height as a function of distance.

To calculate the concentration, we have to obtain the dispersion coefficients, σ_y (m) and σ_z (m). These are calculated as a function of the distance x, from Eq. (C5.26) and the corresponding constants of Table C5.8, for atmospheric stability D. Results are shown in Table C5.20. Hence, from Eq. (C5.25) the concentration C ($\mu g/m^3$) of pollutant NO_2 can now be calculated.

In the diagram of Figure C5.24, it can be seen that the stack and the town hall are in a straight line, while the meteorological station is 200 m away from this line. Hence, having the stack as the emission center, we note that

a) the meteorological station has coordinates $x = 400$ m, and $y = 200$ m, and the concentration is calculated as $C\,(400, 200, 2) = 1.97\ \mu g/m^3$ (z = 2, as the concentration is calculated at the height of a person),

b) the town hall has coordinates $x = 1000$ m and $y = 0$ m, and the concentration is calculated as $C\,(1000, 0, 2) = 388.4\ \mu g/m^3$.

This difference is attributed to the fact that the meteorological station is 200 m away from the straight line between plant and town hall (direction of wind).

Therefore, it is correct that the city inhabitants protest. Moreover, if we examine the values of Table C5.19, we can see that for a continuous exposure for a year the measured limits could be 10 times higher than the established limit (40 $\mu g/m^3$).

❐

C5.8.2. Major Industrial Accidents Caused by Toxic Gas Dispersion

In Table C5.21 an indicative list of major industrial accidents caused by toxic gases dispersion, is given. The list is part of the database MinA - Major Industrial Accidents due to Fires, Explosions and Toxic-Gas Releases [MinA 2006].

Table C5.21. Major Industrial Accidents Caused by Toxic Gases Dispersion.

Date	Country	City/State	D	I	E[a]	Chemical	Unit	M$[b]
2005								
13.11	China	Harbin	5	60	10,000	Benzene	Petrochemicals	
2003								
25.4	USA	Mill Hall, PA		60		PVC	Petrochemicals	
2002								
23.11	Canada	Quebec, QC		65		Ammonia		
13.5	Canada	Montreal, MO		60		Acid		
2001								
10.9	UK	Bridgnorth		23		HCl		
14.7	USA	Detroit, MI	3	10	2,000	Mercaptan	Transport	
8.7	China	Shandong	13	11	48	Ammonia	Fertilizer plant	
7.7	USA	Tulsa, OK		138		Arsenic		
3.4	China	Taichung		100		Acrylnitrile	Transport	
2000								
12.9	Mexico	Salamanca		300		Malathion	Petrochemicals	
9.9	USA	Manhattan, NY		2,000		Anvil	Transport	
1997								
9.2	Russia	Khabarovsk	1	208		Chlorine	Chemicals	
21.1	India	Bhopal		400		Ammonia Nitrate	Transport	
10.1	Pakistan	Lahore	26	900	1,000	Ammonia Nitrate	Transport	
1996								
11.4	USA	Alberton, MO	140		<1,000	Chlorine	Train Wagon	
1995								
1.12	India	Maharashtra		2,000		Ammonia Nitrate	Transport	
9.10	UK	Wilton				Propylene		25
15.7	Iran	Astara	3	200		Chlorine		
28.4	Korea	Taegu	101	104	>10,000	LPG		
20.3	Japan	Tokyo	12	980		Sarin	Metro	

[a] D I E - Number of Deaths, number of Injuries, number of people Evacuated.

[b] M$: Material Damage in million $ (in 2005 prices).

Note: HCl, hydrochloric acid; LPG, liquefied propane gas; PVC, polyvinyl chloride.

- *Continued*

Table C5.21 (cont.). Major Industrial Accidents Caused by Toxic Gases Dispersion.

Date	Country	City/State	D	I	E[a]	Chemical	Unit	M$[b]
1994								
27.6	Japan	Matsumoto	7	54		Sarin	Metro	
1993								
1.11	Vietnam	Nam Khe	47	48		Hydrocarbons		
4.8	Colombia	Remeios	430			Oil	Petrochemicals	
26.7	USA	Richmond, VA		>6,250		Acid		
1.7	Canada	St Terese, MO		36		Dioxins		
2.2	Germany	Griesheim					Reactor	
1992								
27.11	Belgium	Diest				Dioxins	Petrochemicals	
3.10	Germany	Lengrich				Dioxins	Petrochemicals	
10.9	USA	Camden, NJ		12		HCl	Petrochemicals	
1.9	Greece	Elefsina	14	21		LPG	Refinery	
24.3	Senegal	Dakar	>40	>300		Ammonia Nitrate		
23.2	Korea	Kwangju		16	20,000	LPG		
23.1	Germany	Schkopau		186		Chlorine		
1991								
29.12	Mexico	St.Luis Potosi		40		Butane		
3.9	UK	Immingham		127			Fertilizers	
10.8	Taiwan	Kaohsiung		600				
5.1	Switzerland	Nyon			12,000	Chlorine	Chemicals	
1990								
25.7	UK	Birmingham		60	75,000	Phosgene		
22.7	Korea	Ulsan			>10,000	Butane		
4.5	Cuba	Matanzas	3	374	>1,000	Ammonia Nitrate		
18.3	Korea	Daesan		>100	>10,000	HCl		
17.1	Germany	Ahlsfield		>182		Chlorine	Transport	
1989								
22.7	USA	Freeland, MI					Transport	
6.7	UK	Billingham	2			Ammonia	Fertilizers	
20.3	Lithuania	Jonova	7	57		Ammonia	Storage	
5.1	USA	Los Angeles, CA			11,000	Chlorine		
1988								
30.7	USA	Altoona, LA					Train Wagon	
1987								
15.12	Mexico	Minatitlan		200	1,000	Acrylnitrile		
7.7	USSR	Annau	200			Chlorine	Train Wagon	
21.5	UK	Southampton	117			Hydrocarbons		
30.1	USA	Texas City, TX	255		4,000	Acid		

[a] D I E - Number of Deaths, number of Injuries, number of people Evacuated.

[b] M$: Material Damage in million $ (in 2005 prices).

Note: HCl, hydrochloric acid; LPG, liquefied propane gas.

- Continued

Table C5.21 (cont.). Major Industrial Accidents Caused by Toxic Gases Dispersion.

Date	Country	City/State	D	I	E[a]	Chemical	Unit	M$[b]
1986								
1.11	Switzerland	Basel				Pesticides	Storage	
1985								
4.12	India	New Delhi	1	340	>10	Acid		
26.8	USA	S. Charleston, SC		135		HCl		
6.7	USA	Clinton, IA	5	8		Ammonia	Chemicals	19
9.6	USA	Pine Bluff, AR					Train Wagon	
20.3	USA	Plainfield, NJ				HCl		
23.2	USA	Jackson, SC					Train Wagon	
1984								
17.12	Mexico	Matamoros		182	3,000	Ammonia		
25.2	Brazil	Sao Paolo	508	221		Petroleum		
1983								
3.4	USA	Denver, CO				Acid	Train Wagon	
1982								
17.1	France	Moselle River	5				Pipeline	
1981								
21.8	USA	San Francisco, CA		28	7,000		Transport	
1.8	Mexico	Montana	7	280		Chlorine	Train Wagon	
1.6	USA	Geismar, LA		125		Chlorine		
8.4	Mexico	Montanas	29	50		Chlorine		
1980								
20.11	UK	Wealdstone		1		Propane	Storage	
8.10	Mexico	Mexico City	9	28		Ammonia	Distribution	
1979								
10.11	Canada	Toronto, ON		8	215,000	Chlorine	Train Wagon	
18.7	USA	Bayonne, NJ				Chlorine	Train Wagon	
8.4	USA	Crestview, FL		14			Train Wagon	
11.1	Norway	Rafnes				Chlorine	Chemicals	
9.1	Germany	Burghausen				HCl	Chemicals	
1978								
15.4	USA	Chicago, IL	8	29		HCl		
26.2	USA	Youngstown,OH	8	50		Chlorine		
1977								
10.12	Colombia	Pasacabolo	30	22		Ammonia	Fertilizer plant	
9.11	USA	Pensacola, FL				Ammonia	Train Wagon	
2.2	USA	Baton Rouge, LA		10		Chlorine	Chemicals	

[1] D I E - Number of Deaths, number of Injuries, number of people Evacuated.

[2] M$: Material Damage in million $ (in 2005 prices).

Note: HCl, hydrochloric acid.

- *Continued*

Table C5.21 (cont.). Major Industrial Accidents Caused by Toxic Gases Dispersion.

Date	Country	City/State	D	I	E[a]	Chemical	Unit	M$[b]
1976								
10.12	USA	Baton Rouge, LA		3		Chlorine	Storage	
26.9	Italy	Manfredonia		30		Arsenic		
2.9	Germany	Haltern					Storage	
10.7	Italy	Seveso		300		TCDD	Petrochemicals	
11.5	USA	Houston, TX	6	200		Ammonia	Transport	
7.5	USA	Enid, OK				Ammonia	Pipeline	
16.1	Sweden	Landskrona	2			Ammonia	Distribution	
1975								
14.12	USA	Niagara Falls, NY	4	80		Chlorine	Transport	
3.9	USA	Texas City, TX				Ammonia	Pipeline	
2.8	USA	Romulus, MI					Pipeline	
1974								
30.4	Japan	Yokkaichi		521		Chlorine	Transport	
26.4	USA	Chicago, IL	1	160	16,000		Storage	
6.2	USA	Omaha, NE				Chlorine		
1973								
6.12	USA	McPherson, KS				Ammonia	Pipeline	
6.11	USA	Ventura, CA	2	4		LPG	Train Wagon	
13.7	South Africa	Potchefstroom	18	65		Ammonia	Storage	
5.3	Canada	Loos, BC				Chlorine	Train Wagon	
1971								
11.9	USA	Pensacola, FL				Cyclohexane		
5.6	USA	Floral, AR				Ammonia	Pipeline	
24.4	Germany	Emmerich	4	53		Ammonia	Chemicals	
1970								
16.11	USA	Blair, NE				Ammonia	Storage	
19.10	Sweden	Javle				Chlorine	Pipeline	
21.1	USA	Belle, WV		1		Ammonia	Train Wagon	
1969								
11.9	USA	Glendora, MS		1	30,000	VCM	Train Wagon	
8.5	USA	Cleveland, OH	1			Chlorine	Transport	
18.2	USA	Crete, NE	8	20		Ammonia	Transport	
1968								
21.8	France	Lievin	5	35		Ammonia	Transport	105
1967								
14.6	Italy	La Spezia				Chlorine	Train Wagon	

[a] D I E - Number of Deaths, number of Injuries, number of people Evacuated.

[b] M$: Material Damage in million $ (in 2005 prices).

Note: LPG, liquefied propane gas; TCDD, 2,3,7,8-tetrachlorodibenzo-p-dioxin; VCM, vinyl chloride monomer.

- Continued

Table C5.21 (cont.). Major Industrial Accidents Caused by Toxic Gases Dispersion.

Date	Country	City/State	D	I	E[a]	Chemical	Unit	M$[b]
1968								
21.8	France	Lievin	5	35		Ammonia	Transport	105
1967								
14.6	Italy	La Spezia				Chlorine	Train Wagon	
1965								
12.9	USA	Baton Rouge, LA				Chlorine	Transport	
1964								
2.7	USA	Mobile, AL	1			Chlorine	Pipeline	
1963								
9.8	USA	Philadelphia, PA		430		Chlorine	Train Wagon	
28.4	USA	Brandtsville, PA				Chlorine	Train Wagon	
6.3	Netherlands	Amsterdam	8	14		TCDD		
1962								
30.11	Canada	Cornwall, ON		89		Chlorine	Train Wagon	
1961								
23.2	UK	Billingham				Chlorine	Pipeline	
31.1	USA	La Barre, LA	1	114		Chlorine	Train Wagon	
1959								
27.2	USA	Portland, OR				LPG	Transport	
1957								
16.4	USA	Nitro, WV	8			Chlorine		
1956								
10.3	USA	Lake Charles, LA				Chlorine	Storage	
1954								
25.6	Canada	Montreal, QC	1			Chlorine	Natural Gas	
1952								
4.4	Germany	Walsum	7	56		Chlorine	Storage	
1950								
20.7	UK	Billingham				Chlorine	Train Wagon	
16.2	Mexico	Poza Rica	22	320		H_2S	Petrochemicals	
1949								
1.9	USA	Freeport, TX				Chlorine	Pipeline	

[a] D I E - Number of Deaths, number of Injuries, number of people Evacuated.
[b] M$: Material Damage in million $ (in 2005 prices).
Note: H_2S, hydrogen sulfide; LPG, liquefied propane gas; TCDD, 2,3,7,8-tetrachlorodibenzo-p-dioxin.

- Continued

Table C5.21 (cont.). Major Industrial Accidents Caused by Toxic Gases Dispersion.

Date	Country	City/State	D	I	E[a]	Chemical	Unit	M$[b]
1947								
5.11	Finland	Rauma	19	200		Chlorine		
4.2	USA	Chicago, IL				Chlorine	Train Wagon	
1940								
26.1	Norway	Mjodelana	3	34		Chlorine	Train Wagon	
1939								
24.12	Romania	Zarnesti	60			Chlorine	Storage	
1936								
12.11	USA	Johnsonburg, NY	1			Chlorine	Pipeline	
1935								
13.3	USA	Griffith, IN				Chlorine	Train Wagon	
1934								
28.2	USA	Niagara Falls, NY				Chlorine	Train Wagon	
1929								
10.5	USA	Syracuse, NY	1	100		Chlorine	Storage	
1928								
7.7	USA	Asbokan, NY				Chlorine	Train Wagon	
20.5	Germany	Hamburg	10	50		Phosgene	Storage	
1926								
13.12	France	St. Auban	19	105		Chlorine	Storage	
1921								
24.8	UK	Hull				Hydrogen	Transport	

[a] D I E - Number of Deaths, number of Injuries, number of people Evacuated.
[b] M$: Material Damage in million $ (in 2005 prices).

D causes of destruction

D1 Introduction

The previous chapter presented methods for the calculation of the effects from fires, explosions or toxic gas dispersion. The question we will be posing in this chapter is, "How can such an event occur, and what are the usual causes that lead to destruction?"

The most important cause of every accident is ignorance, even if all safety precautions have been taken. The ignorance of hazard, for example, can lead an operator to make a mistake, and ignorance can lead a process designer to select the wrong equipment for a specific process. Hence, all events before an accident can be connected to ignorance, which in turn can only be dealt with in a systematic responsible approach based on experience.

Damages following the explosion on February 8, 2008, at the Imperial Sugar refinery in Port Wentworth, GA, U.S.A. (Reproduced by kind permission of the U.S. Chemical Safety Board.)

Causes of Destruction

The aim of this chapter is to present a series of incidents that took place in the past (and could take place in the future), so that the reader can get a wide picture of the causes of an industrial accident, and thus be able to make contingency plans and take all necessary safety precautions to avoid accidents. Certainly for the proper preparation of such a study, one of the hazard identification techniques discussed in Chapter A ("What-if" analysis, HAZOP analysis, FMEA analysis, etc.), must also be employed. These techniques, however, are directly dependent upon the experiences of the specific engineer, who is called on to apply them. This chapter intervenes at exactly this point aiming both to advise the new engineer, and update the experienced practitioner. Due to space limitations, it is not possible, nor was it our aim, to present all probable causes of destruction, but instead an overview of selected cases. Thus we chose to present a few important and predictable causes of destruction. The work of Kletz [Kletz 1999, Kletz 2003], Lees [Lees 2003] and Sanders [Sanders 1999], was our prime source of literature, along with the Internet and the Major Industrial Accidents Database, MinA (2006).

The chapter is separated into two main sections. In the first section, we show probable causes which can lead to the destruction of equipment or facilities. These are placed in categories according to the equipment involved, e.g., storage tanks, stacks, pipes, etc. Every such case initially includes a description of the technical specifications of the particular equipment. Following that, some possible accidents are discussed.

The second section refers to sources of ignition (where they appear and how they are created) as well as to flammability and explosion limits.

D2

Facilities and Equipment

In this section the possible causes of destruction to facilities and equipment are presented. To help the reader, they are categorized according to the facility or equipment they refer to. The cases discussed are indicative and certainly do not cover all possible causes which can result in an accident.

D2.1. Storage Tanks

Storage tanks can be found in almost all industrial plants (or to that effect, in most areas where engineers are involved). Their size differs greatly, and so differ the material or fluids stored in them. Furthermore, they are fragile and thus easily damaged by overpressure or vacuum. Their widespread use and large size mean that storage tanks figure predominantly amongst major (and minor) industrial accidents. Unfortunately, such tanks are employed for the storage of a very wide variety of fluids, which can be dangerous either directly or indirectly (in contact with other reactive materials), and for this reason they should always be dealt with special caution.

Storage tanks damaged after the disaster at Third Coast plant, Houston, TX, U.S.A. (Reproduced by kind permission of the U.S. Chemical Safety Board.)

Initially we describe the technical characteristics of storage tanks. We also describe, in brief, the steps followed during the design of a storage tank, as given in literature, for the reader who is not so familiar with their design. Following these, a discussion of probable events is presented.

D2.1.1. Technical Specifications

Storage tanks are usually constructed from (a) steel, (b) stainless steel, or (c) aluminum according to specific requirements. They usually consist of large parts welded together and checked at works. Some of the typical design specifications of atmospheric storage tanks include:

- Ability to withstand wind speed of 150 km/h.
- Seismicity of the area.
- 7.5 - 56 mbar design overpressure and 5-7 mbar vacuum design pressure (BS 2654:1989).
- 10° roof inclination to drain rainwater.
- Epoxy-based paint, internally and externally.
- Zinc undercoat for storage tanks near the sea.

D2.1.2. Probable Events

Every incident or accident that happens in the area of a storage tank is characterized by a cause and an effect. In this section we will adopt the following steps: first the effect is described, and more precisely the conditions immediately proceeding the explosion, fire, or dispersion of toxic gas (e.g., the creation of a ground pool from the overfilling of a fuel storage tank) and, following that, the causes that led to this situation.

a) Ground Pool from Overfilling

Although it might seem unbelievable, overfilling is the most common cause of accidents encountered in storage tanks. This usually occurs in open tanks, but as will be shown in the following paragraphs, it can also happen to closed tanks, through the failure of various safety mechanisms, e.g., venting valves.

The consequences of the creation of a pool on the ground from overfilling are obvious, and the probability of fire, explosion or toxic gas dispersion directly depends upon the type of fluid leaked. The reader must always examine the toxicity and the flammability limits of the fluid in question - these will be defined and discussed more extensively in the next section - so that limits are calculated and necessary safety precautions are taken.

Storage tank overfills often occur when the person in charge of the filling procedure is not cautious, or is occupied with something else simultaneously. The December 2, 1999, accident in Sri Racha, Thailand, which destroyed 250,000 barrels of gasoline and caused damages of about US$ 44,000,000 (in 2005 prices) [MinA 2006], was caused by storage tank overfilling. Although the safety instruments that control the filling were activated, the filling continued when the operator closed the wrong valve by mistake.

Storage tanks damaged after the disaster at Third Coast plant, Houston, TX, U.S.A. (Reproduced by kind permission of the U.S. Chemical Safety Board.)

A storage tank can also easily be overfilled when the corresponding control instruments fail. Although every instrument has a specific period of safe operation, maintenance personnel often wait for an instrument to fail before they substitute or repair it. For this reason a second back-up instrument must be available, or alternatively, an operator must be responsible, solely for the proper filling of the storage tank.

There are also cases when the wrong type of instrument or wrong operation manual can cause the storage tank overfilling. Such situations arise when the level control indicator is based on the measurement of weight and not volume for example. This is specifically important when the same storage tank is used for fluids of different density. If the new fluid has lower density, then the operator can allow more material (in volume) to enter in the tank than he should, causing the overfilling of the storage tank. Simply, the operator did not realize that the level indicator measurement was based on weight and not volume.

Sulfuric acid storage tank damaged after explosion at Motiva Enterprises, Delaware City, DE, U.S.A. (Reproduced by kind permission of the U.S. Chemical Safety Board.)

Another cause that leads to overfilling is filling by gravity. Liquid is sometimes transferred from one tank to another by gravity. Overfilling can occur when liquid flows from a tall tank to a shorter one. On one occasion, an overflow occurred when liquid was transferred from one tank to another of the same dimensions several hundred meters away. The operators failed to appreciate the slight slope of the ground that was sufficient to cause the overflow.

Liquid overflow from storage tanks is usually retained in tank dikes built around them for this purpose. Very often, however, drain valves on the dikes - installed so that rainwater can be removed - have been left open, and the spillage is lost to drain. Drain valves should normally be locked shut. In addition, they should be inspected weekly to make sure they are closed and locked.

b) Outflow Because of Over- or Under-Pressure

Atmospheric pressure storage tanks are usually designed to withstand 7.5-56 mbar maximum overpressure. This low pressure makes them specifically fragile. Most storage tanks are designed so that they will burst at the roof/wall weld, so that major spillage is avoided. In many accidents, however, tanks have failed at the base/wall weld, either because the weld was corroded or fatigued, or in some cases where bolts were missing. Corrosion is the most usual cause, and more so when the tank is full of water or water-based solutions. Additionally, constant filling and draining can increase the metal fatigue of the base/wall welds of the storage tank. This can be prevented by leaving about 1 m depth of liquid when the tank is emptied.

The increase in pressure inside a storage tank, over its design pressure, can happen when either gases or liquids are stored. In the case of gases, such phenomena occur easily, as gases are almost always transported and stored under pressure. Furthermore not all operators appreciate how easy it is for a gas under pressure to destroy a storage tank. For example, during an attempt to unblock a level control instrument connected to a storage tank, an operator put through high pressure, 7 bar, air. The compressed air found its way inside the storage tank and eventually blew the roof many hundreds meters away.

In the case of liquids, destruction through overpressure is also possible, but other events must also occur, e.g., venting valves must be closed or not be able to release the extra liquid entering. The bursting overpressure can easily be achieved if the liquid level is higher than 21 cm (about 21 mbar) in any place above the storage tank top, irrespective of quantity, provided it is in contact with the liquid in the tank. In such cases, to avoid bursting, the vent or overflow pipe must always be placed on the vertical wall or near the roof (see Figure D2.1), unless there is a specific reason not to.

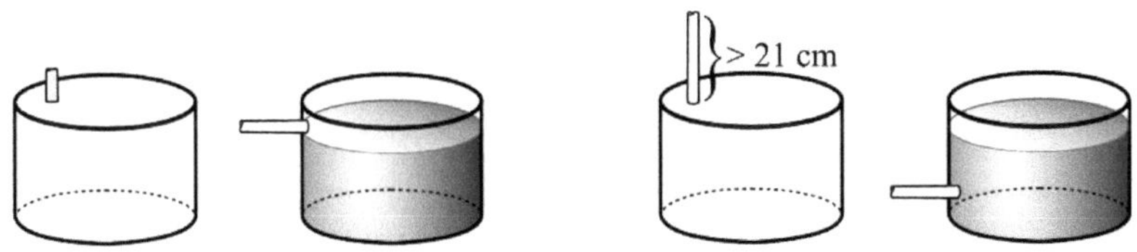

Proper position of vent or overflow — Wrong position of vent or overflow

Figure D2.1. Vent or overflow position.

Although overpressure bursting sounds logical for many people, few appreciate the fact that a storage tank can be destroyed even more easily due to underpressure. Compression is by far the most common way in which tanks are damaged. A common storage tank cannot withstand a vacuum higher than 7 mbar [Lees 2003]. Such a vacuum can easily be reached, even during the transfer of a fluid from a storage tank to another by natural flow, if vents are closed or partially blocked.

The vacuum phenomenon can also appear during the fast cooling of a storage tank, as the decrease in a gas temperature results in the simultaneous decrease in its pressure.

c) Gas Explosive Mixtures

All storage tanks, and especially closed-type storage tanks, comprise a closed space, ideal for the creation of gas explosive mixtures. In this section, we will consider as an "explosive mixture" any gas mixture that can, under proper conditions (see Section D3.2.1), ignite or explode.

Ignition can take place in many different ways, some of which will be described in Section D3, where it will be apparent that the elimination of all ignition sources is almost impossible. Therefore, efforts are concentrated into ways and precautions to reduce the creation of explosive mixtures. Some typical examples where explosive mixtures can be formed include:

- When the storage tank is not completely full, vapors collect in the empty space above the liquid's surface. If air also finds its way through a crack or hole, an explosive mixture can be formed. In floating-roof tanks, vapors usually find their way over the floating roof. These vapors can also form explosive mixtures.
- Vapors of flammable liquid can easily be carried from one tank to another through a partially damaged valve. This is dangerous, as explosive gas mixtures can be formed in tanks where they are least expected.
- Flammable liquids (e.g., gasoline) can be introduced into a tank containing non-flammable fluids, through a pipe that was used in the past for transporting gasoline. The leftover gasoline evaporates and creates flammable vapors.
- Dangerous substances can also be trapped in various places inside a storage tank (such as open pipes, scratches, welds, corroded or dirty spots, etc.) even after its flashing and cleaning.
- When the liquid level is very low, a long time is required for equilibrium to be reached inside the tank. This means that when the liquid level is high, few vapors are formed, but when the level is low the place is full of vapors which are very unstable.
- In some cases, explosive mixtures can be formed even inside a closed tank with non-dangerous substances. It is possible, for example, through biological or chemical reactions to create flammable gases like hydrogen.

The last case verifies what it is said about the design of chemical processes, that every case must be examined separately by itself.

Filling of storage tanks is usually accomplished in two ways, either (a) filling directly from the top, or (b) filling with a pipe through the liquid from the bottom. Both ways hide hazards which must be examined in relation to the liquid in question. As liquid enters from the top of the tank, the momentum created as it falls mixes the liquid already in the tank, while at the same time creates a cloud of droplets over the liquid surface. Such a droplet cloud can have explosive properties, or can store static electricity which in turn can create problems. When filling takes place from the bottom, that is, the filling pipe goes through the existing liquid, then the liquid in the storage tank is not mixed and a flammable (e.g., hydrogen) or toxic gas can be formed through various biological or chemical reactions on its surface.

It is apparent from the aforementioned discussion that it is quite easy for a storage tank to become a trigger that could also ignite other nearby storage tanks into a series of terrible explosions. The reader is advised to check for more information on the subject of storage tank design and the safety precautions that have to be followed in more specialized literature [Cheremisinoff 2000, Walas 1990].

D2.2. Pipes, Valves, and Related Equipment

In contrast to the previous section where accidents in storage tanks are caused by *intrinsic* factors (e.g., overpressure, explosive mixtures), accidents in pipes and related equipment (e.g., valves, measuring, and control instruments) are caused by *extrinsic* factors. In the next section, some typical technical specifications of pipes and valves will be given first and then, the main causes of leaks in these systems will be discussed.

D2.2.1. Technical Specifications

The technical specifications of pipes, valves and related equipment are derived through well-specified design expressions selected according to their operating pressure and the corrosive properties of the fluid to be transported [Parisher & Robert 2001].

Some of the most typical design characteristics are:

- Steel construction.
- Usual length of pipe sections is 2.5, 5 or 10 m.
- Epoxy-based external paint.
- Thermal insulation if required.

D2.2.2. Probable Events

a) Pipe Corrosion

Pipe corrosion occurs when areas of the pipe are weakened or become unable to withstand contact with corrosive substances (see Figure D2.2). This is most often the result of the mistaken selection of technical specifications, and more specifically, of the construction materials used. This way the pipe is in contact with substances that create long- or short-term partial or total rupture. The same failure can occur, even if technical specifications of the pipe were followed, but some other secondary effects like water concentration in pipes' endings, curvatures or other external places had not been considered and compensated for accordingly.

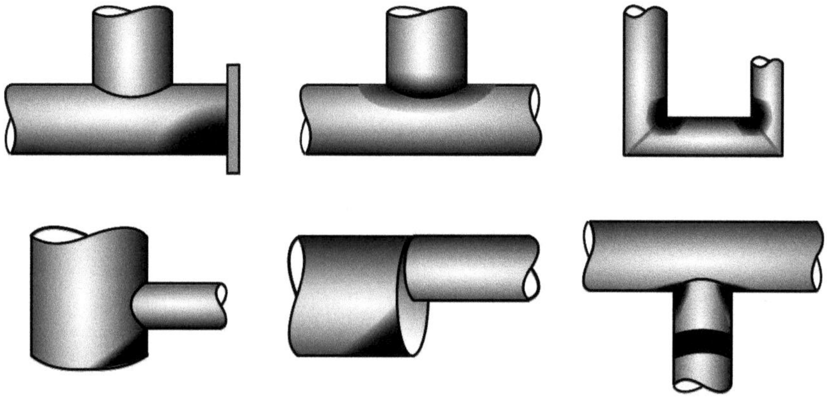

Figure D2.2. Possible places of internal or external corrosion in pipes.

Furthermore, corrosive material can enter a pipeline through an internal leak of another piece of equipment (e.g., a heat exchanger). These substances accumulate with time in specific places and can damage the pipe. These specific places in transport circuits can hide many hazards, which are not obvious to inexperienced technicians. For example, water in small traces can accumulate in one place and cause the failure of the pipe either because of expansion (pipe cooling) or because of sudden instantaneous vaporization. Similar behavior is shown by other substances also (e.g., the explosive decomposition of ethylene oxide because of a rise in temperature).

Rust, in addition to the corrosion of iron pipes, has one more unexpected effect: the density of iron oxide is smaller than that of iron, which means that during its formation, it expands, and can sometimes result in a leaky gasket.

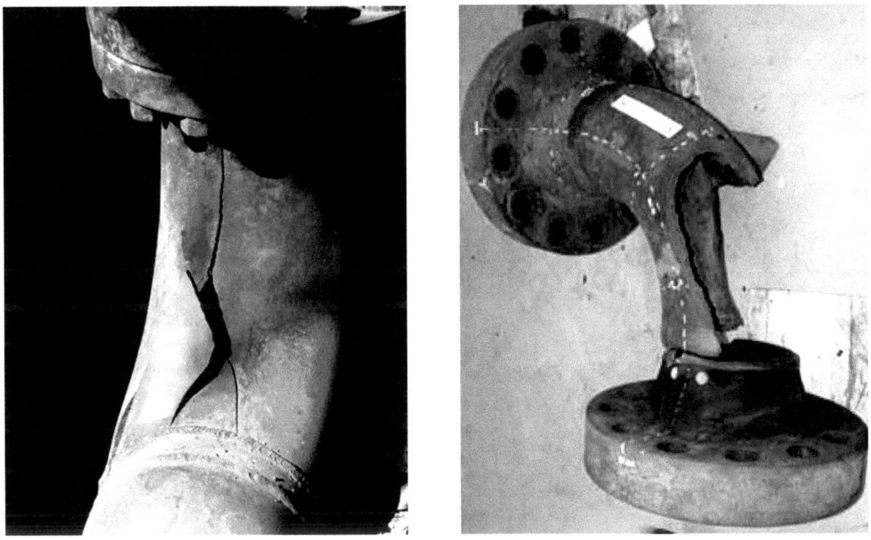

Left: Crack on a 10" pipe that caused a propane fire at Valero McKee refinery in Sunray, TX, U.S.A. on February 16, 2007. Right: Ruptured carbon steel elbow after the incident at BP Refinery in Texas City, TX, U.S.A., on July 28, 2005. (Reproduced by kind permission of the U.S. Chemical Safety Board.)

b) Metal Fatigue

When metallic materials are stressed from repeated or temporally adjustable (dynamic) stresses, they can fail and rupture, even if these stresses are lower than the rupture design stress of the material. This failure is called "metal fatigue." The basic characteristic of the fatigue is that it requires a significant period of operation of the equipment before it appears. Today it is estimated that 90% of all failures associated with mechanical causes are due to metal fatigue.

Metal fatigue is one of the most devious mechanisms of failure, as it appears without any previous warning signs. A failure attributed to metal fatigue can be recognized after its appearance only by examining the damaged surfaces. On April 18, 1982, in the low-density polyethylene production unit in Edmonton, Canada [MinA 2006], a large quantity of high-pressure ethylene leaked, resulting in a flash fire and damages of about US$ 39,000,000 (in 2005 prices). The leak was the result of a rupture of a high-pressure 1/8" pipe connected to a pressure transducer, and was a result of metal fatigue brought about from the vibrations of a nearby compressor. Static electricity in the compressor produced the ignition.

All pipes and related equipment must be properly supported. This happens either by supporting the pipe itself at each end, or with the use of external supports along the conduit. The study of the support of a pipeline is equally as important as the choice of the pipe's material. Each pipe must be properly supported so as to reduce its potential for mechanical fatigue. Another factor which must be studied is the pipe's vibrations, i.e., avoid vibrations with frequencies near the pipe's natural frequency. Something like this can become disastrous in a small time span for any kind of pipe. A dangerous practice is that usually short pipes are left free to vibrate, and get support only after the first time they are damaged.

The support of a pipe must be properly studied and executed by experienced technicians, who must appreciate the fact that welds can very often weaken a pipe or that loose supports can allow the pipe to move.

A typical (and mistaken) practice is the modification of an existing support or part of the piping without performing a full analysis on the effect of this change to the strength of the full, modified system.

Finally, we should mention the effect of temperature on metal fatigue in general, and most importantly, on supports. Exposure to fire, might not have a visible effect, but surely increases the potential for extreme metal fatigue of the equipment.

c) Water Hammer

Water hammer is a pressure surge or wave resulting when a fluid in motion (usually a liquid but sometimes also a gas) is forced to stop or change direction suddenly (momentum change). Water hammer commonly occurs when a valve is closed suddenly at an end of a pipeline system, and a pressure wave propagates in the pipe. The phenomenon of water hammer has disastrous consequences for two main reasons: (a) it incorporates large quantities of energy and (b) it can be carried out to large distances if measures have not been taken, and appear where it is least expected.

A water hammer can appear in the following circumstances:

- If the pipe is suddenly closed at the outlet (downstream), the mass of water before the closure is still moving forward with some velocity, building up a high pressure and shock waves. In domestic plumbing this is experienced as a loud bang resembling a hammering noise. Water hammer can cause pipelines to break if the pressure is high enough.
- In gas pipes, when liquid slugs are suddenly dragged along the flowing gases. This usually appears when condensates are formed inside gas pipes. It is specially dangerous in steam pipes, where steam traps are essential in order to avoid this phenomenon.

When a valve in a pipe is closed, the water downstream from the valve will attempt to continue flowing, creating a vacuum that may cause the pipe to collapse or implode. This problem can be particularly acute if the pipe is on a downhill slope. To prevent this, air and vacuum relief valves, or air vents, are installed just downstream from the valve to allow air to enter the line and prevent such a vacuum from occurring.

d) Pipe Reusability

In industry (specifically during maintenance and repair cycles) there is a strong effort to reduce cost by reusing older pipes that have been withdrawn for various reasons. Sometimes as a result of this action, insufficient attention is given to the quality, condition, or the previous use of the piping to be reused. To start with, the pipe may not be of the correct material for this use in relation to its material of construction, that is for example, the correct type or grade of steel. Additionally, even if the material specifications are correct, many other factors have to be considered, i.e., had the older pipe been used in a corrosive environment, or been used for a very long time, or under conditions which may have increased its metal fatigue, and so on. In general, reusing older pipes should be avoided.

e) Flexible and Rubber Pipes

The use of rubber or flexible metal piping is not common in industrial plants. However, they do exist and can constitute sources of hazard. Rubber pipes are characterized with the same hazards as metallic ones, but with a higher frequency of failure. They are usually employed because of their ease and direct use, which could mean that their strength vis a vis the substance involved, or their connections to other pipelines or vessels, might not have properly been examined. It is also possible to employ the wrong pipe if it is not properly marked for its use. Finally, rubber pipes are easily damaged creating the potential for fluid leakage anywhere and anytime.

Flexible metallic pipes, or bellows, are usually employed for convenience usually during maintenance or when bypassing other equipment for maintenance purposes is required. They require particular attention in their design. The disaster at Flixborough, U.K., in June 1974, which cost US$ 215,000,000 (in 2005 prices) was attributed to the failure of a flexible metallic pipe, which was connected between two cyclohexane reactors when the reactor between them had to be removed for maintenance. The failure was due to bad design.

f) Valves and Other Equipment

Valves are the most common equipment found on pipes; in addition, various measuring and controlling instruments are also used. Each instrument has its own purpose, and each instrument or valve increases the probability of an accident through their existence or their specific operation. If an instrument does not perform its operation as it should and fails, it is certain that a problem will arise. Our discussion, however, will concentrate on how the instrument itself can become a source of leakage.

In any pipeline circuit, the way the pipes, valves and instruments are connected is of prime importance. Any equipment or instrument disrupts the continuum of a pipeline, adding a weak spot to the system. Unfortunately, the

Facilities and Equipment

Storage Tanks

Pipes, Valves and Related Equipment

Stacks

Other Equipment and Procedures

necessary attention to the way equipment and instruments are connected (i.e., whether they should be welded or flanged) is not always given. It is obvious that every method should be applied to specific cases.

The position of the accessories is of equally great importance. There have been cases where the leak from one accessory dropped into another area instead of the designated sewage, resulting in a pool that ignited. In another case, a flammable fluid was collected purposely in an open plastic container, and when the operator went to empty it, it ignited because of static electricity. In a different case, a very toxic liquid leaked by mistake into the sewage line, resulting in the contamination of a whole lake. Great care should be given to the positioning of instruments and accessories on a pipeline, especially when these are related to dangerous substances.

As a typical case, we refer to the destruction of the petrochemical plant on March 7, 1989, in Antwerp, Belgium, that cost US$ 117,000,000 (in 2005 prices) [MinA 2006]. A hairline crack in a welded seam of piping to the level indicator system on the aldehyde column resulted in a minor ethylene oxide leak. As a result of this crack (which was caused by low-cycle fatigue), ethylene oxide escaped near the level indicator and formed polyethylene glycols (PEG) in the mineral wool insulation. It is believed that both the leak and accumulation of PEG occurred over a period of time. During repairs to the level indicator, the metal sheathing of the insulation was removed and air contacted the insulation soaked with PEG. Auto-oxidation of the PEG resulted and the insulating material was ignited. The piping for the level indicator system was heated to such a degree that auto-decomposition of the ethylene oxide within the piping occurred. This auto-decomposition then propagated into the aldehyde column, which subsequently exploded. The force of the explosion destroyed the distillation section of this plant. The large resulting fire and impact of flying debris to other process sections resulted in extensive damage throughout the plant. Because of this incident, this plant was closed for at least 24 months.

In another case, droplets of an extremely flammable liquid leaked from a piece of equipment onto a hot pipe (instead of the intended sewage area where special sensors were installed). This caused the formation of toxic vapors, which traveled through the buildings' ventilation system and collected in the personnel's offices, causing an explosion. Similar cases have been attributed to wrongly placed relief valves, which released dangerous gases into the atmosphere and subsequently spread out over inhabited areas and near potential ignition sources.

D2.3. Stacks

Stacks are a very common sight in industrial plants today, as they are the sole means of releasing and outgassing waste gases. Therefore a stack incorporates all possible hazards which characterize a gas, that is, flammability, explosiveness, and toxicity. In addition, stacks present a higher hazard risk as they have a wide radius of impact. This means that when a gas mass is released from the stack, because of atmospheric dispersion, it can affect many people or even create a flammable mixture that could explode a great distance from the stack. For this reason, it is essential to study not only what happens inside the stack, but also what happens to the released gases upon their exit (atmospheric dispersion - see Section C5).

In the next section, some typical technical specifications of stacks will be given and following that, we will discuss some typical cases where special attention should be given to stack design.

Flare stacks at the Holy refinery in Utah, UT, U.S.A. (Reproduced by kind permission of M. Christensen.)

D2.3.1. Technical Specifications

Taking into consideration the composition of the gas mixture through the stack, the most important parameter in the stack's design is the maximum composition at ground level. Hence the proper calculation model (see Section C5) is employed in order to calculate the necessary stack height to achieve the required dispersion.

Another important parameter is maintaining the temperature inside the stack above the condensation temperature of the vapors (mostly of steam). A possible condensation could result in the incorrect operation of the stack, or the formation of acids, with disastrous consequences. Furthermore, and in special cases, a special mechanism which burns a continuous flame is placed on the stack's exit. The aim of this flame is to "burn" any dangerous substances that should not be released into the atmosphere.

D2.3.2. Probable Events

In order to avoid mistakes which could result in an explosion or a fire in a stack, one must pay extra attention to the following:

- The most common mistake during the design of a stack is the mixing of many different gases, especially those which can, under proper conditions, produce explosive or flammable mixtures. For example, hydrogen must be released through a different stack to reduce the danger of explosion.
- The oxygen in air is the main component of every ignition. Therefore stack designers and production engineers should always check oxygen levels inside the stack, especially when flammable gases are present. Air can enter the stack following the gas flow as a component of exhaust gases or from cracks in the stack, e.g., from joints which are screwed instead of vein welded.
- In many cases, because of the high flammability of the gases involved, an inert gas is employed, e.g., nitrogen, in order to avoid the inflow of air (by increasing the total flow of gases). The flow of the inert gas must always be registered and controlled according to the requirements. There are cases where the flow of the inert gas was interrupted unexpectedly or decreased below the necessary value when the concentration of the other gas components changed, with disastrous effects.
- When toxic or flammable gases are to be released, sometimes it is preferred to burn them at stack exit with a flare - Flare Stack. This method is very useful and practical, but obviously it also incorporates many hazards, if it is not properly applied. The concentrations of the gases inside the stack must never be near the flammability limits, because in that case the fire can travel down the stack and damage the whole plant. Certainly this is valid in all stack cases; however, it is even more so in the case of flare stacks.

On October 12, 1994, in Pasadena, Texas, USA, an explosion in a methanol unit at this facility caused extensive damage to process equipment and buildings, resulting in a loss of US$ 71,000,000 (in 2005 prices) [MinA 2006]. Shrapnel from the explosion also penetrated the outer shell of a large ammonia storage tank, which was nearby. A mixture of combustible gases, which was left in the methanol plant piping and vessels during a shutdown due to incomplete purging, traveled to the flare stack where it ignited. The burning gases flashed back through the flare drum and permitted the burning gases to reach the contact cooler where they exploded. The unit was down for six months.

The aforementioned discussion refers to ways of avoiding fire and explosion. However, because of the nature of stacks, it is relatively easy for a small fire or explosion to occur. Therefore there should be some mechanisms in operation that will minimize the hazard of spreading the disaster. Hence it is necessary, if a fire appears in a stack, to be able to put it out, by interrupting the gas flow, or increasing the flow of inert gases or supply steam, etc. In any case, the aim is not only to stop the fire spreading, but also to avoid the formation of explosive mixtures.

In one case, a fire appeared in a stack following a small explosion. However, as it was raining heavily, it did not spread and died away. Unfortunately, nobody looked for the cause or for any damages, and so they did not appreciate that the flame arrestor inside the stack was damaged by the fire. Some days later the problem reappeared and this time the fire traveled down the stack and a series of explosions destroyed part of the plant.

Demolition of a stack at the Solo Suisice plant, Czech Republic on June 10, 2007. (Reproduced by kind permission of J. Pavlik.)

Facilities and Equipment

Storage Tanks

Pipes, Valves and Related Equipment

Stacks

Other Equipment and Procedures

A flare stack can also sometimes damage nearby facilities, because of the large heat flux it radiates. Temperatures in a stack are generally quite high and thus large quantities of heat are radiated. In this way, materials around the stack can be affected and their life expectancy shortened.

It should also be pointed out that a flare stack can act as an ignition source for flammable gases leaked from other processes. On April 15, 1978, in the industrial area of the city of Abqaiq, Saudi Arabia, a 55-cm gas transmission pipeline operating at 35 bars developed a crack due to corrosion [MinA 2006]. The crack expanded and the line parted, releasing a vapor cloud near a gas processing plant area that covered an 80-m radius area. After about seven minutes, ignition occurred from a flare located 500 m downwind. The jet whipping action of the escaping gas threw a 7-m pipe's section, 130 m away where it struck one of two 10,000-barrel spheroids. A second vapor cloud formed and ignited, developing detonation overpressures of 0.5 bar. Total damages reached US$ 139,000,000 (in 2005 prices).

D2.4. Other Equipment and Procedures

In an industrial plant many and different processes and procedures take place, and each one of them has its own equipment. For the most important ones (important implying that they can be a hazard or cause an accident) a description of possible hazards was presented. In this section we will discuss some additional equipment (e.g., pumps, heat exchangers, etc.) as well as some processes and procedures (e.g., filtering, cleaning, etc.).

D2.4.1. Pumps and Centrifugal Systems

Pumps and other centrifugal systems (and in general equipment that incorporate movable parts through which fluids pass) are susceptible to small or large leaks. This happens as the sealing of different parts (movable and non-movable) is often quite difficult.

Sealing is usually achieved by using o-rings of some rubber-type material. Unfortunately these o-rings wear out because of friction and must be changed often to avoid leaks, especially in cases of flammable or toxic fluids. In more rare cases, the body of the pump itself can be corroded or worn out, and in such cases a mass outflow occurs. In this case, only precautionary measures which may have been taken can prevent fire or explosion.

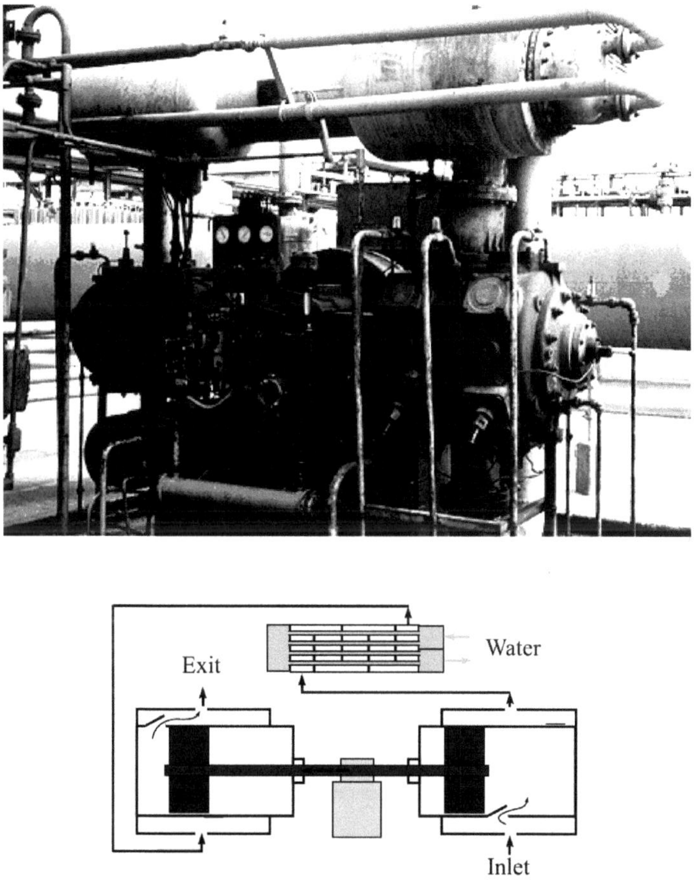

Photograph of a 2-stage pump with intercooler (reproduced by kind permission of N. Zachariadis) and schematic diagram of operation.

Another hazard, which is sometimes associated with pumps, is the partial or total blockage of the exit of the pump. Then the pressure increases inside the pump, and the fluid can auto-ignite. This phenomenon is very similar to the operational principle of the internal-combustion petrol engine.

Finally, pumps, and in general all equipment with movable parts, can be ideal ignition sources, as internal metal parts come in contact with each other and, therefore, can create sparks.

D2.4.2. Heat Exchangers

Heat exchangers in general are fairly hazard-free types of equipment. They are carefully designed for maximum efficiency. The usual accidents that are associated with heat exchangers are of two types:

1) Leakage of fluid from one side to the other, from the hot circuit to the cold or vice versa (see Figure D2.3). Specifically hydrocarbons can enter the steam circuit or the condensate circuit, exit at any point and ignite. It is obvious that such a case cannot be predicted, hence necessary checks must be done to the exchanger on a periodic basis.
2) The flashing of liquefied gases (usually with pressure reduction) results in a corresponding decrease of the temperature. This phenomenon has two consequences. First, if the temperature decrease is significant, it can affect the metals in contact (e.g., shell or tubes) and increase their brittleness. Second, it can cause the freezing of water or steam in the other circuit and cause the exchanger's failure.

Heat exchangers can also be affected by the water-hammer phenomenon (Section D2.2.2., paragraph c), and for this reason, extra precautionary measures must be taken (steam traps, etc.).

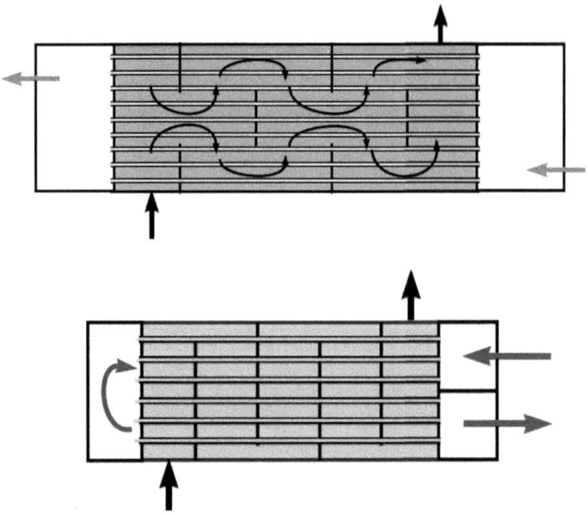

Figure D2.3. Different types of heat exchangers.

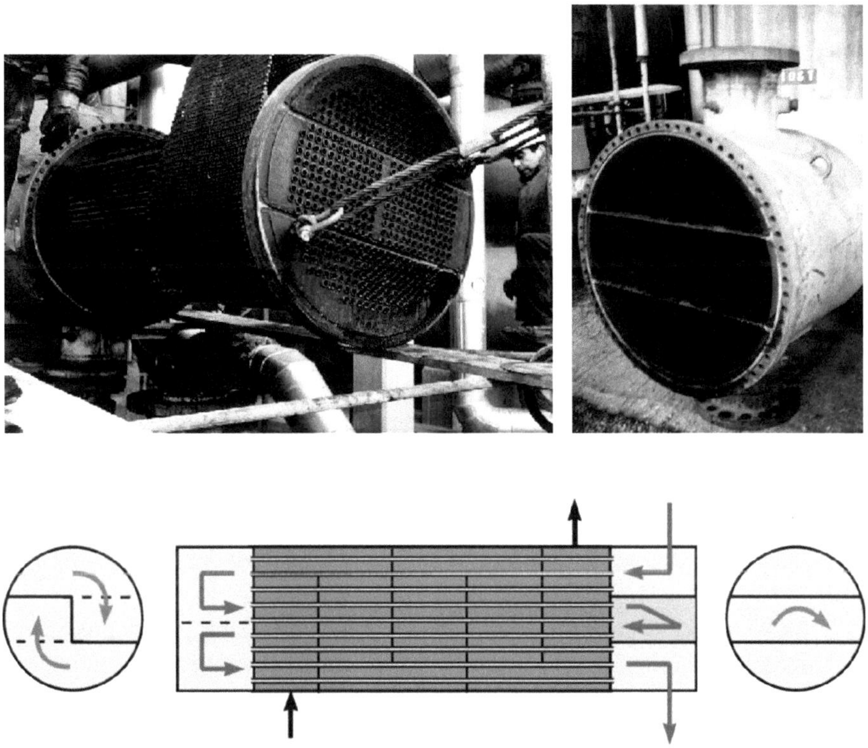

Photograph of a heat exchanger (reproduced by kind permission of N. Zachariadis) and schematic diagram of operation.

D2.4.3. Furnaces

Most accidents that take place in furnaces happen during their start-up, that is, during the stage of ignition. The atmosphere must be checked thoroughly with special instruments before the first ignition flame is lit, as a flammable environment can very easily be formed.

When the fuel is a light oil, an explosive or flammable atmosphere can be created at any moment in time because of the oil's evaporation. In one specific case, a furnace had to be stopped as the flame was interrupted. In the in-between time, however, a small amount of oil was still fed into the main burning chamber. The moment the operator placed the ignition equipment into the burning chamber, a large explosion took place. The cause was the evaporated oil, which went unnoticed during the flame stoppage.

Facilities and Equipment

Storage Tanks

Pipes, Valves and Related Equipment

Stacks

Other Equipment and Procedures

D2.4.4. Maintenance, Cleaning, and Painting

Unfortunately during maintenance and repair works, there is rarely enough time to take all necessary precautions. Improper maintenance is the most common cause of industrial disasters, in conjunction with the human factor and overlooking proper procedures. The disaster of the Piper Alpha offshore oil platform in the Black Sea, on July 6, 1988 [MinA 2006], with a materials loss of US$ 1,503,000,000 (in 2005 prices), was caused by not following proper procedures during the maintenance cycle. On the morning in question, the programmed maintenance of the relief valve in one of the two condensate pumps compressing the gas for transport to the coast took place. When maintenance was over, because there was no crane available to place the pump back in its place, its replacement was postponed for next morning, without informing anyone. During the night, the other pump stopped. In order for the gas supply to the coast to continue, power was switched to the first pump as the maintenance work was assumed to be finished. As the condensate pump was not in place, a large quantity of petrol was released, a vapor cloud was created and exploded. The explosion and the consequent fires caused the death of 165 of the 225 personnel that were working there.

Another very common source of accident is overlooking proper procedures and precautions during cleaning and painting of closed spaces. In order to remove strong remains, sometimes very powerful solvents are employed. Such solvents can ignite relatively easily. The same is valid for paints, which can include flammable substances that easily ignite if they reach the necessary flammability limits. During the cleaning of a butadiene storage tank on March 17, 2000 in the petrochemical plant in Pasadena, Texas, USA [MinA 2006], the tank exploded killing one technician, injuring 69 others, and resulting in material damages of US$ 81,000,000 (in 2005 prices). The explosion occured when technicians tried to remove butadiene remains that were still in the tank.

D3
Ignition

In the previous sections of this chapter, typical causes leading to the leak of flammable gases or liquids, were discussed. For a fire or explosion to appear, the concentration of the fuel mixture must be between its flammability limits and, an ignition source must be present. These concepts will be discussed in this section.

D3.1. Ignition Sources

In the case where the leaked gas is toxic, its consequences must be examined with the use of a dispersion model (see Chapter C5). In all other cases, there is always an additional stage before the final destruction; the ignition of the substance can result in a fire or an explosion or both. In an industrial environment, almost anything can act as an ignition source. In the following subsections some typical ignition sources will be presented.

D3.1.1. Construction and Maintenance Tools

Various tools used for construction or maintenance (e.g., welding equipment, cutting instruments, etc.) constitute probably the most serious ignition source in an industrial plant. This is because such equipment/instruments are used in many different places throughout the plant, in closed spaces, near dangerous equipment, and many times unfortunately are operated by unskilled personnel.

The most characteristic instruments of all are welding equipment which uses either electricity or bare flame. When a bare flame is employed, obviously this constitutes an ignition source. However, even when the welding is taking place with the use of electricity, it is still considered quite a dangerous ignition source because of the high temperatures it creates and the small pieces of burning metal that fly everywhere.

More generally, any equipment that comes into contact with metals, e.g., cutting tools, can create burning metal scrap which can easily cause ignition. For this reason, when they are employed in industrial spaces, all safety precautions must be taken, not only for the equipment to be used, but also for the surrounding equipment or facility. It is also a good practice to place a special cloth cover or a metal plate to collect the burning metal scrap.

The great fire in the crude oil storage tanks in the Jet Oil Depot in Thessaloniki, Greece, on February 24, 1986 was caused by the ignition of a small crude oil leak from sparks generated during the welding of a pipe [MinA 2006]. The fire, and the consequent explosions lasted for 7 days, spread to ten storage tanks (of fixed and of floating roof types) and destroyed 75% of the depot with a cost of US$ 368,000,000 (in 2005 prices).

D3.1.2. Hot and Electric Appliances

Most, if not all, appliances operate by electricity and produce heat. They can easily be a source of ignition when in contact with flammable substances. Electric appliances have cables in contact with electric circuits. If the contacts are not clean and the cable is not properly fixed to the contact, small electric sparks can appear, capable of producing ignition. Furthermore, most of the electric appliances have movable parts. The motion of these parts, if metallic, can also produce sparks capable of ignition. Typical examples are light switches, which must comply to special technical specifications if used in industrial areas.

Appliances or equipment where high temperatures appear can also be sources of ignition. For this to happen, however, temperatures must be quite high, which is rare in controlled environments. Such equipment includes compressors, electric resistances, etc.

A similar type of equipment where high temperatures are noted are catalytic equipment. In these, the temperature of the catalyst can reach hundreds of degrees and thus cause the ignition of nearby flammable substances or materials.

D3.1.3. Natural Phenomena

In nature there are many phenomena that can result in high temperatures or sparks in a degree sufficient to cause ignition. An obvious phenomenon is lightning, which is by itself a very strong electric discharge between sky and earth. Usually their targets are high structures, like stacks. Possible lightning targets must always be protected using proper lightning arrestors. Cases of lightning falling on a stack and igniting its contents are not rare. On May 10, 1998, lightning hit a crude oil storage tank in Ras Gharib, Egypt [MinA 2006], resulting in fire which spread to 16 other storage tanks in the plant. The cost of the destruction rose up to US$ 38,000,000 (in 2005 prices).

Lightning. (Reproduced by M. Coldren at www.PublicDomainPictures.net.)

Another phenomenon which can easily produce sparks is static electricity. Static electricity results from the accumulation of electric charges in two surfaces (in one negative charges and in the other positive ones) that are in constant motion. Typical examples are the flow of a fluid inside a pipe, the movement of droplets or particles in the air, walking, rubbing clothes, etc. Usually these static electric charges find a way through the substances and are carried down to the ground. However, when charges are accumulated on a non-conductive material (e.g., plastic, hydrocarbons), or on a conductive material (e.g., metal, water, etc.) which is not properly grounded, charges can remain on this surface for a certain period of time. If the static electric charge is large, then the moment that it finds a way towards the ground, it forms sparks, strong enough to ignite most flammable gas mixtures.

There are many such examples on how easily a static charge can result in a fire or explosion. A worker tried to clean a pool of gasoline on the floor with a broom that had plastic end pieces. A static electric charge accumulated at the brooms' end, and suddenly the charge was released and the gasoline ignited severely injuring the worker.

Static electricity also constitutes a very serious source of ignition in a dry environment with a lot of dust. On May 6, 1982, an explosion occurred in the drying and grinding area of a chemical plant where fumaric acid was being processed. The ensuing fire spread rapidly through the area, into an adjoining process area and to the 20 m by 40 m warehouse, which was destroyed. Apparently, a static spark from an electric motor ignited dust from the powdered acid and caused the explosion. The cost of destruction was about US$ 26,000,000 (in 2005 prices).

Another natural phenomenon that can cause sparks is the sun's radiation. The sun (especially during the summer days) can easily heat a metal plate to high temperatures. This phenomenon is a serious problem when designing storage tanks, and must be taken into account with the design so it is minimized.

In conclusion, since natural phenomena cannot always be properly predicted, it is best if the uncontrolled transport of flammable mixtures as well as the creation of explosive mixtures is avoided, as a source of ignition will always be present. Special care should also be given to properly grounding high structures to be protected from lightning.

Ignition

Ignition
Sources

Ignition
Characteristics

D3.2. Ignition Characteristics

Burning of a fuel-air mixture can only occur if the fuel concentration is between certain limits (flammability limits) and there exists an ignition source or an increase in the temperature of the mixture. In the first case, the necessary energy is supplied by the external source, e.g., a flame or a spark, while in the second case the gas reaches its autoignition temperature. These ideas will be further discussed in this section.

D3.2.1. Flammability Limits

A flammable gas will only burn as long as the fuel concentration is between specific concentration limits (a minimum and a maximum) - see Figure D3.2. Below the minimum fuel concentration (Lower Flammability Limit – LFL), the fuel-air mixture is considered "poor" and will not ignite. Correspondingly, above the maximum fuel concentration (Upper Flammability Limit – UFL), the fuel-air mixture is considered "too rich" and also will not ignite.

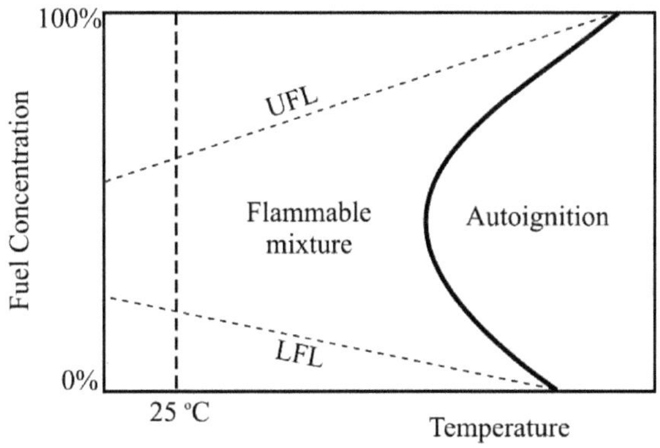

Figure D3.2. Flammability limits.

Flammability limits are influenced mostly by temperature and pressure and are calculated experimentally. As shown in Figure D3.2, the flammable range will widen when the initial temperature is increased. Changes in initial pressure for hydrocarbons in air will not change the LFL significantly, but the UFL will increase. Usually the flammability limits are also employed as the explosion limits of the substance. Table D3.1 shows the flammability limits of some substances that are very common in industry. We note the very wide flammability limits of acetylene, hydrogen or ethylene oxide.

Table D3.1. Flammability Limits,
Autognition and Flash Temperatures [Lees 2003].

	Flammability limits LFL (vol. %)	UFL	Autoignition temperature (°C)	Flash[a] temperature (°C)
Acetone	2.6	13	465	-18
Acetylene	2.5	100	305	-
Ammonia	15	28	651	-
Benzene	1.4	8	562	-11
n-Butane	1.8	8.4	405	-60
Carbon Disulfide	1.3	50	90	-30
Carbon Monoxide	12.5	74	-	-
Cyclohexane	1.3	7.8	245	-20
Ethane	3.0	12.4	515	-135
Ethylene	2.7	36	490	-121
Ethylene Dichloride	6.2	15.9	413	13
Ethylene Oxide	3	100	429	-
Hydrogen	4	75	400	-
Methane	5	15	540	-
Propane	2.1	9.5	450	< -104
Propylene	2.4	11	460	-108
Styrene	1.1	6.1	490	32
Toluene	1.3	7	536	4
Vinyl Chloride	4	22	472	-

[a] Flash temperatures are closed cup values.

D3.2.2. Mixture Flammability Limits

The flammability limits of a gas mixture, composed of many flammable components, can be calculated from the following two equations [Lees 2003]. For the lower flammability limit:

$$LFL_{\mathrm{m}} = \frac{1}{\sum_{n} (y_i / LFL_i)}, \tag{D3.1}$$

where LFL_{m} (vol. %) denotes the lower flammability limit of the mixture, LFL_i (vol. %) the lower flammability limit of each component i, and y_i (-) is the mole fraction of component i in the fuel mixture (without taking into consideration the air).

For the upper flammability limit:

$$UFL_m = \frac{1}{\sum_n (y_i / UFL_i)}, \qquad \text{(D3.2)}$$

where UFL_m (vol. %) denotes the upper flammability limit of the mixture, UFL_i (vol. %) the upper flammability mixture of each component i, and y_i (-) the mole fraction of component i in the fuel mixture (without taking into consideration the air).

D3.2.3. Autoignition Temperature

When the temperature of a flammable mixture is increased uniformly, it reaches a certain temperature where the burning reaction will start spontaneously - see Figure D3.3. For every flammable mixture of fuel-air there is an analogy corresponding to the lowest temperature that the mixture ignites. This temperature is called an "autoignition temperature." Table D3.1 shows also the autoignition temperatures of various substances.

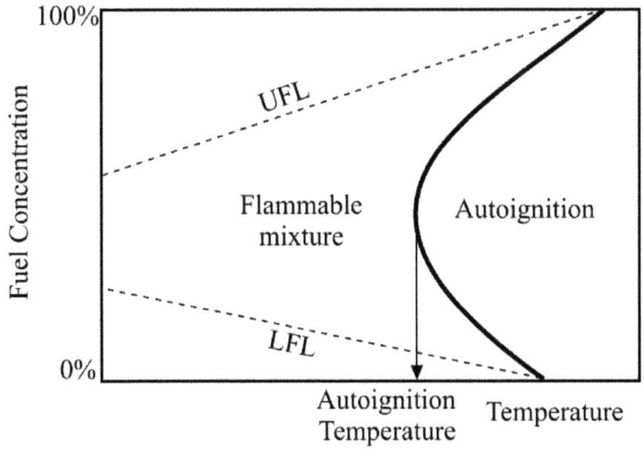

Figure D3.3. Autoignition temperature.

The autoignition temperature is a property that is highly dependent on the hot surfaces that the mixture comes in contact with. It is interesting to note that the autoignition temperature on a clean surface can decrease from 100 to 200°C if the surface has dust or is not flat.

D3.2.4. Flash Temperature

The flash temperature is the minimum temperature at which the fuel gives off sufficient vapor to form a flammable mixture with air, near the surface of the liquid or within the vessel (in all locations) used - see Figure D3.4. Operating at temperatures lower than the flash temperature of the liquid fuel will not lead to a flammable mixture being formed unless a mist cloud (e.g., due to splashing) is generated.

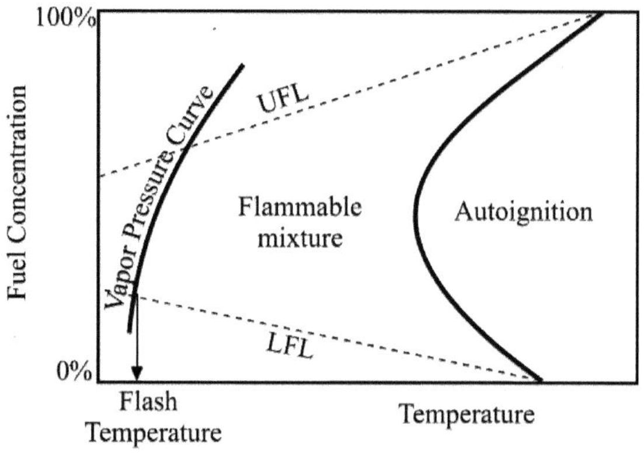

Figure D3.4. Flash temperature.

references

Abramowitz M. and Stegun I.A. (Eds), 1970, "Handbook of Mathematical Functions with Formulas, Graphs, and Mathematical Tables", Dover Publications Inc., New York.

Assael M.J., Trusler J.P. Martin and Tsolakis Th.F., 1996, "Thermophysical Properties of Fluids. An Introduction to their Calculation", Imperial College Press, World Scientific Publications, London.

Arturson G., 1987, "The Tragedy of San Juanico - the most Severe LPG Disaster in History", *Burns* **13**:87.

Babrauskas V., 1983, "Estimating Large Pool Fire Burning Rates", *Fire Techn.* **19**:251-261.

Bagster, D.G. and Pittblado, R.M., 1989, "Thermal Hazards in the Process Industry", *Chemical Engineering Progress*, July, 69-75.

Baker Q.A., Tang M.J., Scheier E.A., and Silva G.J., 1996, "Vapor Cloud Explosion Analysis", *Proc. Safety Prog.* **15**:106-109.

Baker Q.A., Doolittle C.M., Fitzgerald G.A., and Tang M.J., 1998, "Recent Developments in the Baker-Strehlow VCE Analysis Methodology", *Proc. Safety Prog.* **17**:297-301.

Baker W.E., Cox P.A., and Westine P.S., 1983, "Explosion Hazards and Evaluation", Elsevier Scientific Publishing Company, New York.

Beek W.J. and Mutzall K.M.K., 1975, "Transport Phenomena", Wiley, New York.

Bell R.P., 1978, "Isopleth Calculations for Ruptures in Sour-Gas Pipelines", *Energy Processing - Canada* **71**:36-39.

Berg van den A.C., 1985, "The Multi-Energy Method. A Framework for Vapour Cloud Explosion Blast Prediction", *J. Hazard. Mater.* **12**:1-10.

Berg van den A.C, and Lannoy A., 1993, "Methods for Vapor Cloud Explosion Blast Modelling", *J. Hazard. Mater.* **34**:151-171.

Berg van den A.C., Voort van der M.M., Weerheijm J., and Versloot N.H.A., 2004, "Expansion-controlled Evaporation: A Safe Approach to BLEVE Blast", *J. Loss Prev. Proc. Industr.* **17**:397-405.

Birk A.M., Poirier D., and Davison C., 2006, "On the Response of 500 gal Propane Tanks to a 25% Engulfing Fire", *J. Loss Prev. Proc. Industr.* **19**:544-558.

Bjorklund J.R., Bowers J.F., Dodd G.C., and White J.M., 1998, "Open Burn/Open Detonation Dispersion Model (OBODM) User's Guide. Vol. II. Technical Description", DPG Doc. No. DPG-TR-96-008b, West Desert Test Center, U.S. Army Dugway, Utah.

Bowen J.G, Fletcher E.R., and Richmond D.R., 1968, "Estimate of Man's Tolerance to the Direct Effects of Air Blast", Lovelace Foundation for Medical Education and Research, Albuquerque, New Mexico, Report DASA-2113.

Bowers J.R., Bjorklund J.R., and Cheney C.S., 1979, "Industrial Source Complex (ISC) Dispersion Model User's Guide. Volume II", EPA-450/4-79-031, U.S. Environmental Protection Agency, Research Triangle Park, North Carolina, 27711.

Brasie W.C. and Simpson D.W., 1968, "Guidelines for Estimating Damage from Chemical Explosions", *Symp. on Loss Prevention in Process Industries, 63rd AIChE Meeting*, St. Louis, Missouri, USA.

Briggs G.A., 1969, "Plume Rise", USAEC Critical Review Series, TID-25075, National Technical Information Service, Springfield, Virginia, 22161.

Briggs G.A., 1975, "Plume Rise Predications", Lectures on Air Pollution and Environmental Impact Analysis, American Meteorological Society, Boston, Massachusetts, USA.

Brighton P.W.M., Prince A.J., and Webber D.M., 1985, "Determination of Cloud Area and Path from Visual and Concentration Records", *J. Hazard. Mater.* **11**:155-178.

Britter R.E. and McQuaid J., 1988, "Workshop on the Dispersion of Dense Gases", HSE Contract Research Report No. 17.

Burgess D.S. and Herzberg M., 1974, "Advances in Thermal Engineering", Wiley and Sons, Chichester, UK, Chapter 27, 413.

Burgess D.S., Strasser A. and Grumer L., 1961, "Diffusive Burning of Liquid Fuels in Open Trays", *Fire Res. Abstr. Rev.* **3**:177.

Chadwell G.B., Leverenz Jr. F.L., and Rose S.E., 1999, "Contribution of Human Factors to Incidents in the Petroleum Refining Industry", *Process Saf. Prog.* **18**:206-210.

Chamberlain G.A., 1987, "Developments in Design Methods for Predicting Thermal Radiation from Flares", *Chem. Eng. Res. Des.* **65**:299-309.

Cheremisinoff N.P., 2000, "Handbook of Chemical Processing Equipment", Butterworth-Heinemann, Boston, USA.

Chen J.R., Richardson S.M., and Saville G., 1992, "Numerical Solution of Full-bore Ruptures of Pipelines Containing Perfect Gases", *Proc. Safety Environ. Prot.*, **70**:59-69.

Clemens P.L., 1982, "A Compendium of Hazard Identification and Evaluation Techniques for System Safety Applications", *Hazard Prevention*, March/April, 11-18.

Collenbrander G.W., 1980, "A Mathematical Model for the Transient Behaviour of Dense Vapour Clouds", *Loss Prevention and Safety Promotion* **3**:1104.

Directive Council 1999/30/EC, "Relating to Limit Values for Sulphur Dioxide, Nitrogen Dioxide and Oxides of Nitrogen, Particulate Matter and Lead in Ambient Air", Official Journal of the European Communities

Directive Council Proposal 2005/0183 COD, COM 2005 447 final, "On the Ambient Air quality and Cleaner Air for Europe".

Directive Council 2008/50/EC, "On ambient air quality and cleaner air for Europe", Official Journal of the European Communities.

Eisenberg N.A., Lynch C.J. and Breeding R.J., 1975, "Vulnerability Model: A Simulation for Assessing Damage Resulting from Marine Spills (VMI)", US Coast Guard, AD/A-015 245, NTIS Report No. CG-D-137-75.

EPA–454/B–95–003a, 1995, "User's Guide for the Industrial Source Complex ISC3 Dispersion models, Volume I", Environmental Protection Agency, Research Triangle Park, North Carolina, 27111.

Ermak D.L., 1990, "Steady-state SLAB Simple Prediction Heavy-Gas Dispersion Model", UCRL-MA-105607, Lawrence Livermore National Laboratory, University of California.

European Parliament Press release, 2006, "Cleaner air in Europe: more ambitious targets but greater flexibility", REF.: 20060922IPR10875.

Fanning J.T., 1877, "A Practical Treatise on Water-Supply Engineering", Van Nostrand Publ., New York.

Fay J.A. and Zemba S.G., 1986, "Integral Model of Dense Gas Plume Dispersion", *Atmospheric Environment* **20**:1347-1354.

Garrison W.G., 1988, "Major Fires and Explosions Analyzed for 30-year Period", *Hydrocarbon Processing*, **65**:115-120.

Giesbrecht H., Hess K., Leuckel W., and Maurer B., 1981, "Analysis of Explosion Hazards on Spontaneous Release of Inflammable Gases into the Atmosphere", *Ger. Chem. Eng.* **4**:305-314.

Gosselin R.E., Smith R.P., and Hodge H.C., 1984, "Clinical Toxicology of Commercial Products", 5th ed. Baltimore, MD, Williams and Wilkins Company, pp. III-375 to III-376.

Hagglund, B. and Person, L., 1976, "The Heat Radiation from Petroleum Fires", FOA Raport, Forvarets, Forskingsanstalt, Stokholm.

Hanna S.R. and Drivas P.J., 1987, "Guidelines for the Use of Vapor Cloud Dispersion Models", CCPS/AIChE, New York.

Hardee H.C. and Lee D.O., 1975, "Expansion of Clouds from Pressurized Liquids", *Accid. Anal. & Pret.* **7**:91-102.

Hardee H.C., Lee D.O., and Benedict W.B., 1978, "Thermal Hazards from LNG Fireballs", *Comb. Sci. Techn.* **7**:189-197.

Havens J.A., 1985, "The Atmospheric Dispersion of Heavy Gases: An Update", in the *Assessment and Control of Major Hazards*, IChemE Symposium Series No. 93, 143-164.

Heskestad G., 1983, "Luminous Heights of Turbulent Diffusion Flames", *Fire Safety J.* **5**:103-108.

Hirsch F.G., 1968, "Effects of Overpressure on the Ear - A Review", *Ann. N Y Acad. Sci.* **152**:147-162.

Hodge H.C., and Sterner J.H., 1956, "Combined Tabulation of Toxicity Classes", in Spector W.S., ed., "Handbook of Toxicology", Philadelphia, W.B. Saunders Company, Vol. 1.

Hoiset S., Hjertager B.H., Solberg T., and Malo K.A., 2000, "Flixborough Revisited - An Explosion Simulated Approach", *J. Hazard. Mater.* **77**:1-9.

Hottel H.C., 1958, "Review - Certain Laws Governing Diffusive Burning of Liquids, by V.I. Blinov and G.N. Khudiakov", *Fire Res. Abstr. Rev.* **1**:41-44.

HSE, 1975 "The Flixborough Disaster : Report of the Court of Inquiry", Health and Safety Executive, HMSO, London.

HSE, 1981, "Canvey: A Second Report", Health and Safety Executive, HMSO, London.

Hynes I., 1983, "The Physiological and Pathological Effects of Thermal Radiation", U.K. Atomic Energy Authorities, Safety and Reliability Directorate, Culchetch, Washington, SRD R 275.

Kalghatgi G.T., 1983, "The Visible Flame Shape and Size of a Turbulent Hydrocarbon Jet Diffusion Flame in a Crosswind", *Comb. Flame* **52**:91-106.

Kalghatgi G.T., 1984, "Lift-off Heights and Visible Lengths of Vertical Turbulent Jet Diffusion Flames in Still Air", *Comb. Sci. Techn.* **41**:17-29.

Kletz T., 1999, "What Went Wrong: Case Studies of Process Plant Disaster", Butterworth-Heinemann, Boston, USA.

Kletz T., 2003, "Still Going Wrong!: Case Histories of Process Plant Disasters and How They Could Have Been Avoided", Butterworth-Heinemann, Boston, USA.

Lea C.G. and Ledin H.S., 2002, "A Review of the State-of-the-Art in Gas Explosion Modelling", Health & Safety Laboratory Report 02.

Lees F.P., 2003, "Loss Prevention in the Process Industries", Elsevier Science Ltd, Oxford.

Lerdo de T., 1985, "San Juan Ixhuatepec: Memoria de una Emergencia", Report by the State of Mexico.

Luttgens G. and Wilson N., 1997, "Electrostatics Hazards", Butterworth-Heinemann, Boston, USA.

Marsh's Risk Consulting Practice, 2003, "The 100 Largest Losses 1972-2001", Marsh's Risk Consulting Practice Publication.

Melhem G.A. and Groce P.A., 1993, "Advance Consequence Modelling: Emission, Dispersion, Fires and Explosions", Van Nostrand Reinhold Inc., New York.

MinA, Major Industrial Accidents Database, 2006, Laboratory of Thermophysical Properties and Environmental Processes, Chemical Engineering Department, Aristotle University of Thessaloniki, Greece.

Mercx W.P.M., Berg van den A.C., Hayhurst C.J., Robertson N.J., and Moran C.J., 2000, "Developments in Vapour Cloud Explosion Blast Modelling", *J. Hazard. Mater.* **71**:301-319.

Mizner G.A. and Eyre J.A., 1982, "Large Scale LNG and LPG Pool Fires", *I. Chem. Symp.* **71**:147-163.

Moorhouse J. and Pritchard M.J., 1981, "Thermal Radiation Hazards from Large Pool Fires and Fireballs - A Literature Review", *IChemE Symp. Series 71*, "The Assessment of Major Hazards", Manchester, April, 14-16.

Mudan K.S., 1984, "Thermal Radiation Hazards from Hydrocarbon Pool Fires", *Prog. Energy Combust. Sci.* **10**:59-80.

Mudan K.S., 1985, "Thermal Radiation from Hydrocarbon Pool and Vapor Fires", Arthur D. Little Inc. Report 50688, July.

Mudan K.S., 1987, "Geometric View Factors for Thermal Radiation Hazard Assessment", *Fire Safety J.* **12**:89-96.

Mudan K.S. and Croce P.A., 1988, "Fire Hazard Calculations for Large Open Hydrocarbon Fires", Section 2, Chapter 4, "The SFPE Handbook of Fire Protection Engineering", First Edition.

Orbey M. and Vera J.M., 1983, "Correlation for the 3rd Virial Coefficient Using T^C, P^C and ω as Parameters", *A.I.Ch.E. J.* **29**:107-113.

OSHA - U.S. Occupational Safety and Health Administration, 1992, "Process Safety Management of Highly Hazardous Chemicals; Explosives and Blasting Agents", Final Rule 29 CFR 1910.119, February.

Papazoglou I.A. and Aneziris O.N., 1999, "Uncertainty Quantification in the Health Consequences of the Boiling Liquid Expanding Vapor Phenomenon", *J. Hazard. Mater.* **A67**:217-235.

Parisher A.R. and Robert A.R, 2001, "Pipe Drafting and Design", 2nd ed., Gulf Professional Publishing, USA.

Pasquill F., 1961, "The Estimation of the Dispersion of Windborne Material", *Meteor. Mag*, **90,** 33-49.

Picard D.J. and Bishnoi P.R.,1989, "The Importance of Real-Fluid Behaviour in Predicting Release Rates Resulting from High Pressure Sour-Gas Pipeline Ruptures", *Can. J. Eng.* **67**:3-9.

Pierorazio A.J., Thomas J.K., Baker Q.A., and Ketchum D.E., 2004, "An Update to the Baker-Stehlow-Tang Vapor Cloud Explosion Prediction Methodology Flame Speed Table". *Proc. AIChE 2004 Spring Meeting*, April 25-29, New Orleans, Louisiana, USA.

Pietersen C.M., 1988, "Analysis of the LPG - Disaster in Mexico City", *J. Hazard. Mater.* **20**:85.

Pitzer K.S. and Curl R.F., 1958, "Volumetric and Thermodynamic Properties of Fluids - Enthalpy, Free Energy, and Entropy", *Ind. Eng. Chem.* **50**:265.

Planas-Cuchi E., Gasulla N., Ventosa A., and Casal J., 2004, "Explosion of a Road Tanker Containing Liquified Natural Gas", *J. Loss Prev. Proc. Industr.* **17**:315-321.

Pope S.B., 2000, "Turbulent Flows", Cambridge University Press, Cambridge.

Pritchard, M.J. and Binding, T.M., 1992, "FIRE2: A new Approach for Predicting Thermal Radiation Levels from Hydrocarbon Pool Fires", *IChemE Symp.Series* **130**:491-505.

Raj P.P.K., Moussa A.N., and Aravamudan K., 1979, "Experiments Involving Pool and Vapor Fires from Spills of Liquefied Natural Gas on Water", Final Report, NTIS report No. GG-D-55-79, March.

Roberts A.F., 1982, "The Effect of Conditions Prior to Loss of Containment on Fireball Behaviour", *IChemE Symp. Series* **71**:397-429.

Sadee C., Samuels D.E., and O'Brien T.P., 1976/1976, "The Characteristics of the Explosion of Cyclohexane at the NYPRO (UK) Flixborough Plant on 1st June 1974", *J. Occup. Accid.* 203-235.

Sanders R.E., 1999, "Chemical Process Safety: Learning from Case Histories", Butterworth-Heinemann, USA.

Tang M.J. and Baker Q.A., 1999, "A New Set of Blast Curves from Vapor Cloud Explosion", *Proc. Safety Prog.* **18**:235-240.

Tang M.J. and Baker Q.A., 2000, "Comparison of Blast Curves from Vapor Cloud Explosions", *J. Loss Prev. Proc. Industr.* **13**:433-438.

Thomas P.H., 1963, "The Size of Flames of Natural Fires", *Proc. 9th Symp. on Combustion*, Academic Press, New York, 844-859.

TNO, 1989, "Methods for the Determination of Possible Damage to People and Objects Resulting from Releases of Hazardous Materials", Roos A.J. (Ed.), Green book, Report CPR 16E, Chapter 1.

TNO, 2005, "Methods for the Calculation of Physical Effects due to Releases of Hazardous Materials (Liquids and Gases)", van den Bosch C.J.H, Weterings R.A.P.M. (Eds.), Yellow book, Report CPR 14E, 3rd Ed., 2nd Rev.

Tsao C.K. and Perry W.W., 1979, "Modifications to the Vulnerability Model: A Simulation for Assessing Damage Resulting from Marine Spills (VM4)", US Coast Guard, AD/A-075 231, NTIS Report No. CG-D-38-79.

Tsonopoulos C., 1974, "Empirical Correlation of Second Virial Coefficients", *AIChE J.* **20**:263-272.

Turner D.Br., 1994, "Workbook of Atmospheric Dispersion Estimates", CRC Press Inc., Boca Raton, Florida, USA.

Zabetakis M.G. and Burgess D.S., 1961, "Research on the Hazards Associated with the Production and Handling of Liquid Hydrogen", *R.I. 5707*, Bureau of Mines, Pittsburgh.

Walas St.M., 1990, "Chemical Process Equipment", Butterworth-Heinemann, Boston, USA.

Wheatley C.J. and Prince A.J., 1987, "Translational Cloud Speeds in the Thorney Island Trials: Mathematical Modelling and Data Analysis", *J. Hazard. Mater.* **16**:185-199.

Wilson D.J., 1979, "The Release and Dispersion of Gas from Pipeline Ruptures", *Report for Alberta Environment Centre,* 7900686, Dept. of Mech. Engng., Univ. of Alberta, Canada.

World Bank, 1988, "Techniques for Industrial Assessing Hazards - A Manual", World Bank Publications, Technica Ltd.

physical properties A

In this appendix some physical properties of the following selected compounds are given:

- Methane
- Ethane
- Propane
- *n*-Butane
- *n*-Pentane
- *n*-Hexane
- *n*-Heptane
- *n*-Octane
- Gasoline
- Kerosene (or JP-5)
- Diesel

The properties included in the table have been calculated or obtained, and verified by the following literature sources.

REFERENCES

- Assael M.J., Trusler J.P.M., and Tsolakis Th.F., 1996, "Thermophysical Properties of Fluids. An Introduction to their Calculation", Imperial College Press, World Scientific Publications, London.
- S.E.P.E., Handbook of Fire Protection Engineering, 2002, 3rd Ed., NFPA Publications, p.3-26.
- "SUPERTRAPP", FORTRAN Package for the Calculation of the Transport Properties of NonPolar Fluids and their Mixtures, Produced and Supplied by N.I.S.T., Gaithersburg, U.S.A.
- "N.I.S.T. Chemistry WebBook", N.I.S.T. Standard Reference Database No 69, June 2005 Release (http://webbook .nist.gov/).
- International Programme on Chemical Safety, I.P.C.S. (http://www.inchem.org/).
- "ASPEN PROPERTIES", 2001, Ver. 11.1, Aspentech, Canada.

			Methane	Ethane	Propane	n-Butane
Chemical Compound			CH_4	C_2H_6	C_3H_8	C_4H_{10}
CAS Number			74-82-8	74-84-0	74-98-6	106-97-8
Molecular Weight		(g/mol)	16.043	30.069	44.096	58.123
Boiling Temperature	T_b	(K)	111.6	184.6	231.1	272.7
Melting Temperature	T_m	(K)	91	90	86	135
Flash Temperature	T	(K)	84.6		169.5	195.4
Critical Temperature	T^c	(K)	190.55	305.33	369.85	425.25
Critical Pressure	P^c	(MPa)	4.59	4.87	4.25	3.79
Heat of Vaporization[a]	ΔH_v	(kJ/kg)	510	489	385	355
Heat of Combustion	ΔH_c	(kJ/kg)	50,000	47,500	46,010	45,860
Burning Rate	m'	(kg/m^{2}s)	0.078		0.099	0.078
Vapor Phase (@ 298.15 K)						
@ Pressure	P	(MPa)	0.1	0.1	0.1	0.1
Density	ρ_G	(kg/m^3)	0.65	1.22	1.81	2.41
Specific Heat Capacity	C_{PG}	(kJ/kg K)	2.37	1.79	1.70	1.74
Kinematic Viscosity	η_G	(μPa s)	11.1	9.4	8.28	7.50
Thermal Conductivity	λ_G	(mW/m K)	36.1	22.5	18.0	16.1
Liquid Phase (@ 298.15 K)						
@ Pressure	P	(MPa)		4.3	1.0	0.3
Density	ρ_L	(kg/m^3)	-	313	492.6	573.4
Specific Heat Capacity	C_{PL}	(kJ/kg K)	-	7.95	2.75	2.45
Kinematic Viscosity	η_L	(μPa s)	-	37	97	158
Thermal Conductivity	λ_L	(mW/m K)	-	108	93	106
Liquid Phase (@ T_b, 0.1 MPa)						
Density	ρ_L	(kg/m^3)	424.2	545.1	581.9	603.1
Specific Heat Capacity	C_{PL}	(kJ/kg K)	3.79	2.52	2.26	2.30
Kinematic Viscosity	η_L	(μPa s)	117	164	198	201
Thermal Conductivity	λ_L	(mW/m K)	189	164	130	117

[a] Heat of Vaporization @ T_b.

n-Pentane	*n*-Hexane	*n*-Heptane	*n*-Octane			
C_5H_{12}	C_6H_{14}	C_7H_{16}	C_8H_{18}			Chemical Compound
109-66-0	110-54-3	142-82-5	111-65-9			CAS Number
72.15	86.177	100.203	114.23	(g/mol)		Molecular Weight
309.2	341.9	371.6	398.8	(K)	T_b	Boiling Temperature
129	178	182	216	(K)	T_m	Melting Temperature
224	251	269	286	(K)	T	Flash Temperature
469.8	507.9	540.15	568.9	(K)	T^c	Critical Temperature
3.38	3.04	2.74	2.49	(MPa)	P^c	Critical Pressure
358	334	318	303	(kJ/kg)	ΔH_v	Heat of Vaporization[a]
45,000	44,700	44,600	44,400	(kJ/kg)	ΔH_c	Heat of Combustion
	0.074	0.101		(kg/m^{2}s)	m'	Burning Rate
						Vapor Phase (@ 298.15 *K*)
0.06	0.02	0.005	0.0018	(MPa)	P	@ Pressure
1.79	0.70	0.20	0.082	(kg/m^3)	ρ_G	Density
1.72	1.70	1.67	1.64	(kJ/kg K)	C_{PG}	Specific Heat Capacity
6.92	6.54	5.90	5.20	(μPa s)	η_G	Kinematic Viscosity
14.3	13.3	11.7	11.6	(mW/m K)	λ_G	Thermal Conductivity
						Liquid Phase (@ 298.15 K)
0.1	0.1	0.1	0.1	(MPa)	P	@ Pressure
620.8	656.1	678.2	700.2	(kg/m^3)	ρ_L	Density
2.29	2.22	2.16	2.15	(kJ/kg K)	C_{PL}	Specific Heat Capacity
217	294	384	509	(μPa s)	η_L	Kinematic Viscosity
112	119	124	128	(mW/m K)	λ_L	Thermal Conductivity
						Liquid Phase (@ T_b, 0.1 MPa)
610.1	615	612	611	(kg/m^3)	ρ_L	Density
2.35	2.41	2.50	2.58	(kJ/kg K)	C_{PL}	Specific Heat Capacity
197	198	193	215	(μPa s)	η_L	Kinematic Viscosity
109	104	100	98	(mW/m K)	λ_L	Thermal Conductivity

[a] Heat of Vaporization @ T_b.

			Gasoline	Kerosene[c]	Diesel
Chemical Compound[b]			C_4-C_{12}		C_3-C_{25}
CAS Number			86290-81-5	8008-20-6	68476-34-6
Molecular Weight		(g/mol)	~115		~200
Boiling Temperature	T_b	(K)	333-473	333-573	555-611
Melting Temperature	T_m	(K)	253	253	243-255
Flash Temperature	T	(K)	227	310-338	325
Critical Temperature	T^c	(K)	470		
Critical Pressure	P^c	(MPa)			
Heat of Vaporization[a]	ΔH_v	(kJ/kg)	370	250	
Heat of Combustion	ΔH_c	(kJ/kg)	43,700	43,200	44,400
Burning Rate	m'	(kg/m^{2}s)	0.055	0.039	0.045
Vapor Phase (@ 298.15 *K*)					
@ Pressure	P	(MPa)	0.1		
Density	ρ_G	(kg/m^3)	4.8		
Specific Heat Capacity	C_{PG}	(kJ/kg K)	119.5		
Kinematic Viscosity	η_G	(µPa s)	6.6		
Thermal Conductivity	λ_G	(mW/m K)			
Liquid Phase (@ 298.15 K)					
@ Pressure	P	(MPa)	0.1	0.1	
Density	ρ_L	(kg/m^3)	730	810	
Specific Heat Capacity	C_{PL}	(kJ/kg K)	2.23	1.8	
Kinematic Viscosity	η_L	(µPa s)	442		
Thermal Conductivity	λ_L	(mW/m K)			
Liquid Phase (@ T_b, 0.1 MPa)					
Density	ρ_L	(kg/m^3)	693		
Specific Heat Capacity	C_{PL}	(kJ/kg K)	2.32		
Kinematic Viscosity	η_L	(µPa s)			
Thermal Conductivity	λ_L	(mW/m K)			

[a] Heat of Vaporization @ T_b. [b] Number of carbon atoms. [c] Also known as JP-5.

IDLH values B

USA Department of Health and Human Services
Center for Disease Control and Prevention, NTIS Publication No. PB-94-195047.

Documentation for IDLH concentrations:
National Institute for Occupational Safety and Health (NIOSH).

IDLH values (as of 3/1/1995).

Chemical Compound	IDLH value
Acetaldehyde	2,000 ppm
Acetic acid	50 ppm
Acetic anhydride	200 ppm
Acetone	2,500 ppm
Acetonitrile	500 ppm
Acetylene tetrabromide	8 ppm
Acrolein	2 ppm
Acrylamide	60 mg/m^3
Acrylonitrile	85 ppm
Aldrin	25 mg/m^3
Allyl alcohol	20 ppm
Allyl chloride	250 ppm
Allyl glycidyl ether	50 ppm
2 Aminopyridine	5 ppm
Ammonia	300 ppm
Ammonium sulfamate	1,500 mg/m^3
n-Amyl acetate	1,000 ppm
sec-Amyl acetate	1,000 ppm
Aniline	100 ppm
o-Anisidine	50 mg/m^3
p-Anisidine	50 mg/m^3
Antimony compounds	50 mg Sb/m^3
Arsenic	5 mg As/m^3
Arsine	3 ppm
Azinphosmethyl	10 mg/m^3

Chemical Compound	IDLH value
Benzene	500 ppm
Benzoyl peroxide	1,500 mg/m^3
Benzyl chloride	10 ppm
Beryllium compounds	4 mg Be/m^3
Boron oxide	2,000 mg/m^3
Boron trifluoride	25 ppm
Bromine	3 ppm
Bromoform	850 ppm
1,3-Butadiene	2,000 ppm
2-Butanone	3,000 ppm
2-Butoxyethanol	700 ppm
n-Butyl acetate	1,700 ppm
sec-Butyl acetate	1,700 ppm
tert-Butyl acetate	1,500 ppm
n-Butyl alcohol	1,400 ppm
sec-Butyl alcohol	2,000 ppm
tert-Butyl alcohol	1,600 ppm
n-Butylamine	300 ppm
n-Butyl glycidyl ether	250 ppm
n-Butyl mercaptan	500 ppm
p-tert-Butyltoluene	100 ppm
Cadmium dust	9 mg Cd/m^3
Cadmium fume	9 mg Cd/m^3
Calcium arsenate	5 mg As/m^3
Calcium oxide	25 mg/m^3

Chemical Compound	IDLH value
Carbaryl	100 mg/m^3
Carbon black	1,750 mg/m^3
Carbon dioxide	40,000 ppm
Carbon disulfide	500 ppm
Carbon monoxide	1,200 ppm
Carbon tetrachloride	200 ppm
Chlordane	100 mg/m^3
Chlorinated camphene	200 mg/m^3
Chlorine	10 ppm
Chlorine dioxide	5 ppm
Chlorine trifluoride	20 ppm
Chloroacetaldehyde	45 ppm
α-Chloroacetophenone	15 mg/m^3
Chlorobenzene	1,000 ppm
o-Chlorobenzylidene malononitrile	2 mg/m^3
Chlorobromomethane	2,000 ppm
Chloroform	500 ppm
1-Chloro-1-nitropropane	100 ppm
Chloropicrin	2 ppm
beta-Chloroprene	300 ppm
Chromium metal	250 mg Cr/m^3
Coal tar pitch volatiles	80 mg/m^3
Cobalt metal, dust, fume	20 mg Co/m^3
Copper (dusts and mists)	100 mg Cu/m^3
Copper fume	100 mg Cu/m^3
Cotton dust (raw)	100 mg/m^3
Cresol (*o*, *m*, *p* isomers)	250 ppm
Crotonaldehyde	50 ppm
Cumene	900 ppm
Cyanides (as CN)	25 mg/m^3
Cyclohexane	1,300 ppm
Cyclohexanol	400 ppm
Cyclohexanone	700 ppm
Cyclohexene	2,000 ppm
Cyclopentadiene	750 ppm
DDT	500 mg/m^3
Decaborane	15 mg/m^3
Demeton	10 mg/m^3
Diacetone alcohol	1,800 ppm
Diazomethane	2 ppm
Diborane	15 ppm
Dibutyl phthalate	4,000 mg/m^3
o-Dichlorobenzene	200 ppm
p-Dichlorobenzene	150 ppm
Dichlorodifluoromethane	15,000 ppm
1,1-Dichloroethane	3,000 ppm
1,2-Dichloroethylene	1,000 ppm
Dichloroethyl ether	100 ppm
Dichloromonofluoro methane	5,000 ppm
1,1-Dichloro 1-nitroethane	25 ppm
Dichlorotetrafluoroethane	15,000 ppm
Dichlorvos	100 mg/m^3
Dieldrin	50 mg/m^3
Diethylamine	200 ppm
2-Diethylaminoethanol	100 ppm
Difluorodibromomethane	2,000 ppm
Diglycidyl ether	10 ppm
Diisobutyl ketone	500 ppm
Diisopropylamine	200 ppm
Dimethyl acetamide	300 ppm
Dimethylamine	500 ppm
n-Dimethylaniline	100 ppm
Dimethylformamide	500 ppm
1,1-Dimethylhydrazine	15 ppm
Dimethylphthalate	2,000 mg/m^3
Dimethyl sulfate	7 ppm
Dinitrobenzene (*o*, *m*, *p* isomers)	50 mg/m^3
Dinitroocresol	5 mg/m^3
Dinitrotoluene	50 mg/m^3
Di sec-octyl phthalate	5,000 mg/m^3
Dioxane	500 ppm
Diphenyl	100 mg/m^3
Endrin	2 mg/m^3
Epichlorohydrin	75 ppm
EPN	5 mg/m^3
Ethanolamine	30 ppm
2-Ethoxyethanol	500 ppm
2-Ethoxyethyl acetate	500 ppm
Ethyl acetate	2,000 ppm
Ethyl acrylate	300 ppm
Ethyl alcohol	3,300 ppm
Ethyl benzene	800 ppm
Ethyl bromide	2,000 ppm

Chemical Compound	IDLH value
Ethyl butyl ketone	1,000 ppm
Ethyl chloride	3,800 ppm
Ethylene chlorohydrin	7 ppm
Ethylenediamine	1,000 ppm
Ethylene dibromide	100 ppm
Ethylene dichloride	50 ppm
Ethylene glycol dinitrate	75 mg/m^3
Ethyleneimine	100 ppm
Ethylene oxide	800 ppm
Ethyl ether	1,900 ppm
Ethyl formate	1,500 ppm
Ethyl mercaptan	500 ppm
n-Ethylmorpholine	100 ppm
Ethyl silicate	700 ppm
Ferrovanadium dust	500 mg/m^3
Fluorides	250 mg F/m^3
Fluorine	25 ppm
Fluorotrichloromethane	2,000 ppm
Formaldehyde	20 ppm
Formic acid	30 ppm
Furfural	100 ppm
Furfuryl alcohol	75 ppm
Glycidol	150 ppm
Graphite (natural)	1,250 mg/m^3
Hafnium compounds	50 mg Hf/m^3
Heptachlor	35 mg/m^3
n-Heptane	750 ppm
Hexachloroethane	300 ppm
Hexachloronaphthalene	2 mg/m^3
n-Hexane	1,100 ppm
2-Hexanone	1,600 ppm
Hexone	500 ppm
sec-Hexyl acetate	500 ppm
Hydrazine	50 ppm
Hydrogen bromide	30 ppm
Hydrogen chloride	50 ppm
Hydrogen cyanide	50 ppm
Hydrogen fluoride (as F)	30 ppm
Hydrogen peroxide	75 ppm
Hydrogen selenide (as Se)	1 ppm
Iodine	2 ppm
Hydroquinone	50 mg/m^3
Iron oxide dust and fume	2,500 mg Fe/m^3
iso-amyl acetate	1,000 ppm
iso-amyl alcohol (primary and secondary)	500 ppm
iso-butyl acetate	1,300 ppm
iso-butyl alcohol	1,600 ppm
iso-phorone	200 ppm
iso-propyl acetate	1,800 ppm
iso-propyl alcohol	2,000 ppm
iso-propylamine	750 ppm
iso-propyl ether	1,400 ppm
iso-propyl glycidyl ether	400 ppm
Ketene	5 ppm
Lead compounds	100 mg Pb/m^3
Lindane	50 mg/m^3
Lithium hydride	0.5 mg/m^3
L.P.G.	2,000 ppm
Magnesium oxide fume	750 mg/m^3
Malathion	250 mg/m^3
Maleic anhydride	10 mg/m^3
Manganese compounds	500 mg Mn/m^3
Mercury compounds [except (organo) alkyls]	10 mg Hg/m^3
Mercury (organo) alkyl compounds	2 mg Hg/m^3
Mesityl oxide	1,400 ppm
Methoxychlor	5,000 mg/m^3
Methyl acetate	3,100 ppm
Methyl acetylene	1,700 ppm
Methyl acrylate	250 ppm
Methylal	2,200 ppm
Methyl alcohol	6,000 ppm
Methylamine	100 ppm
Methyl (namyl) ketone	800 ppm
Methyl bromide	250 ppm
Methyl Cellosolve (r)	200 ppm
Methyl Cellosolve (r) acetate	200 ppm
Methyl chloride	2,000 ppm
Methyl chloroform	700 ppm
Methylcyclohexanol	500 ppm
o-Methylcyclohexanone	600 ppm
Methylene bisphenyl *iso*-cyanate	75 mg/m^3

Chemical Compound	IDLH value
Methylene chloride	2,300 ppm
Methyl formate	4,500 ppm
5-Methyl 3-heptanone	100 ppm
Methyl hydrazine	20 ppm
Methyl iodide	100 ppm
Methyl isobutyl carbinol	400 ppm
Methyl isocyanate	3 ppm
Methyl mercaptan	150 ppm
Methyl methacrylate	1,000 ppm
Methyl styrene	700 ppm
Mica	1,500 mg/m^3
Molybdenum (insoluble compounds)	5,000 mg Mo/m^3
Molybdenum (soluble compounds)	1,000 mg Mo/m^3
Monomethyl aniline	100 ppm
Morpholine	1,400 ppm
Naphtha (coal tar)	1,000 ppm
Naphthalene	250 ppm
Nickel carbonyl	2 ppm
Nickel metal and other compounds	10 mg Ni/m^3
Nicotine	5 mg/m^3
Nitric acid	25 ppm
Nitric oxide	100 ppm
p-Nitroaniline	300 mg/m^3
Nitrobenzene	200 ppm
p-Nitrochlorobenzene	100 mg/m^3
Nitroethane	1,000 ppm
Nitrogen dioxide	20 ppm
Nitrogen trifluoride	1,000 ppm
Nitroglycerine	75 mg/m^3
Nitromethane	750 ppm
1-Nitropropane	1,000 ppm
2-Nitropropane	100 ppm
Nitrotoluene (*o*, *m*, *p* isomers)	200 ppm
Octane	1,000 ppm
Oil mist (mineral)	2,500 mg/m^3
Oxalic acid	500 mg/m^3
Oxygen difluoride	0.5 ppm
Ozone	5 ppm
Paraquat	1 mg/m^3

Chemical Compound	IDLH value
Parathion	10 mg/m^3
Pentaborane	1 ppm
Pentachlorophenol	2.5 mg/m^3
n-Pentane	1,500 ppm
2-Pentanone	1,500 ppm
Perchloromethyl mercaptan	10 ppm
Perchloryl fluoride	100 ppm
Petroleum distillates (naphtha)	1,100 ppm
Phenol	250 ppm
p-Phenylene diamine	25 mg/m^3
Phenyl ether (vapor)	100 ppm
Phenyl etherbiphenyl mixture (vapor)	10 ppm
Phenyl glycidyl ether	100 ppm
Phenylhydrazine	15 ppm
Phosdrin	4 ppm
Phosgene	2 ppm
Phosphine	50 ppm
Phosphoric acid	1,000 mg/m^3
Phosphorus (yellow)	5 mg/m^3
Phosphorus pentachloride	70 mg/m^3
Phosphorus pentasulfide	250 mg/m^3
Phosphorus trichloride	25 ppm
Phthalic anhydride	60 mg/m^3
Picric acid	75 mg/m^3
Pindone	100 mg/m^3
Platinum (soluble salts)	4 mg Pt/m^3
Portland cement	5,000 mg/m^3
Propane	2,100 ppm
n-Propyl acetate	1,700 ppm
n-Propyl alcohol	800 ppm
Propylene dichloride	400 ppm
Propylene imine	100 ppm
Propylene oxide	400 ppm
n-Propyl nitrate	500 ppm
Pyrethrum	5,000 mg/m^3
Pyridine	1,000 ppm
Quinone	100 mg/m^3
Rhodium (metal fume and insoluble compounds)	100 mg Rh/m^3
Rhodium (soluble compounds)	2 mg Rh/m^3

Chemical Compound	IDLH value
Ronnel	300 mg/m^3
Rotenone	2,500 mg/m^3
Selenium compounds	1 mg Se/m^3
Selenium hexafluoride	2 ppm
Silica, amorphous	3,000 mg/m^3
Silver (metal dust and soluble compounds)	10 mg Ag/m^3
Soapstone	3,000 mg/m^3
Sodium fluoroacetate	2.5 mg/m^3
Sodium hydroxide	10 mg/m^3
Stibine	5 ppm
Stoddard solvent	20,000 mg/m^3
Strychnine	3 mg/m^3
Styrene	700 ppm
Sulfur dioxide	100 ppm
Sulfuric acid	15 mg/m^3
Sulfur monochloride	5 ppm
Sulfur pentafluoride	1 ppm
Sulfuryl fluoride	200 ppm
Talc	1,000 mg/m^3
Tellurium compounds	25 mg Te/m^3
Tellurium hexafluoride	1 ppm
Terphenyl (*o*, *m*, *p* isomers)	500 mg/m^3
1,1,1,2-Tetrachloro 2,2-difluoroethane	2,000 ppm
1,1,2,2-Tetrachloro 1,2-difluoroethane	2,000 ppm
1,1,2,2-Tetrachloroethane	100 ppm
Tetrachloroethylene	150 ppm
Tetraethyl lead	40 mg Pb/m^3
Tetrahydrofuran	2,000 ppm
Tetramethyl lead	40 mg Pb/m^3
Tetranitromethane	4 ppm
Tetryl	750 mg/m^3
Thallium (soluble compounds)	15 mg Tl/m^3
Thiram	100 mg/m^3
Tin (inorganic compounds)	100 mg Sn/m^3
Tin (organic compounds)	25 mg Sn/m^3
Titanium dioxide	5,000 mg/m^3
Toluene	500 ppm
Toluene 2,4-diisocyanate	2.5 ppm

Chemical Compound	IDLH value
o-Toluidine	50 ppm
Tributyl phosphate	30 ppm
1,1,2-Trichloroethane	100 ppm
Trichloroethylene	1,000 ppm
1,2,3-Trichloropropane	100 ppm
1,1,2-Trichloro 1,2,2-trifluoroethane	2,000 ppm
Triethylamine	200 ppm
Trifluorobromomethane	40,000 ppm
2,4,6-Trinitrotoluene	500 mg/m^3
Triorthocresyl phosphate	40 mg/m^3
Triphenyl phosphate	1,000 mg/m^3
Turpentine	800 ppm
Uranium (insoluble compounds)	10 mg U/m^3
Uranium (soluble compounds)	10 mg U/m^3
Vanadium dust	35 mg V/m^3
Vanadium fume	35 mg V/m^3
Vinyl toluene	400 ppm
Warfarin	100 mg/m^3
Xylene (*o*, *m*, *p* isomers)	900 ppm
Xylidine	50 ppm
Yttrium compounds	500 mg Y/m^3
Zinc chloride fume	50 mg/m^3
Zinc oxide	500 mg/m^3
Zirconium compounds	25 mg Zr/m^3

index